Copy No.

A.O. (46) 1.
8th March, 1946.

CHIEFS OF STAFF COMMITTEE

TECHNICAL SUB-COMMITTEE ON AXIS OIL

OIL AS A FACTOR IN THE GERMAN WAR EFFORT, 1933–1945

Offices of the Cabinet and
Minister of Defence, S.W. 1.
8th March, 1946.

30934

Published by Books Express Publishing

ISBN 978-1-78039-905-8

Books Express publications are available from all good retail and online booksellers. For publishing proposals and direct ordering please contact us at: info@books-express.com

THE JOINT CHIEFS OF STAFF

JOINT INTELLIGENCE COMMITTEE

WASHINGTON 25, D.C.

Joint Intelligence
Objectives Agency

JIOA 408

5 August 1946

MEMORANDUM FOR: Commandant,
Command and General Staff School,
Fort Leavenworth, Kansas.

Subject: British Chiefs of Staff Committee Secret Report A.O. (46) 1, dated 8 March 1946, title, "Oil As a Factor in the German War Effort, 1933-1945."

1. The subject report is forwarded herewith for your information in accordance with the request of the British Joint Staff Mission.

2. It is requested that the attached receipt be signed and returned to Joint Intelligence Objectives Agency.

THOMAS J. FORD
Colonel, CWS
Director

JOINT INTELLIGENCE COMMITTEE

Joint Intelligence
Objectives Agency

JIOA 408

5 August 1946

MEMORANDUM FOR: Commandant,
Command and General Staff School,
Fort Leavenworth, Kansas.

Subject: British Chiefs of Staff Committee Secret Report A.O. (46) 1, dated 8 March 1946, title, "Oil As a Factor in the German War Effort, 1933-1945."

1. The subject report is forwarded herewith for your information in accordance with the request of the British Joint Staff Mission.

2. It is requested that the attached receipt be signed and returned to Joint Intelligence Objectives Agency.

THOMAS J. FORD
Colonel, CWS
Director

	Page
FOREWORD	(v)
SECTION I.—PLANNING FOR WAR.	
Lessons of the 1914–18 War	1
The Organisation of the Industry	1
The *WIFO* Organisation	2
The Four-Year Plan	3
SECTION II.—THE POSITION AT THE OUTBREAK OF WAR.	
Stocks	7
Productive Resources	8
Crude Oil	8
Refining Capacity	9
Synthetic Oil	9
Tar Oils, Alcohol and Benzol	9
Consumption Requirements	10
SECTION III.—THE BLOCKADE.	
Restrictive Allocations to Neutrals	11
Holland	12
Spain	12
Portugal	12
Sweden	13
Switzerland	13
Turkey	13
SECTION IV.—ROUMANIA.	
Pre-emption	14
Clandestine Commercial Activity	16
Sabotage	17
Attempts to Block the Danube	17
SECTION V.—THE FIRST YEAR OF WAR.	
The Restriction of Civilian Consumption	19
Oil Consumption in the Conquest of Poland	20
The Western Campaigns	20
Captured Stocks	22
Stocktaking	23
SECTION VI.—OIL IMPORTS AND THE EFFICIENCY OF TRANSPORT.	
German Oil Policy in Roumania	25
Purchases of Hungarian Oil	26
Oil Transport and the Vienna Conference	28
Pipelines	28
Efficiency of Rail Transport	30
SECTION VII.—THE FAILURE OF THE SHORT WAR.	
The Attack upon Russia	31
Stock Losses	32
The Plans for the Caucasus	33
The Revision of Plans	35
Consumption and Stocks in 1942	36
SECTION VIII.—THE OIL REQUIREMENTS OF ITALY.	
Oil Resources	38
Blockade Considerations	38
Naval [illegible]	39
Army [illegible] Air Force Supplies	40
Tanker Sinkings and the North African Campaign	40
SECTION IX.—GERMANY ON THE DEFENSIVE.	
The Increase in Crude Oil Production	42
Adjustment of Refining Capacity	44
The Exploitation of Shale Oil Deposits	45
The Expansion of the Synthetic Oil Industry	45
Substitute Fuels	47
Consumption Requirements	49
Stocks and the Preparations to Resist Invasion	50
SECTION X.—THE PREPARATIONS FOR THE OIL OFFENSIVE.	
The Early Attacks	51
The Casablanca Directive	52
The Combined Bomber Offensive Plan	52
The Low-Level Attack upon Ploesti	53
Inability to Begin the Oil Offensive before 1944	54

SECTION XI.—THE BEGINNING OF THE STRATEGIC BOMBING OFFENSIVE AGAINST OIL

Selection of Plans to Support " Overlord " ... [illegible]
First Phase of the Oil Offensive ... [illegible]
German Reactions ... 57
The Second Phase: July–September 1944 ... 58
Additional Oil Targets ... 61

SECTION XII.—THE CONCLUDING PHASES OF THE OFFENSIVE

The Third Phase: October–November 1944 ... 63
The Achievements of the Plant Repair Organisation ... 64
The December Attacks ... 66
The Final Phase: January–April 1945 ... 67
The Attacks upon Oil Storage ... 69

SECTION XIII.—THE MILITARY EFFECTS OF OIL SHORTAGE.

The Effects of Fuel Shortages on the German Army ... 73
Difficulties in Italy ... 74
The Ardennes Counter-Offensive ... 75
The Offensive in Hungary ... 77

SECTION XIV.—THE CRIPPLING EFFECT OF FUEL SHORTAGES ON THE LUFTWAFFE.

The Depletion of Reserves ... 79
Restrictions upon Training ... 80
The Establishment of Emergency Reserves ... 81
The Restriction of Operations ... 82

SECTION XV.—THE OIL SITUATION AS AFFECTING THE GERMAN NAVY AND GERMAN INDUSTRY.

The Oil Supplies of the Navy... ... 84
The Effect of Oil Shortage on Industrial Output ... 86

ANNEX A.—THE INTELLIGENCE ASSESSMENT OF THE GERMAN OIL POSITION.

The Industrial Intelligence Centre ... 87
The Hankey Committee ... 87
The Lloyd Committee ... 87
The Hartley Committee ... 87
The Enemy Oils and Fuels Committee ... 88
The Combined Strategic Targets Committee ... 88
The First Reports on the Position ... 88
The Air Ministry View of the Position ... 89
Subsequent Reports ... 90
The Assessment of the Oil Target System ... 90
The Estimates of Consumption ... 91
Liaison with Washington ... 92
The Co-ordination of Intelligence ... 93
Commentary upon Target Selection ... 94
The Results Achieved ... 95

APPENDICES.

1. Organisation of the German Oil Administration ... 96
2. Rationing of Liquid Fuels for Civilian Consumption ... 98
3. The Formation and Development of the Wirtschaftliche Forschungsgesellschaft (*WIFO*) ... 99
4. Kontinentale Oel A.G. ... 103
5. The Levant Plan ... 107
6. The Operations of the Goeland Transport and Trading Company, Limited 108
7. The Attempt to Block the Iron Gates ... 110
8. Summary of German Oil and Gas Fields ... 111
9. Development of the Oil Shales of Southern Germany ... 113
10. Composition of the Various Committees ... 115
11. The Working Committee (Oil) of the Combined Strategic Targets Committee ... 117
12. The Reports to Hitler on the Effects of the Attacks ... 123
13. The Results of the Bombing of the Roumanian Oil Refineries ... 137
14. Alternative Target Systems: Lubricants and Ethyl Fluid Production ... 138
15. Vulnerability of Ethyl Fluid Production as a Target System ... 139
16. Effects of Fuel Shortage on *Luftwaffe* Aircrew Replacements ... 143
17. Contracts between Deutsche Erdoel A.G. and the German Navy for Heide-Meldorf Oil ... 145
18. Some German Views upon Allied Strategic Bombing Policy ... 147
19. Note on Oil Plant Defence Measures ... 149
20. Note upon Aiming Error and Weapon Effectiveness in Relation to Oil Plants ... 151
21. The Geilenberg Plan for Plant Dispersal ... 152

ILLUSTRATIONS.

Plate		Facing page
1.	The Leuna Plant of the *I.G. Farbenindustrie* ...	4
2.	The Bruex Synthetic Oil Plant ...	12
8.	The Columbia Aquila Refinery, Ploesti ...	20
3.	The *Deurag-Nerag* Refinery at Misburg, Hanover ...	32
4.	Underground Oil Storage ...	48
5.	The *Rhenania-Ossag* Refinery, Hamburg ...	64
6.	Leuna Under Attack ...	80
7.	A Reconnaissance View of the Misburg Refinery ...	96

FIGURES.

Figure		
1.	Aviation Gasoline, Motor Gasoline and Diesel Oil Stocks, Production and Consumption, 1940–44 ...	8
2.	Production of Oil Products by Process ...	16
3.	Comparison of Production Plans with Actual Production ...	28
4.	Accuracy of Hartley Committee Estimates ...	40
5.	Synthetic Fuel Production by Process ...	56
6.	Plan of an Underground Lubricating Oil Plant ...	72
7.	Aviation Gasoline Position in relation to Attacks ...	88

TABLES.

PRODUCTION.

Table		Page
1.	Production in Greater Germany ...	159
2.	German Crude Oil Production ...	160
3.	Synthetic Oil Plants' Capacities ...	160
4.	Capacities of Refineries ...	161
5.	Statistics of the Roumanian Oil Industry ...	163
6.	Roumanian Oil Exports by Products ...	163
7.	Roumanian Oil Exports by Countries ...	163
8.	Hungarian Crude Oil Production ...	164
9.	Output Achieved by the Geilenberg Dispersal Plants ...	164

CONSUMPTION.

Table		
10.	Allocation of German Civil Consumption in 1938 ...	164
11.	Consumption of Alcohol as Motor Fuel ...	164
12.	Consumption of Principal Oil Products in Germany ...	165
13.	German Civilian Consumption of Automotive Fuels ...	166
14.	Motor Gasoline Allocations to the Civil Economy ...	167
15.	Diesel Oil Allocations to the Civil Economy ...	168
16.	Civil and Industrial Lubricating Oil Consumption ...	169
17.	Generator Fuels as Gasoline Substitutes in Germany ...	170

ATTACK STATISTICS.

Table		
18.	Over-All Attack Data of Combined Strategic Air Forces in the European Theatre of Operations ...	171
19.	Summary of Oil Capacity, Production and Attack Data for Greater Germany ...	172
20.	Attack Data for Oil Targets in Greater Germany by Type of Target ...	172
21.	Chronological Summary of Strategic Attacks on Oil Targets ...	173
22.	Weight of H.E. Bombs Dropped and the Effects on Capacity and Production ...	184

INDEX TO REFERENCES ... 155

INDEX ... 185

MAP ... At end

Foreword.

This Report is an account of Germany's oil economy from 1933 up to the time of her defeat. Its compilation is the result of the work of three groups: the Technical Sub-Committee on Axis Oil, which has been responsible for reporting upon Germany's oil position since 1942; the Oil Committee of the Combined Strategic Targets Committee, which assisted in the direction of the Allied air offensive against oil targets; and the Inter-Departmental Bombing Survey.

The compilers of the report have had access to all the relevant records so far available, have interrogated German officials and have examined many of the bombed targets. The facts thus obtained have provided a record of the efforts made by the Germans to safeguard their oil supplies, of the success of the Allies in destroying the sources of production, and of the extent to which shortage of oil is reported to have influenced military operations.

The Report attempts to summarise the evidence now available on these questions so that those in possession of similar evidence on all the other factors involved may put in its correct perspective the part oil played in the military defeat of Germany.

Acknowledgment.

The Sub-Committee acknowledges its indebtedness in the preparation of this Report to a large number of persons and agencies, both British and American. So many have contributed towards its compilation that to mention a few would omit others no less deserving of our appreciation. A special acknowledgment is, however, due to the members of the United States Strategic Bombing Survey, whose resourceful exploitation of many sources of information in Germany has been of much assistance, and also to Mr. O. F. Thompson for his skill and patience in the difficult task of compiling and editing the report.

(Signed) H. HARTLEY, *Chairman*,
On behalf of the Technical
Sub-Committee on Axis Oil.

SECTION I.

PLANNING FOR WAR.

Although the re-establishment of the German war machine had been progressing for some time it was not until 1933, when the National Socialist Party came into power, that these preparations became intensified. The provision of liquid fuel supplies for the armed forces formed an essential part of these preparations.

2. The Germans were only too well aware of Curzon's dictum that the Allies had floated to victory on a wave of oil. Furthermore, they shared the belief of the statement that "*La victoire des Alliés sur l'Allemagne fut la victoire du camion sur la locomotive*" and it was clear that in the conduct of future wars even more reliance would be placed upon the internal combustion engine. It was also clear that the task of providing adequate supplies of oil for a future war would be a formidable one as the natural oil resources of the *Reich* were at that time providing less than 10 per cent. of Germany's peace-time requirements.

Lessons of the 1914–1918 *War.*

3. In the forming of these plans advantage was taken of the lessons of the 1914–18 War. Although the collapse of 1918 cannot be ascribed directly to the oil position it was an important contributing factor. While most branches of the armed forces had adequate supplies this was only at the expense of other needs and oil was consequently one of the commodities that was basically in short supply. Ludendorff, in his memoirs, stated that the fate of Bulgaria in September 1918, and the threat thereby involved to German control of Roumanian supplies, was one of the reasons that led the German military authorities to their decision to ask for an Armistice.

4. The first consideration of Hitler's oil planning staff was therefore to expand the sources of supply and to ensure that consumption was kept within reasonable limits. Crude oil production had to be subsidised, a synthetic oil industry had to be developed, and such steps as might be considered necessary had to be taken to provide for the maximum import of oil from other countries. It was also considered necessary to proceed with the construction of underground storage facilities.

5. An appropriate administration had also to be set up to co-ordinate the whole process of the supply and consumption of oil. In 1918 the lack of such co-ordination had precipitated the economic collapse. At that time the three services had bickered amongst themselves for the available supplies resulting in both inequality in distribution in the armed forces(1) and also in industry being deprived of supplies necessary to maintain the delivery of munitions.

The Organisation of the Industry.(2)

6. The oil industry in Germany had for many years been relatively closely integrated although before 1933 this integration was of a commercial nature with a minimum of Government interference. However, in view of the vital importance of the industry from a war point of view the Nazi Government lost no time in ensuring that it should come under Government control.

7. The process of mobilisation was gradual and maximum use was made of industrial federations in a consultative and executive capacity. In 1934 a start was made with the foundation of the *Ueberwachungstelle fuer Mineraloel*, an official organisation charged with the duty of controlling the import, storage, distribution and consumption of all mineral oil products. Similar supervisory

NOTE.—Throughout the footnotes to this Report the sources of information, whether persons, agencies or documents, are given in abbreviated form. The details of these abbreviations are given in the Index of References, immediately preceding the Tables at the end of the Report.

(1) At the time of the Armistice there were substantial stocks for submarines and aircraft, but there was a crippling shortage of gasoline for motor transport. "*Das Erdoel im Weltkrieg*," Friedensburg, 1939.

(2) Further details are given in Appendix 1.

bodies were set up for other commodities. It is doubtful whether at this time any of the members of these federations were aware that they were in any way participating in any preparations for war.

8. The absence of direct control over oil production, as opposed to stocks and distribution, is noteworthy. The Government did not have such a firm grip on oil output as they had on the production of other commodities. The influence of the *I.G. Farbenindustrie* was to some extent responsible for preventing the industry from becoming deprived of all freedom of action. Moreover, the industry was competently operated and any direction of its activities by the Government was probably unnecessary.

9. During this time the industry was being assisted by Government subsidies for exploratory drilling programmes and tax relief was also being afforded to home-produced fuels.

10. The rapid development of the synthetic oil industry was initiated in **1935** when a number of power, chemical, oil and mining concerns were forcibly directed to establish the *Braunkohlen Benzin A.G.*, a company which, exclusive of the original *I.G. Farbenindustrie* plant at Leuna, formed the basis of all synthetic oil production from the brown coal resources of Central Germany.

11. Meanwhile the co-operation of the industrial federations was deemed essential and, in order that this co-operation should be complete, membership of the *Wirtschaftsgruppe Kraftstoffindustrie* was made compulsory and a politically reliable president, Dr. E. R. Fischer of *I.G. Farbenindustrie*, was appointed.

12. In addition to these groups there was the influence asserted in **1936** by the Four-Year Plan. One section of the Four-Year Plan consisted of two closely co-operating offices, the *Rohstoffamt* and the *Planungsamt*, which were respectively responsible for administering and importing, and making the best use of Germany's strategically valuable raw materials. There was, in addition, a section dealing especially with oil (*Abteilung* 2). Under this section crude oil production problems were delegated to Dr. Bentz with the official title of *Beauftragete fuer die Erdoelgewinnung*. Synthetic oil came under the direction of Dr. Krauch, another politically reliable *I.G. Farbenindustrie* official, as *Beauftragte fuer Sonderfragen der Chemische Erzeugung*.

13. Not long before the outbreak of war the *Ueberwachungstelle fuer Mineraloel* was embodied in the *Reichswirtschaftsministerium*, and its name was changed to the *Reichstelle fuer Mineraloel* to bring it in line with other *Reichstellen*. Production, however, continued to be outside its competence and remained within that of the Four-Year Plan Organisation.

14. Distribution was handled through three organisations. There was firstly the *WIFO* organisation, the purpose and operations of which are described below, and which was principally engaged in supplying the *Luftwaffe*. There was secondly the *Zentralbuero* which was formed in September **1939** and which was a combine of all distributors and dealers in oil products. It was responsible for the supply of liquid fuels, exclusive of aviation spirit, to the armed forces and to industry. Finally, there was the *ASV* (*Arbeitsgemeinschaft Schmierstoff Verteilung, G.m.b.H.*), which co-ordinated the distribution of lubricants to all consumers other than the armed forces.

15. In spite of this comparatively formidable State direction, the integration of the German oil industry was not completely achieved by the time the War started. The trade federation was still the most efficient unifying organisation. It may well be that the omission to co-ordinate closely the planning of production and consumption, as well as the belief that the War would be of brief duration, were to some extent responsible for Germany having insufficient oil stocks and productive capacity at the outbreak of war to sustain active operations for more than a short time.

The WIFO Organisation.(3)

16. The earliest discernible sign of the National Socialist Government's determination that a lack of oil stocks should not impair Germany's war potential can be seen in the formation in the Autumn of **1934** of the *Wirtschaftliche Forschungsgesellschaft m.b.H.* (Economic Research Co.). This innocent title

(3) See Appendix 3.

and the equally innocent articles of association(4) of the company camouflaged an organisation of which the primary purpose was the construction of strategic oil storage and the acquisition of oil stocks as a war reserve.

17. The organisation was formed at the request of the German War Office and Air Ministry and was sponsored by the *Reichswirtschaftsministerium*. A proposal that the oil industry should be directed to execute the project was overruled on security grounds and the *Wirtschaftliche Forschungsges. m.b.H.* was therefore founded with a capital of RM.20,000, by the *Deutsche Ges. fuer öffentliche Arbeiten A.G.* and by *I.G. Farbenindustrie*, the former subscribing RM.16,000 and *I.G.* the remainder. This capital was nominal and the funds required for the projects constructed by the company and for their operation were provided by the Government. *WIFO* thus acted as the trustee of these properties for the State. In April 1943 these expenditures were capitalised to the extent of raising the capital of *WIFO* to RM.100 million. The remaining funds which had been provided were converted into a loan of RM.670 million.

18. Direction of the undertaking was entrusted to *Regierungs-Baurat* Franz Wehling, whose capabilities had been proved in similar engineering projects during 1914–18 and in the inter-war period, and he was subject on broad questions of policy to a Technical Commission on which were represented the Ministry of Economics, the Air Ministry and the *Oberkommando der Wehrmacht*. *WIFO* never had any relationship with the Navy except for occasionally supplying small quantities of special oil products.

19. The primary functions of *WIFO* were to construct oil storage installations, to acquire oil, and to deliver oil both in bulk and packed, to the Army and Air Force. It was also responsible for oil transport both by rail and pipeline, for pipeline construction, and for the construction of chemical plants to produce sulphur and sulphuric acid, nitric acid and toluol. These chemical plants were built by *WIFO* but were leased to and operated by chemical companies, principally by the *I.G. Farbenindustrie*.

20. The total planned capacity of oil storage was about 2 million tons, of which 900,000 tons capacity had been constructed by September 1939.(5) Stocks in hand at *WIFO* depots on the outbreak of war amounted to 640,000 tons. These stocks had been acquired partly from German production but also by direct, but surreptitious, purchases from producers in Roumania, Mexico and the United States. Notwithstanding the size of these purchases and the fact that *WIFO* sometimes had to lease commercial storage, a considerable measure of success was achieved in safeguarding the security of the activities of the organisation, a subject on which great stress was laid in instructions issued from time to time by the management.(6)

21. As the means of safely storing the fuel required for the *Luftwaffe* and the Army,(7) the *WIFO* organisation was an important item in Germany's war plans. As will be seen later in this Report, this organisation maintained the efficient distribution of the supplies available for as long as transport conditions permitted. Furthermore the heavy expenditure that was incurred in the protection of the installations was to pay rich dividends in that none of them was considered by the Allied Air Forces to be a vulnerable objective until the concluding stages of the War.

The Four-Year Plan.

22. The Four-Year Plan was announced in 1936. It came at a time when commercial interests were unwilling to plunge into the enormous capital costs

(4) "The erection and maintenance of industrial trading and manufacturing undertakings and plants, especially the erection and maintenance of research and development plants for the purpose of the advancement of the relevant industrial activities."

(5) Throughout this Report "tons" are "metric tons" of 2,204·6 lbs. except in the case of bomb tonnages which are recorded in short tons of 2,000 lbs.

(6) *WIFO* Circular dated 13.1.39—

> "Once again special attention is drawn to the duty of reticence in the presence of strangers, in which category are to be included employees of contracting firms working in our plants and visitors to our plants who come by the authority of the management. Our security officers tell us that the Military Attachés of foreign Powers have expressed surprise at the significant information they have obtained in German plants. Such a happening must be impossible in our installations."

(7) *WIFO* was responsible for supplying only a part of the Army's requirements. Other supplies were obtained by requisition direct to the Central Office for Mineral Oil Production. The Navy also looked after its own requirements and the fuel storage capacity provided was in excess of 2 million tons.

and uneconomic operation of synthetic oil plants. The development of indigenous oil production by State aid was consequently one of the principal objects of the Plan.

23. The oil provisions of the Four-Year Plan, which by 1939 had accounted for RM.574,000 million, or 42 per cent. of the total investment for all plants constructed under the Plan, foresaw the production of the following quantities of oil from German raw materials :—(8)

(In 1,000 of tons.)

	Benzol.	*Crude Oil.*	*Hydrogenation Process.*	*Fischer Tropsch Process.*	*Other.*	*Total.*
1936 ...	400	319(9)	622	8	1,159	2,508
1937 ...	400	319	857	117	1,161	2,854
1938 ...	400	319	953	537	1,214	3,423(10)
1939 ...	400	319	1,898	761	1,287	4,665(10)
1940 ...	400	319	2,707	786	1,342	5,554(10)

24. When the Plan was inaugurated the responsibility for the oil developments rested with an official named Keppler, who proved unequal to the task. He was replaced in 1937 by Krauch, a director of the *I.G. Farbenindustrie*, who had been responsible for the development of the original large-scale hydrogenation plant at Leuna. Krauch soon found that the rapid provision of synthetic oil production capacity was fraught with many difficulties. There was firstly a background of unwillingness on the part of industry to support wholeheartedly the synthetic production of oil, which has been likened, in the commercial sense, to growing bananas in greenhouses.(11) In addition, too much had been planned for execution at one time. There was not enough steel to meet all requirements, a shortage that was aggravated by the demand of *Autobahnen* bridges and by the naval programme. Skilled labour was also in short supply.

25. In consequence a revision of the Plan became inevitable and in June 1938 a conference was summoned by Goering which resulted in the New Four-Year Plan (the so-called Karinhall Plan). The revised objective, which was to be reached in 1942–43, was to provide for an annual production of 13,835,000 tons of oil products a year. It was agreed that to attain this target the steel supplies that were being allocated to the oil industry should be increased from 60,000 tons a month to 110,000 tons a month.(12) The resultant products were to be as follows :—

	Tons per annum.
Aviation spirit	3,000,000
Motor gasoline	4,000,000
Diesel oil	2,000,000
Fuel oil	4,000,000
Lubricants	835,000

26. The target date of 1942–43 coincided with all the other ancillary planning in connection with the main plan of oil production. It was estimated, for instance, that at least 18,000 white products tank cars would be required in the event of war and satisfactorily shown that the existing tank car construction programme would ensure that this number was available by April 1942.

(8) Krauch files (8). Note of 27.5.37. Also Hettlage. (U.S.S.B.S. Interview 12A.)

(9) Crude oil production in Germany in 1936 was 440,000 tons. The figures are evidently net production after refining losses.

(10) Plus 314,000, 484,500, 484,500 tons respectively of liquid products from brown coal carbonisation.

(11) The initiative for making proposals for new plants rested with private industry. To stimulate the erection of plants and to co-ordinate the work Krauch fostered the formation of a company known as the *Mineraloelbaugesellschaft*, which was jointly controlled by the *I.G. Farben, Union Rheinische Braunkohle, Gelsenberg, Stinnes* and *Ruhrchemie* interests. The capital was nominal, originally RM.300,000, later increased to RM.500,000. (Martin. U.S.S.B.S. Interview No. 47.)

(12) Under the revised Four-Year Plan of 1.1.39 (Karinhall Plan) the steel requirements of the oil programme were put at 4½ million tons in the four years ending December 1943. This figure disregards any steel already installed in plants partly finished in 1938 and is probably exclusive of steel required in the mining of coal and brown coal from which the oil is produced. This amount of steel would have been sufficient to build a battle fleet about three and a half times the size of the British Navy as of the 1st January, 1940. (U.S.S.B.S.)

THE LEUNA PLANT OF THE *I.G. FARBENINDUSTRIE.*

This photograph of the plant of the *Ammoniakwerke Merseburg G.m.b.H.* (a subsidiary of the *I.G. Farbenindustrie*) gives an indication of the size of this synthetic oil and chemicals plant. The photograph was taken in May 1945. A total of 85,079 bombs were released on this target, of which 8,643 hit the plant area.

In the period between the first heavy attack (12th May, 1944) and the last attack (4th April, 1945) production was reduced to less than 10 per cent. of normal. Although totally shut-down after many of the attacks, energetic repair measures resulted in a resumption of operations after remarkably short intervals. The length of the shut-down period varied with the severity of bombing and the rate of recovery decreased as each attack added its effect to that of the preceding ones.

The immediate cause of all shut-downs was the loss of utilities through damage to water, steam or electric power supplies.

Up to, but not including the last attack, the damage was estimated at RM.280,582,000.

PHOTOGRAPH BY U.S.S.B.S.

[PLATE I.

Likewise the target date for the completion of the reserve stock programme was to be April 1943 and the quantities in stock by that date were to be :—

	Tons.
Aviation spirit	1,500,000
Motor gasoline	1,500,000
Diesel oil	1,000,000
Fuel oil	1,800,000
Lubricants	100,000

27. These stocks were to be held as follows :—

	Aviation. Spirit.	*Motor Gasoline.*	*Diesel Oil.*	*Fuel Oil.*	*Lubricants.*
Government Storage ...	1,500,000	500,000	750,000	...	100,000
Commercial Storage ...	...	1,000,000	250,000	...	...
Naval Storage	...	...	...	1,800,000	...

28. To enable this level of reserves to be reached, Government storage, already planned at 2·5 million tons, was to be increased to 4 million tons. Industry was to be forced by law to increase its stocks of motor gasoline from 400,000 to 1 million tons; similarly it was to be compelled to maintain diesel oil stocks at 250,000 tons.

29. These, in outline, were the German oil plans for war as formulated in the late summer of 1938 and they provide an interesting background to what actually took place.

30. So far as the production plans were concerned, these were subject to a constant process of revision as it became apparent that even the revised (Karinhall) plan was impossible of attainment. There were continuous difficulties which hindered the realisation of these projects : Technical difficulties which are illustrated by the fact that planned Fischer Tropsch production from the nine plants erected under the Plan was to be 786,000 tons a year, whereas in fact this production never exceeded 474,000 tons a year; difficulties in raw material supply, which can be seen in the constant demands for an increased steel allocation, and above all difficulties of labour supply.

31. An outstanding fact in all this planning is that war was not expected to come before 1942–43. And this is the explanation for Germany's unreadiness for war in 1939 in so far as oil production capacity and oil stocks are concerned. This also accounts for the grandiose programme for the creation of oil storage capacity, which was never completed and never could be fully used. It also explains the dismay of the planners when they found that the large reservoir of productive capacity scheduled under the Four-Year Plan for construction in Eastern Germany and for completion in 1943 would have to be built under the stress of war and subject to all the delays occasioned by the competitive demands upon industry under these conditions.

SECTION II.

THE POSITION AT THE OUTBREAK OF WAR.

On the next page is a translation of a document that was drawn up by the *Wirtschaftsruestungsamt* in July 1939. It is a summary of the oil resources for war, and these resources are given as at the 1st August and the 1st October, 1939.

2. These figures confirm that, at the outbreak of War, Germany had no oil reserves in terms of a substantial stockpile that would safeguard military requirements in the event of a long war or the loss of productive resources. The total stock of 2,134,000 tons was but little more than the country would have normally carried for peacetime commercial purposes, and represented approximately four months' peacetime consumption. General Thomas, who was head of the Economic Office of the German Supreme Command, is reported to have expressed concern in July 1938 at the inadequacy of the stock position for a long war, and in March 1939 he referred to the "absolute folly of even contemplating going to war" while supplies were so short.(1) However, Hitler had planned on a short war. He disregarded whatever advice he may have been given on the inadequacy of the reserves set aside, and he evidently thought that these were sufficient to attain his purpose.

3. The low level of German oil stocks at the outbreak of war is shown by a comparison with those in Great Britain, which at that time amounted to 6,700,000 tons. While the requirements of the United Kingdom were of a different order to those of Germany, the disparity in these figures serves to reflect the sparsity of Germany's oil economy in relation to British requirements.(2)

4. Some of the figures in this German document call for special comment. Stocks of motor gasoline were equivalent to only two months' consumption; under mobilisation it was anticipated that, together with indigenous production, they would suffice for from four to five months. Stocks at filling stations are not included in the total. In a footnote to these statistics it is stated that stocks had substantially decreased since May of that year on account of the heavy seasonal demand.

5. The aviation fuel stocks, which were equivalent to about three months' consumption, are of particular interest. The quantity accumulated, which included 87,000 tons surreptitiously imported in 1938, represented the partial completion of a plan to remedy Germany's deficiency in aviation supplies. In 1938 the planners were talking in what at that time were big figures. It was intended by the end of 1941 that domestic production should be approaching a level of 200,000 tons a month. Meanwhile, purchases from abroad were to be steadily increased; imports of aviation fuel in 1941 were to be double the total world production of aviation fuel of 1938, and were to be obtainable at a cost in foreign exchange of half a milliard marks.(3) These projects clearly did not envisage a state of readiness before 1942 at the earliest.

6. There was sufficient Naval diesel oil to maintain active submarine warfare for at least two years. Stocks of Naval furnace oil are recorded as only equal to two months' consumption, although, in the light of later events, requirements were overestimated. No provision is made for consumption by the Mercantile Marine. However, in a footnote to the table it is explained that the 92,000 tons per month of oil normally required by the Merchant Marine "has not been considered as it is assumed that this would be ineffective in a war against England."

7. Stocks of tetra-ethyl-lead fluid were dangerously low, and the plan for safeguarding supplies of this vital commodity by the erection of an additional plant to the one at Gapel-Doeberitz had not yet reached fruition. Consumption requirements were over 300 tons per month, and the situation was extremely precarious until France was overrun and some production could be obtained from Paimbœuf and until the new plant at Frose, near Magdeburg, came into operation.(4)

(1) I.C.F./284, 1.6.39.

(2) The paucity of Germany's oil economy is shown by the fact that in 1938 her total production and imports amounted to about 7 million tons. In the same year the United States produced 164 million tons of crude oil and Russia 29 million tons.

(3) A.D.I. (K) Report No. 391A/1945.

(4) Interrogation of Ahrens, Head of Supply Section, Mineral Oil Department, R.L.M. Further information on tetra-ethyl-lead is given in Appendices 14 and 15.

Oil Position on Mobilization, July **1939**.

Products.	War-time Needs.			Production in Germany except West fortification zone.		Stocks available in Germany except West fortification zone.				Probable duration of full covering of the war-time demands by domestic production and available stocks.	
						On 1st August, 1939.		On 1st October, 1939.			
	Services.	Commerce.	Total.	Tons Per Month.	Percentage of Mobilization Demand.	Total.	Of this National Reserve Stock.	Total.	Of this National Reserve Stocks.	On 1st August, 1939.	On 1st October, 1939.
	Tons Per Month.	Tons Per Month.	Tons Per Month.			Tons.	Tons.	Tons.	Tons.	Months.	Months.
Aviation fuel	152,500	...	152,500	46,200	30	480,000	385,000	450,000	355,000	4·8	4·5
Tetra-ethyl-lead for aviation and—	430	...	430	158([5])	32	860	615	800	550	2·6	2·7
other motor fuels	43	...	43	193([6])	39						
	473	...	473								
Other motor fuels	115,000	70,000	185,000	117,000	63	350,000	...	300,000	...	5·2	4·4
Aviation diesel fuel	3,300	...	3,300	...	...	48,000	40,000	48,000	40,000	14·5	14·5
Diesel fuel, including tractor diesel fuel ...	39,000	103,000	142,000	45,000	32	308,000	93,000	308,000	93,000	3·2	3·2
Fuel oil for economy	...	25,000	25,000	6,000	24	50,000	...	50,000	...	2·6	2·6
Fuel oil for the Navy	137,000	...	137,000	32,000	24	260,000	...	260,000			
Ship diesel fuel for the Navy	20,000	...	20,000	6,300	32	500,000	...	500,000			
	157,000	...	157,000	38,300	24	760,000	...	760,000	...	6·4	6·4
Aviation motor oil	9,500	...	9,500	2,400	25	50,000	33,000	50,000	33,000	7·0	7·0
Other motor oils	8,500	7,000	15,500	8,800	57	88,000	30,000	88,000	30,000	13·0	13·0
Total (excluding T.E.L.)	484,000	205,000	689,800	263,700	...	2,134,000	581,000	2,084,000	551,000	...	...

([5]) = 1st August. ([6]) = 1st October.

8. The overall oil stock position was, in fact, so low that there was considerable apprehension amongst Hitler's professional advisers at the lack of provision for substantial reserves of liquid fuels. These apprehensions were, however, soon dispelled, firstly by the unexpected speed and economy of the first two campaigns, and, secondly, by the capture of large quantities of oil in Western Europe.

Productive Resources.

9. In peacetime Germany met her oil requirements of some 670,000 tons per month very largely by imports which amounted in 1938 to 5,250,000 tons or 437,000 tons a month, the remainder coming from crude oil production, synthetic oil production and the production of coal tar oils, alcohol and benzol.(7)

10. On the outbreak of war all imports except those from Roumania and Russia were stopped by the blockade, and German resources were limited to an oil income of approximately 455,000 tons per month made up as follows :—

Crude oil from German and Austrian fields ...	80,000
Synthetic oil	120,000
Tar oils, alcohol and benzol	75,000
Imports from Roumania—averaging	120,000
Imports from Russia—averaging	60,000

Crude Oil.

11. On the outbreak of war German crude oil production was running at a rate of some 700,000 tons a year. The following table shows the steadily increasing trend of production since 1930 :—

	Tons.		*Tons.*
1930	174,000	1935	429,000
1931	229,000	1936	444,000
1932	230,000	1937	453,000
1933	238,000	1938	552,000
1934	314,000	1939	700,000

12. Bent upon the total mobilisation of all domestic resources, the National Socialists, from the moment of their seizure of power, did everything possible to expand crude oil production. The Mining Laws were immediately altered to permit large-scale exploration and a comprehensive geophysical survey was set on foot. A large programme was worked out for increased drilling and exploration with public funds and, in addition, industry was compelled to spend large sums of its own for the same purpose. As a result of these measures total drilling increased from 62,000 metres in 1932 to 220,000 metres in 1938, the exploration drilling in virgin areas increasing from 13,000 to 100,000 metres. The effect was a 150 per cent. growth in crude oil output. But whereas almost all the output in 1933 had come from the older producing districts, the figure for 1936 included over 130,000 tons extracted from newly-discovered areas.

13. From 1933 until the outbreak of war about a dozen new fields had been discovered in association with salt dome formations in North Germany. Some of these were extensions of older pools, and a number of others were highly productive in the initial stages of their development, thus giving promise of a substantial underground reserve ready for producing when war demands arose. The geological survey of Germany had located 68 favourable structures in the oil district of Hannover alone. At the outbreak of war there were some 170 drilling rigs working in the *Altreich*.

14. While the immediate prospects of a significant increase in production from the German fields were good, there were equally good prospects, though slightly more distant, of a considerable increase in the production from the Austrian fields in the Zistersdorf area. In 1937 Austrian production was only 30,000 tons, but much exploratory work had been done, mainly by foreign companies, and prospects were known to be extremely promising. Production, in fact, reached a figure of over 20,000 tons per month in the last quarter of 1939 and thereafter continued to increase.

(7) See Table 1 on page 159.

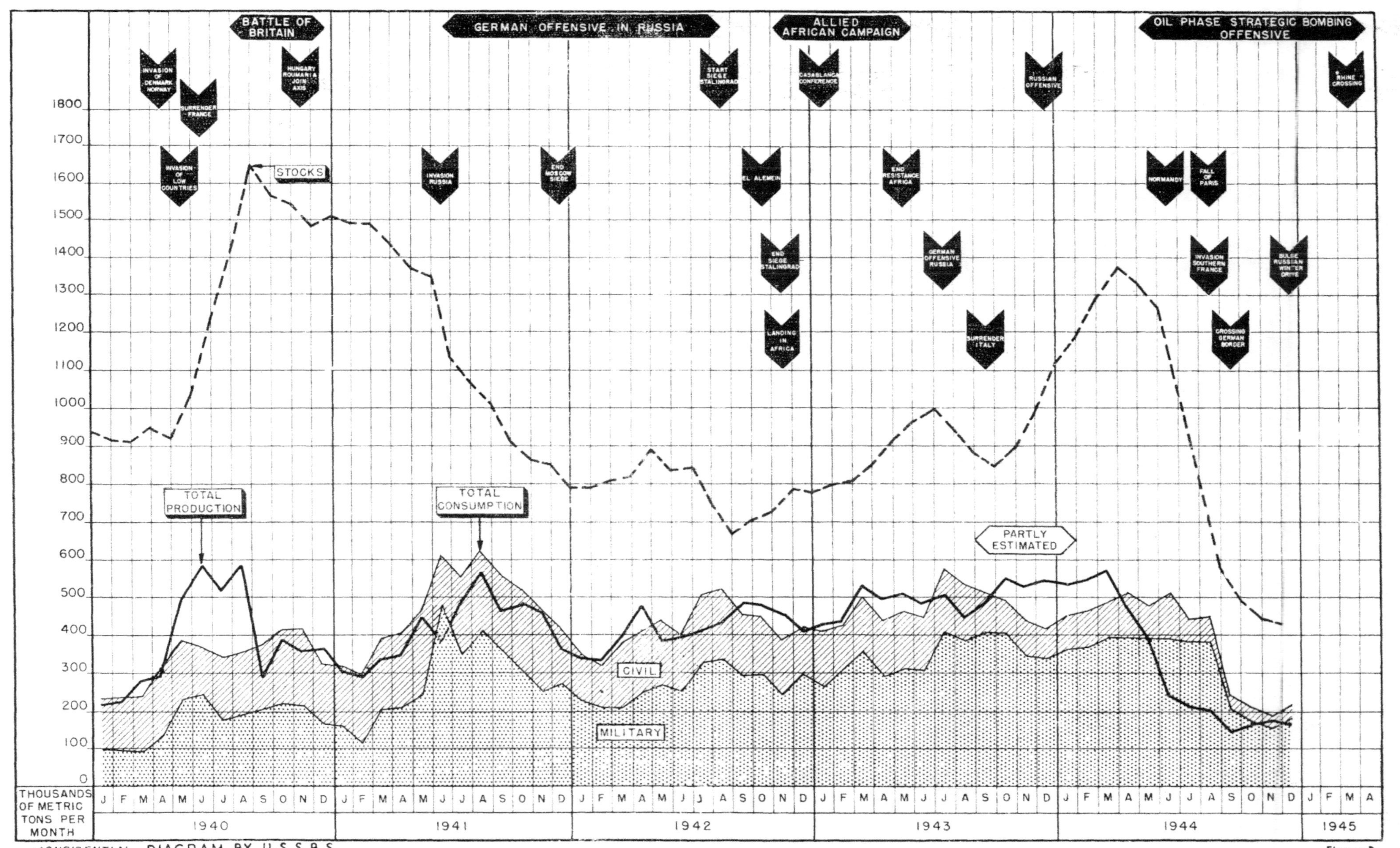
TOTAL AVIATION GASOLINE, MOTOR GASOLINE AND DIESEL OIL
GERMAN STOCKS — PRODUCTION — CONSUMPTION
PRODUCTION INCLUDES IMPORTS & CAPTURED STOCKS — CIVIL CONSUMPTION INCLUDES EXPORTS
BATTLE OF BRITAIN
GERMAN OFFENSIVE IN RUSSIA
ALLIED AFRICAN CAMPAIGN
OIL PHASE STRATEGIC BOMBING OFFENSIVE
INVASION OF DENMARK NORWAY
SURRENDER FRANCE
INVASION OF LOW COUNTRIES
HUNGARY ROUMANIA JOIN AXIS
INVASION RUSSIA
END MOSCOW SIEGE
START SIEGE STALINGRAD
EL ALEMEIN
END SIEGE STALINGRAD
LANDING IN AFRICA
CASABLANCA CONFERENCE
END RESISTANCE AFRICA
GERMAN OFFENSIVE RUSSIA
SURRENDER ITALY
RUSSIAN OFFENSIVE
NORMANDY
FALL OF PARIS
INVASION SOUTHERN FRANCE
CROSSING GERMAN BORDER
BULGE RUSSIAN WINTER DRIVE
RHINE CROSSING
STOCKS
TOTAL PRODUCTION
TOTAL CONSUMPTION
PARTLY ESTIMATED
CIVIL
MILITARY
1800
1700
1600
1500
1400
1300
1200
1100
1000
900
800
700
600
500
400
300
200
100
0
THOUSANDS OF METRIC TONS PER MONTH
J F M A M J J A S O N D
1940
1941
1942
1943
1944
J F M A
1945
CONFIDENTIAL DIAGRAM BY U.S.S.B.S.
Figure 1

Refining Capacity.

15. There was ample refining capacity for both the German and Austrian crude oil and especially as the lack of imports released those refineries which were normally engaged on processing imported crude. The total throughput capacity available was, in fact, in excess of 3,000,000 tons a year, which provided an ample margin to cover anticipated increases in domestic production. Later, the occupation of Poland, the invasion of the West, and the entry of Italy into the War were each in turn to add more refineries to those already available to Germany.

16. In one aspect, however, this large amount of processing capacity did not entirely suit the production programme. German and Austrian crudes yield by distillation only a small percentage of gasoline. This yield could only be increased by cracking, and the total cracking capacity provided by these refineries was small. This deficiency was later to prove a serious handicap.

Synthetic Oil.

17. The original discoveries in the synthesis of oil from coal were made shortly before 1914, and in 1916, under the stress of war, a small experimental plant was built at Mannheim. In 1925 motor spirit was successfully produced from brown coal tar at Ludwigshafen-Oppau. By this time technical development was far enough advanced for a start to be made on the construction of a plant to begin production on a commercial scale, and in 1927, at Leuna, large-scale production of synthetic oil by the hydrogenation of coal started. Between 1925 and 1936 work had proceeded on a second process, the synthesis of liquid fuel from gas by the Fischer–Tropsch method, and by the later date commercial production was under way.

18. By the Autumn of 1939 synthetic production had reached a monthly rate of 120,000 tons. The output was coming from seven hydrogenation plants and seven Fischer–Tropsch plants.(⁸) Although this production was only meeting a small proportion of requirements the line on the graph of synthetic production was firmly set upwards.

19. The future increase in production was mostly to come from the hydrogenation process, there being only two additional Fischer–Tropsch plants in course of erection at that time. Altogether eleven new hydrogenation plants were planned, although several of these were a special type designed for only a small output. The largest projects were the enormous plant at Bruex in Czechoslovakia and three plants in Upper Silesia, all strategically situated at long range from air attack from the West. The Bruex plant was planned for an output of 600,000 tons a year of liquid fuel, a rate which was achieved in one month only, in May 1944; it was primarily designed for the production of high-quality diesel fuel for submarines.

20. The plants in Upper Silesia consisted of two *I.G. Farbenindustrie* projects, Heydebreck (Blechhammer South) and Auschwitz (Oswiecim), and one plant sponsored by the *Herman Goering Werke* at Blechhammer. The Heydebreck plant was of a comparable size to Bruex and the Blechhammer plant was almost as large. Auschwitz was a methanol plant designed for the production of synthetic rubber, high octane aviation fuel and other synthetic products. There was much delay in the completion of these plants and none of them had achieved more than a part of their designed output by the time they became targets for bombing attacks.

21. The large increase in the production of synthetic oil, which attained a peak of 350,000 tons in the month of April 1944, was principally due to Bruex and to the extensions made to a number of pre-war plants.

Tar Oils, Alcohol and Benzol.

22. From coke ovens, gas works, low temperature carbonisation and tar distillation plants, and from alcohol distilleries, Germany was obtaining nearly a millon tons of fuel or fuel components in the year before war broke out. In general these supplies were not susceptible of any great increase. Wartime requirements brought a greatly increased demand for alcohol for industrial

(⁸) See Table 3 on page 160.

purposes, especially for explosives, and even before the War the growth of the movement towards increased home food production had already tended to bring into disfavour the conversion of agricultural products into alcohol. No extensive use of alcohol for blending into motor fuel was attempted until the culminating stages of the War, at which time the effects of depriving the chemical industry and food production had to be ignored on account of the imperative need for motor fuel, whatever its source or quality.

23. The rate of production of benzol from coke ovens and domestic gas plants in 1939 was approximately 550,000 tons, of which slightly over 400,000 tons was available, after other requirements had been met, for use as an automotive fuel. Since benzol production is at the proportion of a little over 1 per cent. of coke output, there was little scope for bringing about any sharp increase in production. Until coke ovens were attacked, Germany could therefore count on this quantity, but on no great increase on it.

24. In contrast to these two auxiliary fuels, the fuels that could be obtained from the low temperature carbonisation of lignite could be produced in growing volume. Although the greater part of the tars thus produced were used as feed-stocks for hydrogenation plants, the output surplus to this requirement formed a valuable contribution to supplies of both furnace oil and diesel oil.

25. Detailed figures of production are lacking, but it is probable that at the outbreak of war Germany counted on at least half a million tons a year of finished fuels (exclusive of benzol) derived from tar, which by 1944 had been nearly doubled by plant construction and expansion.

Consumption Requirements.

26. In the table given on page 7 the *Wirtschaftsruestungsamt* estimated that the consumption requirements[9] of the Armed Forces and the national economy would be as follows :—

	Tons per month.		
	Military.	*Civil Economy.*	
Aviation spirit	152,500	...	
Other motor fuels	115,000	70,000	
Aviation diesel oil	3,300	...	
Diesel oil	39,000	103,000	
Industrial fuel oil	...	25,000	
Naval fuel oil	137,000	...	
Marine diesel	20,000	...	
Aviation lubricants	9,500	...	
Other lubricants	8,500	7,000	
	484,800	205,000	
			689,800

27. With domestic production amounting to about 275,000 tons, and with the continuation of imports from Russia and Roumania of about 180,000 tons, this estimate indicated that the outbreak of war would involve a monthly withdrawal of stocks of upwards of 200,000 tons. As events turned out, for the first year of the war Germany, operated her war and armament machine on almost exactly the same quantity of oil as had sufficed for the nominally peaceful year of 1938.

[9] It is of interest to compare these estimated war-time requirements with the actual peace-time consumption of 1938 :—

	Tons per Month.		
	Aviation and Motor Gasoline.	*Diesel Oil (all grades).*	*Fuel Oil.*
1938	286,000	142,000	92,000
War-time	337,000	165,000	162,000

SECTION III.

THE BLOCKADE.

On the outbreak of war it was immediately clear that Germany would be compelled to supplement her internal resources of liquid fuels if she were to prosecute the war on an active scale. Her peace-time oil economy had been dependent to the extent of over 70 per cent. upon imported supplies amounting to nearly 5 million tons annually, of which nearly 1 million tons came from Roumania and the remainder, apart from small quantities from Russia and as re-exports from other European countries, from overseas.

2. The successful imposition of an oil blockade therefore depended upon the cutting of overseas supply routes, the prevention of re-exports from neutral countries and the maximum possible reduction of exports to Germany from producing countries in Europe and principally from Roumania.(¹)

3. The immediate imposition of a naval blockade and the application of the "navicert" system prevented any direct shipments of oil to Germany. No reliable figures are available of the leakages(²) through the blockade, but it is almost certain that since May 1940 only infinitesimal quantities of oil can have reached Germany from overseas. Until that date it was estimated that the maximum quantity of contraband oil obtained by Germany was 290,000 tons. It is certain that the policy of restricting supplies to neutrals, besides being a bargaining counter of great value in our dealings with them, involved the Germans in the export of substantial quantities of oil which they could ill afford.

Restrictive Allocations to Neutrals.

4. Once the physical machinery for the blockade had been established there remained the more difficult problem of preventing re-exports of oil products to Germany from neutral countries. These re-exports might be derived from current imports or from reserve stocks amassed in neutral countries to meet the impending emergency.

5. It was also recognised that the existence of unusually large stocks in a neutral country might provide additional bait for German aggression. Several countries, and notably Holland and Belgium, had supplies on hand that were out of all proportion to a reserve that might be needed to cover any delays in receiving regular supplies.

6. As an inducement to the voluntary reduction of these stocks His Majesty's Government had to promise that essential supplies would be allowed through the blockade to these countries. In the first months of the war it was not, however, found possible to restrict these imports drastically, one reason being that it was decided, on politico-military grounds, that no such sanction should be applied against Italy and in consequence other neutrals had at first to be accorded equal treatment.

7. Moreover, the continued provision of oil to these countries provided a bargaining counter of some importance. There was, however, a real danger that oil supplied for use in a neutral country might be re-exported, and that the blockade might thereby be circumvented. Such breaches of the blockade were thwarted partly by the action of His Majesty's representatives in neutral countries, who could apply the economic sanction of "black-listing," but mainly by a policy of limiting imports into neutral countries to the bare quantities which were essential to maintain a minimum of industry and transport for the continued existence of the country. These quantities were in general fixed by negotiation in the countries concerned, although when leakages were suspected it was possible to reduce supplies and so apply an economic sanction to ensure future good behaviour.

(¹) A detailed account of the Blockade is given in "The Official History of Economic Warfare, 1939–1945."

(²) A leakage, which for a long time proved a thorn in the side of the Admiralty and the Ministry of Economic Warfare, was the periodical movement of two small Italian tankers, the "Albaro" and the "Celeno," which carried oil from Constanza to Piræus. It was not until 1943 that these movements were stopped. However, the perils of the voyage caused both insobriety amongst the ships' officers and disaffection amongst the crews and the sailings were consequently infrequent.

8. An additional control was provided by the agreement of the British oil companies trading in these countries to conduct their operations so as to ensure as far as possible that the products supplied were consumed without undue delay in neutral territory. Early in 1940 an embargo was placed on American oil imports into Holland, an action which caused a certain amount of criticism in the States. As similar action had not at that time been taken with any other neutral a case could not be made out for the continuance of this embargo. However, willing co-operation was given by a number of American oil companies,(3) which agreed to restrict their exports to the 1936–38 average to ensure, in so far as they were able, that none of the oil supplied would reach Germany.

9. Some indication of the delicate nature of the discussions with the neutrals in the restriction of their oil imports is provided by the following summary of the negotiations with the principal countries concerned. An account of the negotiations with Italy is given later in this report.

Holland.

10. The negotiations with Holland were extremely complicated. In September 1939 a war trade agreement was proposed and this was to include a guarantee against the re-export of vital commodities. As His Majesty's Government was not at that time in a position to enforce the rationing of oil in Holland the Dutch were not averse to delaying matters in order to build up stocks and force us into various concessions. Until the end of that year they did everything possible to increase the amount of oil in storage. Stocks at the beginning of November were estimated at 700,000 tons and it was estimated that they would reach 800,000 tons by January.

11. There was also a "counter-blockade." The Pernis refinery, near Rotterdam, was under contract to supply the British Air Ministry with at least 50,000 tons of 100 octane aviation fuel. This was one of the principal British sources of supply at the time and of considerable importance for the maintenance of full operations by the Royal Air Force. In this and other cases (for example, cocoa-butter) we sent the raw material to Holland for processing and the Dutch could either delay returning the finished product, or argue that if they processed for us they must, to preserve their neutral position, make the same terms with Germany. Ultimately a not very satisfactory compromise was reached on both points, but before it could come into effect Holland was invaded.

Spain.

12. Negotiations with Spain were particularly difficult. With her own tankers and with the assistance of chartered neutral tonnage, Spain began accumulating a substantial stock of oil in the early stages of the war which could well have become subject to German acquisitiveness. The protracted discussions upon the limitation of these stocks were made more difficult by officially inspired Spanish propaganda, which laid the blame for the plight of the country on Great Britain's restrictions upon oil imports. The Germans also caused reports to be circulated that, in dealing with the domestic question of restrictions, the British were meddling with the internal affairs of Spain.

13. In September 1940 an agreement was signed which limited importations to the extent that stocks would not exceed 160,000 tons, which were equivalent to about two and a half months' consumption. This agreement, which was periodically reviewed, successfully prevented the re-export of oil to Germany. In later years the threat of reducing these supplies was a useful factor in forcing Spain to curb her exports to Germany of wolfram and other valuable war supplies.

Portugal.

14. Portugal likewise received all oil supplies from Allied sources in such quantities as her basic requirements justified and any re-export to the enemy was successfully prevented.

(3) Agreements were signed by the following companies, or by their subsidiaries:—

Texas Oil Co.
Gulf Oil Co.
Socony-Vacuum Oil Co.
Standard Oil Co. of New Jersey
Tidewater Associated Oil Co.

THE SYNTHETIC OIL PLANT AT BRUEX, CZECHOSLOVAKIA.

This photograph was taken during the attack on 12th May, 1944, by the United States 8th Air Force. The construction of this very large plant for the hydrogenation of low-temperature tar was begun shortly before the outbreak of war. Sponsored by the Four-Year Plan, it was originally intended primarily for the production of high-grade diesel oil for submarines, although in the course of the War a full range of products, including aviation fuel, were manufactured. The liquid fuel production capacity of the plant was to have been 600,000 tons a year, equal to the Leuna and Poelitz plants. Production was begun in 1942, but full output was attained in one month only, in May 1944. The plant was put totally out of commission by the Allied bombing attacks.

Sweden.

15. A different problem was presented in the case of Sweden. Both ourselves and the Germans were anxious to safeguard those economic or political advantages which the continuance of good relations with Sweden could offer. In order to obtain certain materials of major importance, notably iron ore and ball bearings, the Germans were ready to export oil products and particularly lubricating oils. They also permitted the through shipment by rail of oil from Roumania and Hungary. On the other hand, the Allies, equally anxious to maintain a foothold, permitted restricted quantities of liquid fuels, including aviation spirit, to pass through the blockade.

16. After the occupation of Norway the Germans succeeded for a time in using Swedish rail transit facilities for the movement of oil to the forces in Northern Norway. After pressure had been put upon the Swedish Government it was agreed that the traffic was classifiable as war material and these movements were stopped.

Switzerland.

17. Switzerland was completely surrounded by enemy Powers and her provision with oil became an Axis liability. The country was therefore a small but continuous drain upon the enemy's resources. As in the case of Sweden the Allies were successful, in the later stages of the war, in getting the Swiss Government to classify oil as a war material and this put a stop to the movement through Switzerland of Wehrmacht oil supply trains to Italy.

Turkey.

18. In peace-time Turkey had been largely dependent upon imports of oil from Roumania and both Allied and German inclination was that Roumania should continue to supply most of Turkey's requirements. The Allies were glad that there should be this drain on Roumanian supplies which would otherwise have gone to Germany, and the Germans countenanced the traffic on the grounds that it financially assisted the import into the Axis sphere of Turkish cotton, copper and other valuable war materials.

SECTION IV.

ROUMANIA.

The strategic importance of Roumania's oil had been shown in the 1914–18 War when the sabotage of the principal producing wells had far-reaching effects.(1) Consequently none of the belligerents under-rated the importance of the output of the Ploesti refineries, and some time before the War began in 1939 Bucharest had already become the centre of a struggle in which the contestants endeavoured to make skilful intrigue a substitute for actual weapons. While the Germans were starting their economic encroachment into the Roumanian oil industry,(2) His Majesty's Government was devising ways and means for the denial of the industry to the enemy upon the outbreak of war.

2. The restriction of the quantity of oil which Germany could obtain by importation from Roumania proved a task of great difficulty. The problem of imposing such restrictions was complicated by the following factors :—

(1) Oil exports to Germany had been a long-standing feature of Roumanian trade.
(2) German political pressure was more easily exerted and particularly on account of the delivery of armaments which Roumania needed in return for oil exports.
(3) Inflation in Roumania made it increasingly more difficult to conduct normal commercial trading outside the European Axis zone which was becoming more and more " closed."
(4) The closing of the Mediterranean in June 1940 prevented the normal carriage of oil by sea from Constanza to markets of Allied selection.

3. The policy of the War Cabinet, laid down in the early months of the War, for denying Roumanian oil supplies to Germany, may be summed up as follows :—

(*a*) To buy as much Roumanian oil as possible.
(*b*) To aim at preventing Germany from obtaining more than 1 million tons a year of Roumanian oil.
(*c*) To purchase or charter means of oil transport such as barges, tugs and rail tank cars.

4. The achievement of these ends was to be by means of normal commercial activity with diplomatic assistance and co-ordination, by clandestine commercial activity and by sabotage.

Pre-emption.

5. As soon as the policy of purchase was decided upon in September 1939, the Mines Department (Petroleum Division) concerted preliminary arrangements with the oil companies with British capital in Roumania. The companies were the *Astra Romana*, a subsidiary of the Shell Group, *Unirea*, a subsidiary of Phœnix Oil and Transport Co., Ltd., *Dacia Romana*, and *Steaua Romana*, of which 49 per cent. of the capital was held by British and French interests.

6. Generally the executive instrument of the policy for purchase was the Asiatic Petroleum Co., Ltd. (Shell Group), acting through *Astra Romana*. Negotiations were also taken up with the Petroleum Board in the United Kingdom for the disposal of Roumanian oil shipped to the United Kingdom. Close contact was also maintained with the French authorities in order that similar action for the purchase of oil could be taken by French oil interests in Roumania, and, eventually, complete agreement was reached, including the sharing of costs.

7. A Committee(3) under the War Cabinet was set up, under the Chairmanship of Lord Hankey, to devise and co-ordinate means to frustrate German plans and promote the policy of the War Cabinet. The first meeting was held on the 17th October, 1939. As the purchase of oil was hampered under the provisional

(1) " The action of a few British officers represented without doubt one of the most significant achievements of the World War, one which in its consequence was more effective than most of the great battles fought throughout the whole period." " *Das Erdoel im Weltkrieg,*" Friedensburg.
(2) Details of the activities of *Kontinentale Oel A.G.* are given in Appendix 4.
(3) Committee on Preventing Oil from reaching Germany,

arrangements by financial and political difficulties, the Committee arranged in November that the Oil Adviser([4]) to the Ministry of Economic Warfare and a representative([5]) of the Treasury should proceed on a temporary mission to Roumania with the following instructions:—

(a) To restrict to the lowest possible figure the quantity of oil which Germany could obtain from Roumania and, in pursuance of this object, to procure the purchase, for the account of this country, of the maximum amount of Roumanian oil on the most reasonable terms.

(b) To obtain an immediate reply to proposals already made, which were intended to stimulate the export of oil to this country, and to recommend how these proposals should be modified to obtain satisfactory results.

8. The Committee further arranged that this mission should be followed in December 1939 by an official,([6]) to be attached permanently to the Legation in Bucharest and to represent the Ministry of Economic Warfare and the Mines Department, in order to arrange and co-ordinate plans for the future purchase of oil.

9. Great difficulties were encountered from the outset as the Roumanian Government was subject to the strongest pressure from the Germans, not only to increase supplies of oil to Germany, but also to put every possible difficulty in the way of the Allied plans. The Roumanians proposed to impose a system of quotas and, in consequence, to withhold approval of long-term contracts to be entered into by the oil companies. The Roumanian authorities also withheld export licences for contracts already entered into. The deliberations of Lord Hankey's Committee([7]) during the early months of 1940 were concerned with such obstacles set up by the Roumanian authorities in the way of the British and French policy for the export of oil. In particular they decided—

(a) That the quota system should be strongly resisted as it might guarantee to Germany shipments in excess of 1 million tons a year (an amount of 1,300,000 tons was contemplated).

(b) Attempts by the Roumanian Government to force companies under British and French capital control to sell oil to quarters objected to by the companies were to be strongly resisted. Freedom of export at commercial prices was to be the objective.

(c) The normal Roumanian markets in the Balkans and Mediterranean should be maintained even if oil was sold at a loss.

10. As a result of the unsatisfactory attitude of the Roumanian Government on the oil question, a temporary embargo was placed towards the end of January 1940 on shipments of goods from the United Kingdom subject to export licences, on the ground that trade facilities could only continue on condition that, and so long as, Roumania did not impose restrictions and unreasonable conditions on the export of oil by the companies to the desired destinations. About a month later, when the Roumanians showed a more accommodating attitude, the embargo was lifted.

11. Subsequently, as Roumania had already submitted to German pressure to the extent of granting a quota of 130,000 tons a month (representing a rate 50 per cent. higher than the amounts to which it was hoped to restrict exports to Germany), the Committee directed their efforts to securing agreement from the Roumanians that this arrangement with Germany would be implemented in ways least unsatisfactory to the Allied cause. In particular the quota should not be increased or anticipated in any way, there should be no discrimination in Germany's favour in regard to prices and no increase in storage or transport facilities should be made available to the Germans.

12. Despite the difficulties under which the policy of pre-emptive purchase was carried out, the full co-operation of the companies ensured that 680,000 tons of oil were delivered to the United Kingdom and 325,000 tons to France and French territories in the Mediterranean up to the time when the closing of the Mediterranean on the entry of Italy into the war rendered it impossible to continue export of oil to these destinations.([8]) On the other hand, the average

([4]) The Hon. M. R. Bridgeman, C.B.E.

([5]) Mr. E. L. Hall-Patch, C.M.G.

([6]) Mr. E. A. Berthoud, C.M.G.

([7]) See page 87.

([8]) This policy was assisted by the "Levant Plan," for details of which see Appendix 5.

quantity of oil secured by the Germans during the first nine months of the war did not exceed 80,000 tons a month, or about the monthly rate contemplated by the War Cabinet.

13. It cannot be stated with certainty how far, in fact, the policy of pre-emption reduced the quantities of oil drawn from Roumania by Germany, for the reason that in the early months of the war it was transport rather than the availability of the oil which formed the bottleneck. The denial to the enemy of transport facilities by the Goeland Company's operations, which were part of the War Cabinet plans, contributed towards this bottleneck.(9) In any case, the competition thus set up had the effect of making Germany pay heavily for such oil as she secured, and reduced to that extent her power to buy other raw materials. The policy further benefited Roumanian economy in a way to encourage co-operation with the Allies. It was one of the few ways in which a stand could be made against German pressure on Roumania and there is little doubt that it had the effect of causing the Roumanians to procrastinate in the face of German demands for various facilities. It is indeed significant that, in the autumn of 1940, when the Germans had secured fuller control of Roumanian production and transport, the export of oil to Germany rose to 200,000 tons a month.

Clandestine Commercial Activity.

14. The carrying out of the third part of War Cabinet policy, to purchase or charter means of oil transport such as barges, tugs and rail tank cars, was first put into practice in the spring of 1939.

15. The enforcement of a strict blockade of Germany's ocean-born imports at the outset of the war left open only the Baltic, and the Black Sea and Danube waterways for her imports. The oil, food and mineral resources of the Black Sea and Danube areas made the Danube route of vital importance. It was therefore to be expected that the Germans would need and actively seek to acquire control of Danube shipping, and that, as there had been no extensive building of Danube vessels in preparation for the war and the river would be needed to carry greatly increased quantities, any diminution in the number of vessels available would directly reduce the amount of goods carried to Central Europe. Consequently a policy of reducing the numbers of vessels was clearly a valuable measure of economic warfare.

16. In the spring of 1939 the Anglo-Danubian Transport Company appealed to His Majesty's Government for financial assistance on the ground that otherwise they would be compelled to sell their fleet of tugs and tankers to German interests. In August, His Majesty's Government was obliged to exercise complete control over this company's vessels to prevent their sale to Germany. In November 1939 a limited liability company was formed in London to carry out Danube policy. This company, the Goeland Transport and Trading Company, Limited,(10) is totally Government-owned and was formed with a view to furthering the Government's plans for denying the use of craft on the Danube to the enemy by means of purchase, charter, laying up or hindering river traffic. The organisation was also a cloak for more clandestine activities. The company was registered, with offices in London, in February 1940. Its entire share capital of £750,000 was subscribed by the Ministry of War Transport and its directors were originally representatives of the Ministry of War Transport, the Foreign Office, the Treasury and the Ministry of Economic Warfare.

17. The main activity of the company was to charter or purchase vessels in order to reduce the tonnage of shipping on the Danube, not only to prevent them trading with the enemy but to manipulate their use in trade with neutral countries. It was impossible to lay up too large a number of vessels for fear that the Roumanian Government would requisition them. A show had to be made of carrying some cargoes, which was done by taking uneconomic transports such as firewood, coal and stone. In the course of doing this it was possible to keep agents and observers in the Iron Gates region, the most vulnerable stretch of the river.

(9) See Appendix 6. The freezing of the Danube for an unusually long period early in 1940 also contributed towards this bottleneck.

(10) A summary of the reports of the Goeland Company during its first two years of operation is given in Appendix 6. The full reports of the company are held by the Ministry of Transport.

GERMAN PRODUCTION OF OIL PRODUCTS
BY PROCESS

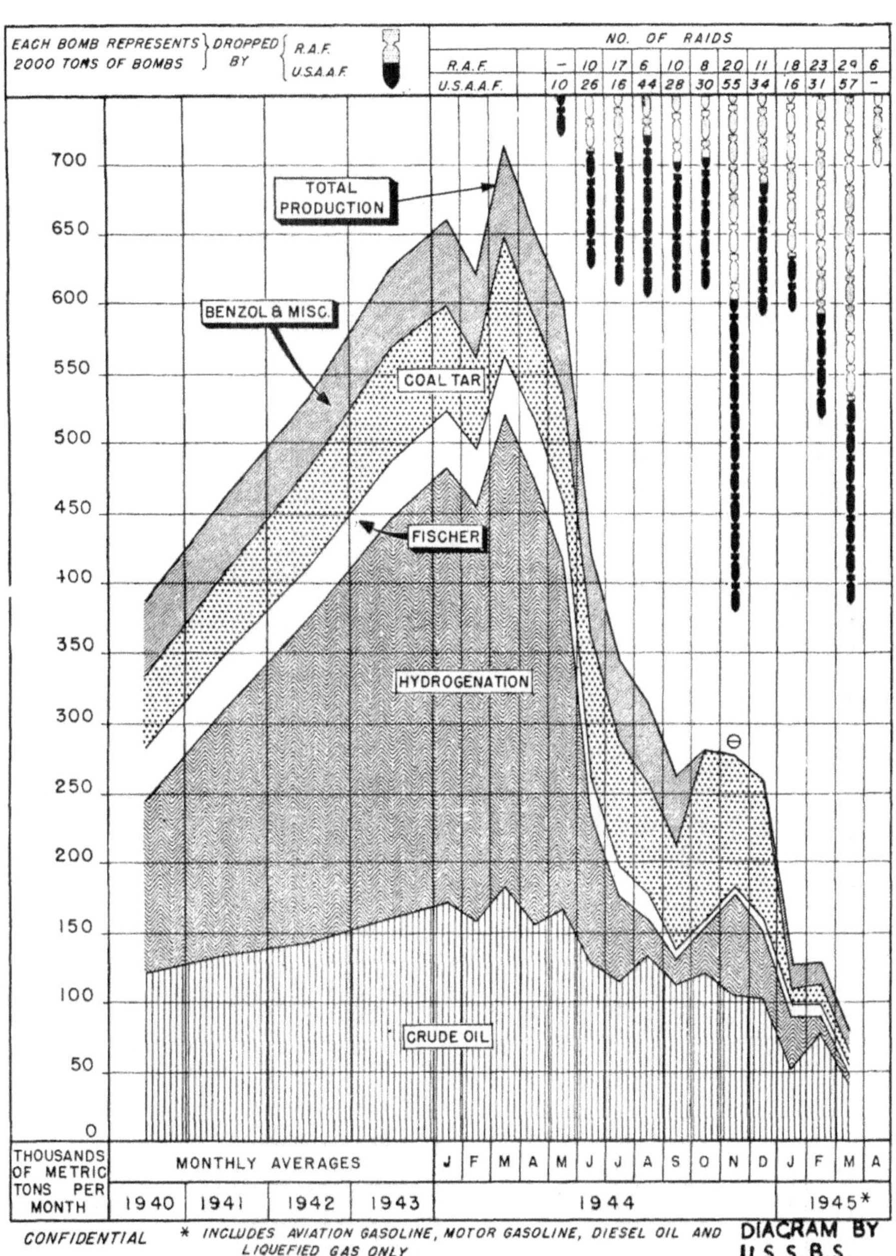

CONFIDENTIAL

* INCLUDES AVIATION GASOLINE, MOTOR GASOLINE, DIESEL OIL AND LIQUEFIED GAS ONLY

⊖ INCLUDED WITH COAL TAR, OCT.–DEC., 1944

DIAGRAM BY U.S.S.B.S.

Figure 2

18. Including a Yugoslav shipping company, which came under the Goeland Company's partial control, the company in December 1940 controlled 270 barges, 50 tank barges, 31 motor tank barges and 51 tugs. A number of these vessels were subsequently evacuated to Turkey and were thus completely denied to the enemy.

19. Among the other activities of the company was a scheme to divert oil shipments to Switzerland from the sea to the river route, the withdrawing of lower Danube River pilots from German service, which was done in the name of the Goeland Company by Naval personnel acting under the Naval attaché, and seducing Iron Gates' pilots from their employment thereby causing a shortage of the highly skilled pilots who were qualified to navigate this particularly difficult 15-mile stretch of the river. The services of the company were also freely used for Naval and other service activities, particularly in the plan for active hostilities and sabotage on the Danube, the enemy's discovery of which caused great embarrassment and difficulty to the Company in its other less violent but equally clandestine activities.

Sabotage.

20. Simultaneously with other plans to deny Roumanian oil to Germany it was decided in the summer of 1939 that in the event of war, which by then appeared inevitable, the oil resources of Roumania must not be allowed to fall intact into the hands of the enemy.

21. The proposal that the oil-fields should be "scorched" was first put forward by the Admiralty and a representative(11) was sent to Roumania in June 1939 to discuss with the Roumanian General Staff the possibility of British help in the work of destruction should the Germans invade. At the same time the War Office sent an intelligence officer(12) to reconnoitre the vulnerable points in the Roumanian communication system.

22. By mid-August, when the project had been taken over by the War Office, it had been agreed that when action became necessary a Field Company of the Royal Engineers should be flown in to Ploesti. This company meanwhile started on special training in the Western Desert, a small reconnaissance party being meanwhile sent to Roumania to study the objectives. This party prepared detailed plans, in co-operation with the French, for the destruction of the seven largest Roumanian refineries and the three most productive oilfields. The French were to deal with the remaining fields and with six other refineries.

23. The political situation took an unfavourable turn after the assassination of Calinescu on the 21st September, 1939, and though a minimum of these plans had been committed to paper there were fears that his successor might reveal them to the Germans. In the following year the Franco-German armistice split the French in Roumania into two factions and, in view of the political uncertainty of some of the French who were aware of the plans, the British mission had suddenly to undertake the task without French assistance.

24. On the 3rd July, 1940, without warning and obviously acting under German orders, the Roumanian police expelled from the country about twenty-five British employees of the oil companies, giving them twenty-four hours in which to leave. They were replaced by Roumanian "commissars." This meant the end of the oilfields plan, since the services of the local British residents were essential to the scheme. At the same time, the placing of strong forces of Roumanian troops around the oil installations put an end to any possibility of large-scale sabotage. On the 24th July the German press published an account of the plans to destroy the Iron Gates and scorch the oilfields, giving the plans in some detail together with names of those who were to take part. It appears likely that this information fell into German hands during the occupation of Paris.

Attempts to Block the Danube.

25. An important part of these plans had been the blockage of the Danube at the narrows formed by the Iron Gates. During the winter of 1939 British representatives in Yugoslavia had arranged for the cliffs above the river at the Iron Gates to be prepared for explosive charges which were to blow the rock into

(11) Lt.-Cmdr. R. D. Watson, R.N.
(12) Major V. Davidson-Houston, R.E.

the stream and so block the narrow passage. In April 1940, a small fleet of merchant vessels entered the Danube. The crews were personnel of the Royal Navy, attired in appropriate disguises, and the cargoes included the explosives required for the task. This force proceeded upstream to Giurgiu, where, in the course of re-fuelling, a lapse occurred in the security arrangements. The German minister insisted that the ships be inspected by customs officials, as a result of which suspicions were aroused and the Roumanian Government insisted upon the immediate withdrawal of the ships from the Danube.(13)

26. One more attempt was made to block the Iron Gates. In April 1941 an endeavour was made to sink a number of barges laden with stones in the narrows. While the barges were manœuvring into position fire was opened upon them from the shore and, although some barges were sunk in the positions intended, the channel was not effectively blocked.

27. Mention should also be made of another and more ambitious plan for blocking the Danube. The site chosen was some distance West of the Iron Gates and the arrangements for the blockage were ready by November 1939. As, however, the execution of the plan might have caused complications with more than one neutral country the project had to be abandoned.

28. On the 15th February, 1941, the British Legation party left Bucharest and with their departure there ended any chance of effective action against the Roumanian oil industry from within the country. The Petroleum Adviser to the Legation(14) proceeded to Cairo where, at a conference with the Foreign Secretary and Chief of Staff, it was decided to press for a policy decision to bomb Roumanian oil targets without delay. Roumanian targets were, however, out of range to the then available bombers from Middle East bases, and permission to use Greek airfields was not forthcoming. When, after the invasion of Greece, these became available for use, there were no bombers which could be spared for these operations.

29. On the 22nd June, however, Russian aircraft made an attack upon Ploesti. In July the Petroleum Adviser proceeded to Moscow and thence to Sevastopol to assist in briefing pilots of the Russian Fleet Air Arm who were preparing further attacks. The results of these attacks were negligible owing to the smallness of the forces employed and to inadequate navigational aids.(15)

---

(13) A German account of this episode is given in Appendix 7.

(14) Mr. E. A. Berthoud, C.M.G.

(15) Although no further direct action was taken against Roumanian oil until the attacks of the strategic bombing forces, one other operation was planned. It was proposed that a unit of the Long Range Desert Group, together with several British officers with a knowledge of the oilfields, should be dropped by parachute in a wooded zone in the vicinity of the Tintea oilfields. The intention was to destroy the high-pressure wells in this area, which would have substantially curtailed the country's output of oil. The party was then to withdraw to the Black Sea coast to be picked up by a Russian submarine. The scene of operations was intimately known to the officers who had volunteered for the task and the scheme had good chances of success if the necessary facilities could have been spared for the purpose. It was also considered likely that the operation would have been of value in bolstering the morale of the Balkans in favour of the Allied cause. Unfortunately the demands upon men and aircraft in other theatres precluded the plan from being carried out.

SECTION V.

THE FIRST YEAR OF WAR.

Towards the end of 1940 the British public gained the impression that there were prospects of a German collapse on account of oil shortage. Whereas, in fact, the enemy's position was far from critical there were a number of reasons that contributed towards this unwarranted optimism.

2. It had been correctly foreseen that Germany's consumption in the first stages of the War could not be covered by production and imports and had therefore entailed some depletion of the very limited initial stocks. Furthermore, it was known that Roumanian production was declining and that imports into Germany had been interrupted by the freezing of the Danube. In addition, there had been the optimistic announcements of the Air Ministry, reporting serious damage to the German oil plants. In these circumstances, the Minister of Economic Warfare(1) and his advisers believed there was justification for the statement that "an oil crisis was likely to occur in a period to be measured by months rather than years."

3. While the British public was avidly noting any report that tended to lend brightness to an outlook that was otherwise gloomy in the extreme, those responsible for the administration of German oil supplies were congratulating themselves on how successfully the position had been improved since 1939. The following account of these twelve months shows how non-military consumption was kept at a low level, how the military campaigns attained their objectives with a minimum expenditure in liquid fuel and, finally, how the very large stocks of gasoline that were captured(2) restored the stock position to the extent that the starting of new military conquests was unhampered by any liquid fuel shortage.

The Restriction of Civilian Consumption.

4. Germany's peace-time oil consumption per head of population in 1938 was less than half that prevailing in the United Kingdom. There was in Germany one motor-driven vehicle per 41 persons as compared with one per 19 persons in the United Kingdom and per 4 persons in the United States. Since a lower rate of consumption per head indicates that oil consumption was being largely confined to essential purposes, and the lower vehicle density similarly indicates that the recreational use of motor transport was relatively limited, the possibility of reducing consumption of oil products by restricting their non-essential use was clearly less in Germany than in other countries with a greater consumption rate and vehicle density. An exceptionally good highway system was, in fact, carrying less than 3 per cent. of the freight traffic of the country.

5. The Germans, however, did succeed in making very considerable savings in civilian consumption by the application of severe rationing regulations which were imposed immediately on the outbreak of war.

6. The means of obtaining these savings were the immediate suppression of all non-essential use of road transport, with a very rigid interpretation of "essential," the close scrutiny by all trade groups of the consumption requirements of their members and the cutting of all supplies of oil to industries not engaged in war-essential activities, and the restriction of household oil requirements. The essential needs of agriculture were met, although the withholding of oil supplies was used both as a weapon against the inefficient farmer and to encourage conversion to substitute fuels. The following figures show the successive reductions in consumption as the war progressed:—(3)

Average Monthly Civilian Consumption.
(In 1,000's of tons.)

	1938.	*1939.*	*1940.*	*1941.*	*1942.*	*1943.*	*1944.*
Motor Spirit	213	192	71	53	29	25	24
Diesel Oil	142	105	85	79	54	47	34
Fuel Oil (excluding bunkers)	33	52	n.a.	14	10	6	4·5

(1) The Rt. Hon. Hugh Dalton.

(2) It was some months before intelligence reports indicated the full extent of this booty. The first reports were of the successful denial of much of these stocks and the erroneous impression given added to the belief in London that the German position was worse than it really was.

(3) In the United Kingdom non-military consumption throughout the war never dropped below 70 per cent. of peace-time requirements. The following figures show the changes in the proportionate consumption by military and non-military users in the United Kingdom:—

	In 1939. *Per cent.*	*From 1940 onwards.* *Per cent.*
Civilian	95	30
Military	5	70

7. Lubricating oil was not susceptible of reduction to the same extent as other products in that it was mainly used in industry, the tempo of which was inevitably to increase with the expansion of war production. Until almost the final stages of the war a high and efficient level of production was maintained which ensured that essential requirements were met. The demand, however, exceeded the supply and a crisis occurred in 1943 which resulted in a rationing scheme being put into effect on the 1st August of that year.

Oil Consumption in the Conquest of Poland.

8. The invasion and occupation of Western Poland started on the 1st September, 1939, and was completed twenty-seven days later when Warsaw surrendered. On the 28th September Poland was partitioned between Russia and Germany.

9. The troops engaged in the operation consisted of 10 armoured, 4 motorised and 37 infantry divisions. The amount of fuel oil used by these divisions is not known but the total consumption of the campaign is not likely to have been in excess of 75,000 tons. The amount of liquid fuel consumed by the Air Force can also only be deduced, in the absence of official statistics, and the total is likely to have been to the order of 20,000 tons. Total oil consumption during September for the German Army and Air Force was approximately 155,000 tons of which under 100,000 tons were expended in the Polish campaign.

10. For this very moderate oil expenditure Germany obtained control of Western Poland among the resources of which were oilfields and refineries which, in one year, would more than repay the oil consumed in their capture.

11. Finely balanced as was the German oil supply position on the outbreak of war, the oil expenditure involved in the Polish campaign must have appeared even to them of minor quantitative importance.

12. The main significance of the campaign so far as oil was concerned was that it was the first occasion on which the new mechanised spearhead of the German army was in action. It showed the German High Command that, regardless of how large the oil thirst of a mechanised army might be, its adequate provision with liquid fuels for a lightning campaign was capable of paying enormous military and territorial dividends.

13. The speed of the German advance into Western Poland precluded the possibility of the destruction of the oilfields with the result that from the end of September the Germans had the full benefit of Western Polish crude oil production.

14. Polish crude oil production in 1939 amounted to some 400,000 tons, of which two-thirds came from fields in Eastern Poland, which are on the Russian side of the line as set up by the Russo-German partition of Poland of the 28th September, 1939, and approximately one-third came within what was to become the German zone. Polish crude oil production had been declining for some years but appeared to have been stabilised at about 400,000 tons a year. As a result of the greater production of former years there was ample if somewhat old-fashioned refining capacity available to treat this Western Polish production. As against their consumption of under 100,000 tons of liquid fuel in the Polish campaign therefore, the Germans got an oil dividend of 130,000 tons of crude oil per year, a quantity which was increased to upwards of 400,000 tons when, in June 1941, Germany attacked Russia and annexed the East Polish fields which were also captured comparatively intact. In addition the capture of all Poland made available a valuable second rail route from Roumania to Germany and added to the already large pool of surplus refining capacity at German disposal.

The Western Campaigns.

15. After the lightning success in Poland, Germany had a period of some seven months in which to complete preparations of the next campaigns, the invasion of Denmark and Norway, to be followed by the occupation of the Low Countries.

16. There was little that could be done in such a short period greatly to increase available oil supplies. A small contribution of gasoline could now be derived from the Western Polish fields, but it was not more than a few thousand tons a month. German crude oil production was stepped up by about 50 per cent.,

THE COLUMBIA AQUILA REFINERY, PLOESTI.

During this devastating low-level attack, Liberators of the United States Strategic Air Force destroyed many storage and process tanks and severely damaged the crude oil distillation plant, cracking plant, and stabiliser installation.

yielding over one million tons in 1940, although the gasoline yield was small. The time table of planned synthetic production did not allow of any sudden increase from this source.([4])

17. The success of the new tactics in Poland and the modest oil expenditure that had been involved must indeed have strengthened the belief of the Germans that their current production, aided by withdrawals from stock, would suffice for the achievement of their plans in the West.([5])

18. The campaign into Denmark and Norway was launched on the 9th April, 1940, and lasted until the 9th June when the Norwegians were ordered to cease hostilities. The German troops engaged consisted of eight divisions, none of them armoured nor mechanised, and the total oil cost of the campaign is estimated at not more than 15,000 tons. No authentic figures of consumption in this campaign are available and the overall consumption figures for the armed forces give no indication of any abnormal rise in consumption except in May when the campaign was overlapping with that into France and the Low Countries.

19. There is therefore no reason to doubt that this estimate is substantially correct, and that the whole campaign was fought and won on a quantity of oil which was insignificant as regards its effect on the German position as a whole, and which may well have been largely offset by such stocks as the Germans were able to capture intact during the course of the campaign.

20. The campaign into France and the Low Countries opened on the 10th May and lasted forty-four days. The most rapid movement occurred in the first eleven days during which fifteen of the seventeen mechanised and armoured divisions advanced an average of 25 miles per day. Latterly the advance was less rapid and some divisions, those for example opposite the Maginot Line, hardly moved at all during the campaign.

21. There were engaged 10 armoured, 7 mechanised, and 91 infantry divisions and total Army consumption for the period is estimated at 185,000 tons.([6])

([4]) At a meeting held at Karinhall, under the chairmanship of Goering, on 27th March, 1940, apprehensions were expressed concerning the future oil position. Thomas assessed the monthly demand for carburettor fuel at 230,000 tons, or a minimum of 200,000 tons, and supplies were calculated at only 140,000 to 150,000 tons. Krauch, Bentz, Fischer and von Schell were respectively urged to expedite plans in connection with synthetic plants, crude oil production, imports, and producer gas. (FD. 4809/45.)

([5]) Hitler duly took the oil factor into consideration when planning the invasion of the Low Countries. The following is an extract from a Directive on the Conduct of the War in the West dated 9th October, 1939. Although unsigned it was undoubtedly written by Hitler. The original document was marked for the personal attention of C.-in-C., Army (then von Brauchitsch), C.-in-C., Navy (Raeder), C.-in-C., *Luftwaffe* (Goering), and Chief of Staff, *Wehrmacht* (Keitel):—

> "The *Luftwaffe* cannot succeed in efficient operations against the industrial centre of England and the South and S.W. parts, which have increased in importance in war-time, until it is no longer compelled to operate offensively from our present small North Sea coast, by extremely devious routes involving long flights. If the Dutch–Belgian area were to fall into the hands of the English and French, then the enemy air force would be able to strike at the industrial heart of Germany and would need to cover barely a sixth of the distance required by the German bomber to reach really important targets. If we were in possession of Holland, Belgium or even the Pas de Calais as jumping-off bases for German aircraft there, without a doubt, Great Britain would be struck a mortal blow, even if the strongest reprisals were attempted.
>
> Such a shortening of air routes would be all the more important to Germany because of our difficulties in fuel supply.
>
> Even 1,000 kg. of fuel saved is not only an asset to our national economy, but means that 1,000 kg. more of explosive can be carried in the aircraft; that is to say, a 1,000 kg. of fuel would become 1,000 kg. of bombs. And this also leads to economy in aircraft, in mechanical wear and tear of the machines, and above all in valuable airmen's lives.
>
> These very facts are reasons for England and France to secure for themselves these regions in all circumstances, just as they compel us, on the other hand, to prevent such an occupation on the part of France and England."

In the concluding part of this directive, which details both the strategy and the tactics to be followed in the War in the West, there is the following final sentence:—

> "The brutal employment of the *Luftwaffe* against the heart of the British will-to-resist can and will follow at the given moment." (W.O.I.R., No. 105, 5.10.45.)

([6]) Thus only 16 per cent. of the divisions engaged in this highly mobile warfare were mechanised. This illustrates the relatively small proportion of the German Army that relied on motor transport. Although there was later some increase in the proportion of mechanised vehicles used, the locomotive and the horse maintained their importance throughout the war.

22. The additional military consumption involved in the campaign into France is clearly shown in the consumption statistics of German armed forces. During the first four months of 1940, for instance, the gasoline consumption of the *Wehrmacht* had averaged 44,000 tons per month; in May and June it averaged 107,000 tons. Diesel oil consumption rose from an average of 16,500 tons to 35,500 tons and aviation spirit from 45,000 to 95,000 tons. Taking the first four months of the year as "normal," the additional consumption involved in May and June was therefore to the order of 265,000 tons, of which 100,000 tons was aviation spirit. Total actual consumption in May and June for the three services, excluding naval fuel oil, was 476,000 tons.

23. The immense territorial acquisitions of Norway, Denmark, the Low Countries and France were thus made for a total oil consumption of under 500,000 tons or an additional consumption over the "normal" level of under 300,000 tons.

24. Although the victorious course of the German Army was stopped when it reached the North Sea, the Channel Coast and the Bay of Biscay, the *Luftwaffe* was not subject to this limitation and, after a period of regrouping, launched its attack on the British Isles in early August.

25. The operations of the *Luftwaffe* in the Battle of Britain and in the intensive bombing of Southern England, particularly London, from September onwards, involved an expenditure of aviation spirit at a rate of 85–90,000 tons per month until December when the rate of consumption was reduced. During this period aviation spirit stocks, which had increased to nearly 700,000 tons by the capture of a quarter of a million tons in France, were reduced by withdrawals from stock to maintain the air offensive by nearly 100,000 tons. In spite of these withdrawals, and a constant over-expenditure during the whole year, aviation spirit stocks at the end of 1940 were over 100,000 tons higher than they had been at the outbreak of war.

26. The changes in the stock position of the three principal products were as follows:—(7)

Stocks on—	*Aviation Spirit.*	*Motor Gasoline.*	*Diesel Oil.*
1st September, 1939 ...	492,000	311,000	(not available)
1st January, 1940	511,000	280,000	138,000
31st December, 1940 ...	613,000	626,000	296,000

Captured Stocks.

27. The speed of the advance through the neutral and allied countries of Western Europe enabled the Germans to capture intact much of the stocks of oil which had been amassed by the neutrals to tide them over the war period, and by France to meet the very menace which, by their capture, they helped to nourish.

28. Notwithstanding the attempts of His Majesty's Government to induce those countries that were neutral for the first nine months of the War not to lay in large reserves of oil products, as it was feared and foreseen that these would be an additional inducement to Germany to invade, there were considerable stocks of oil in each of these countries and a substantial part of them became German booty.

29. The quantity of oil captured was not less than 1½ million tons and the total may have amounted to as much as 2 million tons. Over 1¼ million tons consisted of gasoline and diesel oil.

30. A precise calculation of the disposition of this booty is not possible. Some of it was consumed by the German forces in the course of their operations. Nearly half a million tons were placed at the disposal of transport services and industry in France, and quantities had likewise to be apportioned to the other occupied countries to maintain the civil economy. Whereas the statistical allocation of this windfall cannot be fully recorded, the fact remains that it provided for the essential requirements of the occupied countries and thus deferred for a considerable time the economic decay of these areas. Of no less importance, it also provided, after other demands had been met, a bulk addition of approximately

(7) The stocks are products in main reserve storage installations. They do not include products in transit nor products in course of distribution.

half a million tons to Germany's depleted reserves of aviation fuel, motor gasoline and diesel oil.(8)

31. The failure to destroy this oil had far-reaching consequences.(9) Without it the Germans would have been unable either to have maintained the civil economies of the occupied countries or to have had sufficient reserves to have met military requirements for an attack upon Russia in June of the following year.

Stock Taking.

32. At the end of 1940 the oil position as a whole was in a more healthy condition than at the beginning of that year. The campaigns in the West had been satisfactorily achieved and at a very moderate cost in oil, a cost so modest that with the help of the loot gained and with the rising trend of home production, stocks were higher at the end of the year than at the beginning.

	Aviation Spirit.	*Motor Gasoline.*	*Diesel Oil.*
Stocks on 1st January, 1940 ...	511,000	280,000	138,000
Stocks on 1st January, 1941 ...	613,000	626,000	296,000
Average Monthly Consumption ...	72,000	150,000	111,000
Average Monthly Production and Imports	60,000	150,000	107,000
Loot (included in above stock figure for 1st January, 1941) ...	245,000	309,000	200,000

33. The following rough balance sheet for 1940, which covers only the three products in the table above, shows that it was only the captured stocks that had kept the position in equilibrium.(10) The importance of the purchases from Russia are also to be noted :—

(In 1,000,000's of tons.)

Supply.		*Consumption.*	
Production	3·5	Armed Forces	3·1
Imports—		Industry and Civil ...	2·4
Roumania	·8	Occupied Territories ...	·1
Russia	·6	Balance added to Stocks ...	·3
Sundry	·2		
Captured Oil	·8		5·9
	5·9		

34. In the words of one authority on the subject "the position had been changed from the pretty hopeless situation of 1936 to the admittedly always anxious but generally sufficient position of to-day."(11) A main source of anxiety was, no doubt, the fact that Germany was dependent upon imports for about 40 per cent. of her requirements.

35. Moreover, an undercurrent of uneasiness could be detected in High Command directives. On the 1st July, immediately after the conclusion of the campaign in France, the High Command ordered(12) that the contraction in the construction programme of oil facilities (due to priority of munitions production ordered in the spring of 1940) should be reversed and increased production of oil was to be initiated immediately and that all construction work for this purpose should be put in hand forthwith. Special emphasis was laid on the production of aviation spirit and long-term naval fuel oil production.

(8) The aviation fuel was particularly valuable to the Germans at this time. According to Ahrens this comprised 160,000 tons captured in France and 80,000 tons captured in Holland. All the operations of the *Luftwaffe* against Great Britain in the summer and winter of 1940 were made on this fuel. (A.D.I.(K) 399/1945.)

(9) Numerous attempts to deny oil stocks during the course of the war have proved that such denials can only be successfully carried out after careful planning and methodical execution. Oil is not as destructible as may at first sight appear.

(10) From a lecture by Krauch to the *Generalrat*, June 1941.

(11) "Safeguarding of German Raw Materials by the Four-Year Plan" dated 20.6.40 (Krauch files.)

(12) Krauch files 20 (*d*).

36. Nevertheless, a[illegible] the end of 1940 the stock position was never again to be as secure as it was at that time. It would also seem that, in view of the apparent generosity with which allocations were made for consumption in the occupied countries, no apprehensions about future oil supplies swayed the Germans in their preparations for the invasion of the British Isles.([13])

37. At the same time the planners were thinking ahead. Four days after the signing of the Franco-German Armistice the Office of the Four-Year Plan produced a "Petroleum Plan for Europe." This plan foresaw a Continental oil deficit, exclusive of the requirements of Great Britain and Russia, of 18,500,000 tons a year. This deficit was to be met by 18,200,000 tons of oil from the Middle East. It is interesting that this plan made no allowance for use, either by purchase or capture, of Caucasus oil. Nor did it allow for any consumption by Great Britain which was presumably envisaged as either standing in unconquerable isolation or as a vassal State no longer worthy of the benefits of oil.

([13]) Oil is unlikely to have been a factor in the decision that stopped the preparations for the invasion of Britain. Had the assembly of the necessary number of vessels been completed, the stocks of fuel at that time, and especially of diesel oil, would have been adequate to provision the operations.

"Before the collapse of France, itself most unexpected, no plan has existed for the invasion of England. After the Germans entered Paris, they felt they ought to exploit their advantage, but the outline plan which they then evolved was never developed. The first requirement was absolute air supremacy, and when this proved impossible to obtain, the scheme was dropped. The naval plans for ferrying the assault troops in barges were about as good as those made by Julius Cæsar." Jodl. (W.I.R., 10.8.45.)

SECTION VI.

Oil Imports and the Efficiency of Transport.

In August 1938 the Germans had calculated that, in the event of war, Roumanian oil was essential if Germany was to be adequately supplied. They planned to obtain from Roumania the following percentages of her total exports of oil :—

	Per cent.		Per cent.
1939	98	1942	52
1940	83	1943	39
1941	65		

Thus Roumania was to provide Germany with almost all her exports in 1939 and then on a diminishing scale as the projects of the Four-Year Plan came into fruition. The measure of success which was achieved in this direction is illustrated in the following table :—

Roumanian Oil Exports.(¹)

(In thousands of tons.)

Year.	Crude Prodn.	Total Exports.	Per cent.	Exports to Germany.	Exports to Italy.	Per cent.	Per cent.	Roumanian. Consn.	Per cent.
	A	B	B : A	C	D	C : B	C+D : B	E	E : A
1938	6,610	4,495	68	999	560	22	35	1,674	25
1939	6,240	4,178	67	1,285	629	30	45	1,785	29
1940	5,810	3,493	60	1,430	343	41	50	1,862	32
1941	5,577	4,072	73(²)	2,920	762	72	78	1,811	32(²)
1942	5,665	3,374	59	2,192	862	65	90	2,098	37
1943	5,266	3,150	60	2,511	391	79	92	2,007	38

German Oil Policy in Roumania.

2. German oil policy(³) in Roumania was necessarily directed towards three main objectives :—

(1) To increase production.
(2) To reduce Roumanian internal consumption.
(3) To obtain the maximum share of exports.

3. In regard to the first objective, the Germans gave appropriate priority to the supply of oilfield and refining equipment to Roumania. These measures did not, however, suffice to prevent the declining trend in production that had begun a number of years before the War. In the case of the refineries the Germans provided, after the entry of Roumania into the War, fire fighting personnel and both ground and air defences which did much to mitigate the effects of Allied attacks.

4. As will be seen in the table given above, the endeavours to reduce Roumanian domestic consumption were without avail. The economy of the country was vitally dependent upon oil, and war requirements brought about an inevitable rise in consumption. Any drastic reduction in this consumption would have had the effect of impairing Roumania's ability to wage war. Except for one year, Roumanian internal consumption showed a steady increase in spite of an equally steady decline in production and in spite of the cession of territory in 1941 to Hungary and Bulgaria, which reduced the area and population of Roumania by about 15 per cent. Throughout the War there was this phenomenon of Roumania, paralleled to a lesser extent by Hungary, alone among the nations

(¹) Statistics of Roumanian production and exports are given in Tables 5, 6 and 7 on page 163.
(²) Excess of consumption *plus* exports represents export of stocks.
(³) The documents covering German diplomatic and commercial relations with Roumania and Hungary are not available. This section of the Report is therefore based principally upon intelligence and deductions from intelligence.

of Europe continuing to consume oil at a rate in excess of her pre-war requirements in a continent where there was a universal famine of oil for non-military purposes.

5. The third objective, to obtain the lion's share of exports, was considerably easier to attain, and especially as Germany was able to apply pressure on Roumania from more than one direction. There were, firstly, the results to be gained by their policy of commercial infiltration, which had been largely achieved by means of the activities of the Kontinentale Oel A.G.(4) A financial interest in various oil companies ensured that their management policy was favourable to the German interest. Another factor on the side of the Germans was their stranglehold on European communications.(5)

6. There was also the exercise of diplomatic pressure and no opportunity was lost for stressing Axis solidarity and the need for the Roumanians to support it with all their resources.

7. This was the line taken in September 1942 when a special delegation went to Bucharest to urge upon Antonescu the imperative German need for Roumanian oil. The Roumanians were told that they must be prepared to make sacrifices, that these would entail suffering, but not so much as if both countries were defeated. They would have to reduce road transport, prohibit the use of fuel oil for domestic heating purposes, and find other fuels for plants using fuel oil.

8. A slight reduction in internal consumption is all that can be seen as the result of these representations. Steps were taken to make use of natural gas instead of fuel oil, and German material and technical assistance were forthcoming to this end. A gasoline rationing scheme was introduced, but this unwelcome innovation was too much for the Roumanians who continued to acquire their gasoline as freely as before.

9. In spite of this failure to arrest the decline in production and to reduce internal consumption, the Germans were successful, from 1941 onwards, in obtaining for themselves or their Allies practically all Roumanian oil exports. The success of the British counter-measures before 1941 are reflected in the table given on page 25. For example, in 1940 the Germans only secured 41 per cent. of the exports as compared with 83 per cent. for that year as envisaged under the 1938 plan. When the brake represented by Allied counter-measures was taken off, exports to the Axis immediately reflected the almost unfettered control Germany had over the distribution of Roumania's exportable oil surplus.

10. Germany had little to offer Roumania by way of payment for these exports. Foremost there were armaments which, in the days of Anglo-German competition in Roumania, had been such a powerful bargaining factor, and in addition there was machinery and equipment, and small quantities of manufactured goods. The balance of trade, however, was heavily in Roumania's favour and, in spite of an artificial low rate of exchange and negotiated prices for oil products considerably below the market level, Germany could only finance her oil imports by continual additions to the credit of Roumania in the Clearing Agreement.

Purchases of Hungarian Oil.

11. In April 1941 Germany demanded passage for her troops through Hungary on their way to Yugoslavia. This was at first refused, but later granted, and British diplomatic relations with Hungary were immediately broken off.

(4) See Appendix 4.

(5) The possibility of a British move to interrupt communications with Roumania was continuously appreciated in Berlin. When Molotov had conversations with Hitler in November 1940 Hitler betrayed apprehensions of the establishment of a British base at Salonika and of a possible thrust Northwards to deny the Roumanian oil supplies.

It would appear that these apprehensions were one reason for the attack on Greece. According to Jodl—

> "The planned attack on Greece from the North was not executed merely as an operation in the aid of an ally. Its real purpose was to prevent the British from gaining a foothold in Greece and from menacing our Roumanian oil area from that country."
>
> (Lecture to German *Gau*-leaders, 7.11.43. W.O.I.R., 21.11.45.)

This strategy resulted in the denial to the Allies of air bases in Greece that would have put the Roumanian oil refineries within the range of Wellington bombers. However, our air forces at that stage of the War would have been quite inadequate for the task.

In June 1941, when Germany attacked Russia, a "Russian" air raid on a Hungarian town was staged by the Germans, and Hungary on this pretext declared war on Russia. Finally, in December 1941, Great Britain and the United States declared war on Hungary.

12. This in bare outline is a record of how Hungary came into the War. At the time when it happened the Germans may still have had faith in the success of their lightning tactics which had had such striking success in the Polish and Western European campaigns, and may not have taken into full account the valuable economic resources which Hungary's adherence to the Axis cause would bring. Of these resources oil was to prove among the most valuable.

13. In 1937 oil was struck in the Lispe area at Budafapuszta and in 1940 at Lovaszi. The producing company, known as *MAORT*, was a wholly owned subsidiary of the Standard Oil Company of New Jersey. The total production for Hungary, nearly all of which came from these two areas, was as follows :—

	Tons.		*Tons.*
1937	2,200	1941	421,700
1938	42,800	1942	665,200
1939	143,200	1943	837,710
1940	251,400	1944	809,970

14. The Germans meanwhile had not been inactive, for at the end of 1940 a German syndicate known as *MANAT* had obtained a concession covering all south-eastern Hungary. The syndicate was a combine of the major German oil-producing companies with *Wintershall A.G.* as the operating company. *MANAT*, however, obtained no significant production in this area.

15. It was not until 1941 that Hungary had any exportable oil surplus, and before that time the Allies had made every endeavour to induce Hungary to continue her purchases of Roumanian oil so that Roumanian oil exports to Germany might thereby be lessened. In 1941 there was an exportable surplus of some 150,000 tons of Hungarian production after home consumption requirements had been met, and thereafter Germany, and to a lesser extent Italy, counted on the Hungarian surplus as an addition to their own supplies.(6)

16. The Hungarian Government, however, retained considerable freedom of action in the management and operation of the oil industry. The refining capacity of the country was modernised and extended to enable it to handle the bulk of Hungarian crude production. Facilities for the production of lubricating oil were added to the Shell and Vacuum refineries at Csepel and Almasfuzito and a new Government-sponsored refinery was constructed at Szoeny on the Danube. In this way the Hungarian Government was able to benefit by the financial advantages of exporting finished products rather than crude oil. From the German point of view, the arrangement had the advantage of the creation of a pool of modern refining capacity in an area which, until the Allies obtained bases in Italy, was out of range of air attack.

17. Hungarian independence was also manifested in their refusal to dissipate their crude oil reserves by producing at a wasteful and uneconomic rate to meet a sudden emergency demand. Thus, although it is probable that production at a rate well over a million tons could have been obtained in 1944, the rate was limited to ensure the rational and economic depletion of the producing fields.

18. In 1944 Hungarian oil took on an added importance in the planning of Germany's oil plant dispersal programme. Hungarian crude had always been of particular value in view of its 30 per cent. gasoline content as compared with an average gasoline content of German and Austrian crude oils of not more than 10 per cent.

19. In the Geilenberg dispersal programme(7) an important part in gasoline production was to be played by the plants designated as "*Ofen*," which were primitive topping plants. Over twenty of these plants were planned and their most efficient feedstock would have been Hungarian crude by the use of which the gasoline offtake of these plants would have been three times as large as the use of German crudes would have allowed.

(6) In 1943 Germany imported 302,000 tons of Hungarian oil. (Annual Report of the *Planungsamt*, 29.6.44. F.I.A.T. Final Report No. 403.)

(7) See pages 65 and 152.

20. In addition, greatly as Hungarian refining capacity had been damaged by air attack, repairs were resolutely carried out and, up to the last, important quantities of gasoline and diesel oil were being supplied to the German forces in south-east Europe. This continued operation was assisted by the fact that for some time no attacks could be made on these refineries. This was due to lack of information upon the position of the Soviet forces, notwithstanding the efforts made to improve liaison arrangements with them. In the last three months of 1944, for instance, Hungarian refineries produced 35,000 tons of gasoline and 8,000 tons of diesel oil. At least until the end of February 1945, crude oil production was being maintained at a rate of 60,300 tons a month.

Oil Transport and the Vienna Conference.

21. German control over the destination of Roumanian and Hungarian oil exports underlined the importance of the transport question. Roumanian oil had previously been exported by one of two routes: either up the valley of the Danube by rail or river, a route in which the river port of Giurgiu played an important part, or by ocean-going tanker via Constanza.

22. The lack of shipping and the Allied blockade precluded any substantial shipments moving by way of the Mediterranean and, in July 1940, there was convened at Vienna a conference of the authorities of the railway systems of Germany, Italy, Roumania and of other countries to study the problems of oil transport. The programme arranged by this conference covered the export of over 550,000 tons of oil each month to Germany and Italy, although at that time there was a maximum exportable surplus of 300,000 tons per month. Moreover, on account of the decline in Roumanian exports, even this lesser figure was never attained. If 550,000 tons of oil had been available, there would have been no serious difficulty in moving this quantity by the joint use of the Danube and the Balkan railway system. If the Danube had frozen it is probable that the railways could have carried upwards of 400,000 tons per month without undue difficulty.[8]

23. It was, however, necessary to add to the river fleet on the Danube, to ease a shortage that had doubtless been aggravated by the activities of the Goeland Company. To meet this need for additional vessels over one hundred tanker barges were moved from the Rhine, Elbe and Oder. The transfer was carried out both by rail, using special wagons, and by road, the *Reichsautobahn* permitting the loading clearances necessary. The vessels had to be stripped above the deck-line before movement was possible.[9]

24. The carriage of oil from Roumania and Hungary was the most difficult transport problem, so far as oil was concerned, which the Germans had to meet, other sources of supply and production being much nearer the main areas of consumption. Up to the time of the Allied attacks in 1944 there is no evidence that shipments of oil from Roumania were at any time unduly delayed or that the rail and river systems were ever unable to keep pace with the quantities of oil awaiting transport.

25. Although the Balkan transport system was never called upon to carry the quantities of oil contemplated, the ultimate success of German plans for the conquest of the Caucasus might have resulted in these routes having to cope with greatly increased oil shipments.

Pipelines.

26. The manifold advantages obtained by the Allies by the use of pipelines for oil transport draw attention to the comparative neglect by the Germans of this form of transport.

27. The Germans themselves have given no unanimous nor particularly cogent reason for this. One source[10] has stated that "pipelines were not used because of shortages. There was a critical bottleneck in welding fabrication and this was one of the major limiting factors. There was also a critical shortage of pumps inasmuch as the submarine programme was taking the bulk of pump production."

[8] To expedite movements on the Danube plans were made early in 1942 for locks to be constructed at the Iron Gates. There is, however, no evidence that this project was proceeded with. (Hitler Conferences, Vol. 11, 13.5.42. FD. 3353/45.)

[9] The whole operation was supervised by the *Deutsch-Amerikanische Petroleum G.m.b.H.*, which company should have records of the ownership and destination of these vessels.

[10] Fischer. (U.S.S.B.S. Interview No. 67.)

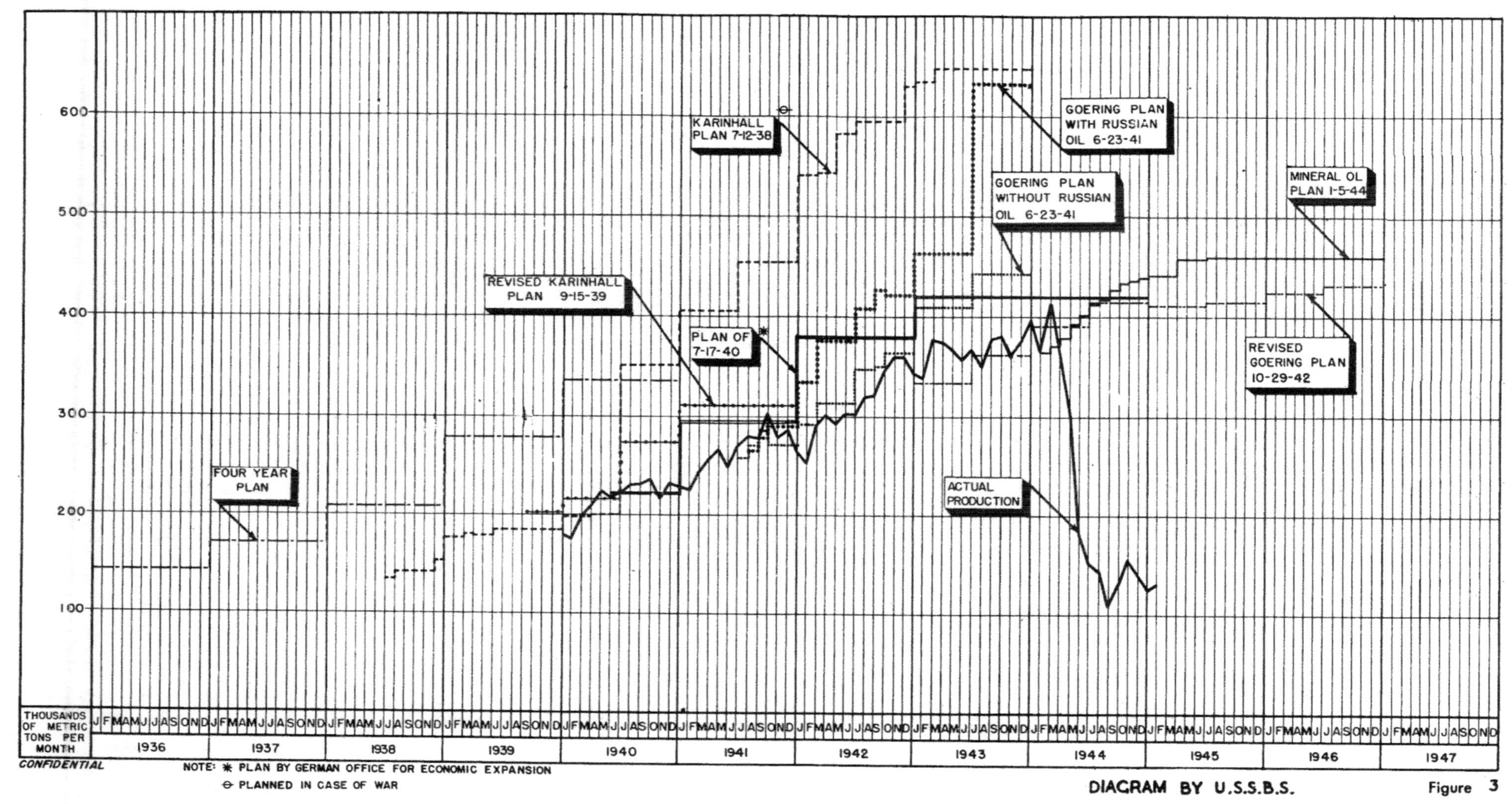
COMPARISON OF GERMAN PRODUCTION PLANS
WITH ACTUAL PRODUCTION
(INCLUDES AVIATION GASOLINE , MOTOR GASOLINE AND DIESEL OIL)
600
500
400
300
200
100
THOUSANDS OF METRIC TONS PER MONTH
1936
1937
1938
1939
1940
1941
1942
1943
1944
1945
1946
1947
FOUR YEAR PLAN
REVISED KARINHALL PLAN 9-15-39
KARINHALL PLAN 7-12-38
PLAN OF 7-17-40
GOERING PLAN WITH RUSSIAN OIL 6-23-41
GOERING PLAN WITHOUT RUSSIAN OIL 6-23-41
ACTUAL PRODUCTION
MINERAL OL PLAN 1-5-44
REVISED GOERING PLAN 10-29-42
CONFIDENTIAL
NOTE: * PLAN BY GERMAN OFFICE FOR ECONOMIC EXPANSION
⊖ PLANNED IN CASE OF WAR
DIAGRAM BY U.S.S.B.S.
Figure 3

28. Another source,([11]) who was in charge of the construction of the few pipelines which were built, attributed the failure to adopt this form of transport to the opposition of the German oil companies, who looked askance at such a threat to their monopoly of distribution, especially as it would be in the hands of the *WIFO* organisation, which was already trespassing upon their preserves. This opposition was continued by the *Zentralbuero fuer Mineraloel* which had the same commercial viewpoint. In the later stages of the War, when commercial opposition could have been more easily over-ruled, lack of materials prevented pipeline construction except for some small tactical pipelines laid across the Rhine.

29. The first German pipeline project was the laying of two additional 10″ lines from Ploesti to the Danube port of Giurgiu. The material for these lines was part of the pipeline captured in France which had been intended to connect Donges and Montargis. This material was supplied and partly laid by the Germans, but the lines were owned and operated, as part of their monopoly, by the Roumanian State Railways. The two extra lines more than doubled the pipeline capacity between Ploesti and Giurgiu, and greatly eased the problems of rail transport of oil from Ploesti by permitting increased despatches by Danube barges and by rail from Giurgiu.

30. The next pipeline projects to be considered were those connected with the plans for the exploitation of Caucasus oil. The main route for this oil was to be up the Danube waterway to Vienna, thence by pipeline to the Elbe, where it could be further distributed by Germany's system of inland waterways.

31. This route entailed the modification of the east stage of the existing Ploesti–Constanza pipeline. The flow of this section of this line between Constanza and Cernavoda was made reversible so that tankers could dock in Constanza and their contents be pumped overland to Cernavoda, where Danube tank barges could load and proceed upstream to Vienna. The reversal of this line was carried out by the *WIFO* organisation.

32. The Danube–Elbe pipelines had their starting-point at the large underground *WIFO* storage installation adjacent to the new refinery that had been built at Lobau, near Vienna. The lines terminated at Roudnice, where a concealed installation, similar to the one at Lobau, had been constructed. The pipelines, both of 12″ diameter, were originally intended for white products, and principally those refined from Russian crude, and later it was proposed to use these lines for crude oil. However, although these lines were duly completed, they never came into operation. These lines are reported to have a connection with the Vacuum Oil Company Refinery on the Elbe at Kolin.

33. A comparatively short line was built from the Bruex hydrogenation plant to Roudnice and was used for the purpose of shipping products from Bruex. Roudnice was designed to become an important distribution and transhipment centre, but events prevented it ever attaining full operation.

34. A more ambitious project, and intended to be an alternative system to the Danube–Elbe route, was to be a pipeline from Odessa to Upper Silesia, where it would have tied in with the River Oder.([12]) There is no evidence of any attempt to put this plan into effect. Another pipeline which was completed on the drawing-board but not constructed, was to carry crude oil from Ploesti to Apahida on the Hungarian frontier.

35. The only pipelines constructed for the direct supply of military operations comprised five pairs of lines (one for motor gasoline and the other for diesel oil) that were laid by the *WIFO* organisation under the Rhine in the winter of 1944–45. These were located near Wesel, Remagen, Mainz, Speyer and Breisach. Each pair had a total capacity of about 7,000 tons a month. The lines at three sites were ready to operate by March, but no fuel was available to put through them.

36. During 1944 a plan was approved in principle for a new central reserve to be established underground in the Harz mountain area, which was to be connected by a pipeline system with the major producing plants and to a number of distribution points. Work, however, was never started on this project.

([11]) Wehling, Director of *WIFO*.

([12]) Record of the 15th Meeting of the *Zentrale Planung*, 20.10.42. Milch was reported as being in favour of this project. The route suggested by *WIFO* for this line was Odessa, Tarnopol, Lemberg, Przemysl, Jaroslau, Rzeszow, Tarnow, Bochnia, Krakau, Kattowitz and Gleiwitz, a total distance of 1,300 kilometres. Twin lines, each of 16 inches diameter, were proposed and the project was to take two years to complete.

Efficiency of Rail Transport.

37. There was a good reason why so little use was made of pipeline transportation. Until air attacks began to hamper seriously rail and water communications, these systems met requirements admirably. Oil represented less than 2 per cent. of the total rail freight traffic and, as oil trains were given appropriate priority, distribution was maintained until deliveries no longer became possible. Movements were facilitated by an abundance of tank cars, resulting from German conquests, and even after considerable losses of rolling-stock in Russia there were more than enough tank cars to meet overall requirements.

38. The production of oil was not seriously affected by the dislocation of communications, a fact which was no doubt largely due to the raw materials being mostly *in situ* at the source of production.(13)

39. Although the delivery of the finished products became impeded when rail transport approached total dislocation, deliveries to combat areas were well maintained until that time. The oil trains were efficiently moved to the railheads except when movements were interrupted by air attacks. One effect of these attacks was to cause diversions over circuitous routes causing delays in delivery, these delays being serious during periods of intense fighting. Another effect, and of no less importance, was the added inconvenience to front line units caused by railheads being located increasingly far back as the air operations of the Allies became intensified. This resulted in the need to substitute road transport for rail, thus increasing the demand for liquid fuel. The amount of liquid fuel that was lost in rail transit by fire or leakage on account of air attacks was very small.(14)

40. One consequence of the efficiency of the transport system in so far as oil was concerned was that it obviated the need for establishing large stocks in forward areas. This permitted the policy of maintaining strategic reserves in the underground storage depots of Central Germany and deliveries into consumption were well maintained as long as transport was possible.

---

(13) Kehrl, amongst others, has confirmed that damage to communications did not seriously affect oil production.

> "In the petroleum industry despatch was fully maintained right to the end. In the synthetic oil industry also, almost all the products manufactured were despatched, because tank trucks were given priority over all other transport." (FD. 4550/45.)

The disruption of the railway system was, however, a retarding factor in the execution of the Geilenberg programme.

(14) Fuel trains comprised a relatively limited proportion of total military traffic during operations in the West. Adequate statistics are not available of train movements in the Summer of 1944, but the following figures indicate the movements to the Western front in the last quarter of the year:—

October	728 supply trains, of which 93 were fuel trains.
November	782 supply trains, of which 95 were fuel trains.
December	830 supply trains, of which 161 were fuel trains (which trains were reduced in size from the standard train of 30 tank wagons).

These same statistics suggest that damage directly inflicted by air attack was slight:—

October	10 wagons destroyed.
November	7 wagons destroyed.
December	16 wagons destroyed.

It is likely, however, that the ratio of loss from air attack was higher in both the preceding and succeeding months when the weather conditions were much more favourable for air attacks.

On the other hand, deliveries in the West were frequently subject to long delays. This was in large part due to the almost excessive precautions which were taken in moving fuel trains. Except when unavoidable they were never kept on sidings in major rail centres, they were directed as far as possible from areas of likely attack and they were unloaded with extreme caution, mainly at night. During the fighting in Normandy cases were frequently reported of fuel trains having to be unloaded on the East bank of the Seine because of the difficulty of rail movement nearer the battle area.

SECTION VII.

THE FAILURE OF THE SHORT WAR.

After the conclusion of the campaign in Western Europe the German army was not engaged in any active operations until April 1941 when Yugoslavia and Greece were invaded. In the air, however, the *Luftwaffe* maintained its attacks on the British Isles, first by day and then by night, involving a rate of consumption of aviation spirit substantially higher than current production. It was clear that production would have to be replanned on a more substantial scale. Goering's position as head of the *Luftwaffe* and dictator of the Four-Year Plan enabled him to carry out this replanning with a minimum of interference.

2. On the 27th April Athens was entered. On the 19th May the Battle of Crete was joined and on the 1st June Allied evacuation from Crete was complete. The oil consumption of the 20 divisions which took part in the Balkan campaign is estimated at approximately 100,000 tons, a quantity of minor importance when compared with the strategic and political advantages which had been gained.

3. From the point of view of oil supplies, perhaps the most important result of these conquests was the domination gained over the Danube waterway and over Hungary and Roumania. In an official German memorandum written in August 1938(1) it was admitted to be imperative for the success of the Four-Year Plan for oil that the south-eastern Balkan industrial area should be kept open for Germany and that oil imports from Roumania should be unimpeded. Although Hungary and Roumania were not fighting on Germany's side until December 1941, German ends had already been attained by the beginning of June.

The Attack upon Russia.

4. One year to the day after the signing of the Franco-German armistice Germany attacked Russia on the 22nd June, 1941. The troops engaged consisted of 159 divisions of which 35 were armoured or mechanised.

5. The successes in Poland, France and the Low Countries, and in the Balkans gave expectations that the Russian campaign would be almost as short-lived. It was true that the distances involved were much greater and that railway communications were more sparse and would at first be inconvenient to work on account of the difference of gauge. Nevertheless, Hitler and the political leaders were convinced that they could win the Russian war in three months.(2) One German official,(3) who was aware of the plans to attack Russia in January 1941, has confirmed that " The High Command calculated on a short war. Everyone thought it would be over by October 1941. Normal production rates were sufficient to equip fully all necessary divisions for this type of conflict." The planners were prepared for a rate of oil consumption twice as high as normal needs for the short time the campaign was expected to last.

6. Based upon these optimistic expectations the hopes that sufficient liquid fuels would be provided for the forces in Russia appeared to be well grounded. Before the campaign began current production of motor gasoline and diesel oil was running at a rate little below that of estimated consumption and was backed by reserve stocks that were in excess of those available in 1939. Gasoline production and imports were averaging 190,000 tons a month and consumption 208,000 tons a month; with 1940 opening stocks of nearly 600,000 tons the deficit between consumption and production could be met for many months without uneasiness. The position with regard to diesel oil was as satisfactory so long as the synthetic plants were engaged, as they were until June 1941, in a maximum output of this product.

(1) Krauch files.
(2) Koller. (U.S.S.B.S. Interview No. 8.)
The views of Jodl are also of interest:—

" It was easy to start the Russian campaign because the campaigns in Poland and France had been ended with really ridiculously small losses. In addition, we had secured large and precious booty in Czechoslovakia, Poland, France, &c., so that all was plentiful for the Russian War. We started preparing for the Russian War in November or December 1940."

(FD. 4472/45.)

(3) Saur. (U.S.S.B.S. Interview No. 9.)

7. The gap between aviation spirit production and consumption was, however, much greater. Average production was 75,000 tons a month as compared with an average demand of over 105,000 tons a month. Although stocks were over 600,000 tons, which would permit of a withdrawal of 30,000 tons a month for some time, the supplies were unbalanced in that there was a deficiency of high octane fighter fuel. Three months before the attack on Russia, Goering had been apprehensive lest the operations of the Messerschmidt 109-F would be restricted by a lack of this fuel, and he had ordered that production was to be increased as rapidly as possible.(4)

8. A report of the *Zentrale Planung* shows that total oil consumption of the armed forces in 1941 averaged about 400,000 tons a month. *Luftwaffe* consumption was at a level of about 100,000 tons a month with a peak figure of 137,000 tons in July. Naval consumption remained steady at about 100,000 tons a month.(5)

9. This document also estimated the optimum requirements of the army in 1941 at 240,000 tons a month but admits that the army only received an average of 185,000 tons a month. However, the difference between the desirable and actual receipts of the German Army was only in part due to the deficit between production and consumption; supply and transport difficulties due to the depth and length of the Russian front were at least as much responsible. "By the end of 1941, however, the vast distances which had been covered made operations increasingly difficult. Hitler did not establish proper tactical reserves of either aircraft or ground forces. Instead he threw everything into the Moscow attacks and came to grief."(6)

10. There is no evidence that a shortage of oil supplies was responsible for the failure to force a decision in the 1941 Russian campaign. In spite of their readiness on other occasions to attribute their failures to political or economic factors, the German General Staff, so far as is known, have not attempted to suggest that with more generous supplies of oil they would have been able to achieve victory by the end of 1941.

Stock Losses.

11. A summary of the position in 1941, as presented to the *Zentrale Planung* in 1942, was as follows:—(7)

		In millions of tons.
Production—		
Supplies available to German controlled Europe of which from 1·2 to 1·5 were obtained at the expense of depleting stocks		12·7
Consumption—		
Civilian economy, German Europe ...		8·0
of which		
Agriculture—Germany	·17	
Civilian economy—Germany ...	3·70	
Eastern Area	·15	
Supplied to Italy	2·0	
Wehrmacht		4·8
		12·8

12. By the end of 1941 it had become clear that the Russian campaign would not be as short as was hoped. It was also obvious that withdrawals from stocks at the rate of one and a quarter to one and a half million tons a year was an experience that could not be repeated. The consensus of Allied intelligence confirmed the difficulties of the position. At a meeting of the War Cabinet on

(4) A.D.I.(K) Report No. 391A/1945.
(5) Shorthand notes of meeting of *Zentrale Planung*, October 1942.
(6) Saur and Buhle.
(7) Shorthand notes of meeting of *Zentrale Planung*, October 1942.

PHOTOGRAPH BY U.S.S.B.S.

THE *DEURAG-NERAG* REFINERY AT MISBURG, HANOVER.

One of the principal German refineries, it was engaged in processing German and Austrian crudes. During the nine months between 18th June, 1944, and 17th March, 1945, the refinery was attacked on fourteen occasions, the total bombs carried amounting to 4,711 tons. Approximately 1,650 high explosive bombs fell in the plant area, destroying 90 per cent. of the tank storage capacity, 50 per cent. of the processing structures and 80 per cent. of all other buildings. As relatively small damage was done to machinery and processing equipment only 35 per cent. of the investment value of the refinery was destroyed. The attacks were frequently impeded by a smoke screen and, in addition, some near-by cement plants were attacked more than once in error.

The refinery was never put totally out of action for any appreciable length of time. It was possible for individual units to be started up shortly after each attack. These resumptions were accurately assessed by aerial reconnaissance and the attacks were well timed. Although the refinery was out of operation at the end of the War it was working at 53 per cent. of normal capacity by 1st July, 1945.

the 22nd December it was reported that detailed investigations had led to the conclusion that Germany's oil position was now at a crucial stage.(8)

13. The year 1941 had not been an easy one for the German oil planners. In February Goering had ordered that oil production was to be given every priority and that previously planned production for 1942 was to be increased by 25 per cent. Production officials regarded this as impossible of attainment owing to shortages of steel and labour; in fact, they continually had to whittle down production plans as construction difficulties postponed plant completion dates.

14. In June the imperative need for a higher production of aviation spirit made it necessary to change the processing programmes of the synthetic plants, which had been geared for a maximum production of diesel oil, so that more high octane gasoline would be produced at the expense of other liquid fuels.

15. According to the revised Four-Year Plan aviation spirit production was to be at the rate of 1·44 million tons a year by the end of 1942 with a final target of 2·04 million tons a year later. The plan was now amended to 2·4 million tons a year by the end of 1942 with a maximum of 3·6 million tons by the end of 1943. This plan was never achieved and the production experts realised that the only hope of achievement was that aviation spirit should be given first priority and if 1½ million tons of Russian oil was made available as from the spring of 1942.(9) The need to obtain the oil of the Caucasus therefore became of greater urgency.

16. These difficulties were largely attributable to faulty planning. Production instructions were given without sufficient study of the availability of labour and materials, and programme modifications were made without regard to the dislocations they would cause and apparently without a full calculation as to their necessity. An example is to be found in the fact that the planned aviation spirit output of 300,000 tons a month in 1943 was far above the monthly quantity the *Luftwaffe* ever consumed, and this production could only have been achieved by the sacrifice of a greater quantity of less highly refined fuel.

The Plans for the Caucasus.

17. Although the attack on Russia was decided upon for political and military reasons, there were also economic motives. These motives were primarily the acquisition of food and oil.(10)

18. No attempt was made in German official circles to deny their aspirations for the oil of the Caucasus or the importance of obtaining it. Shortly before the occupation of the Maikop oilfield Keitel is reported to have stated to Admiral Robertelli, who was in Berlin begging for oil for the oilbound Italian Navy, that if Maikop were not occupied the German oil situation would be "tragic." Goering's aviation spirit plan was dependent upon Caucasus oil and preparations were being made for the conversion of the German hydrogenation plants from coal to Russian crude oil feedstocks which would have greatly increased their output capacity.(11)

(8) W.M. (41) 133rd. Based upon the Eighth Report of the Committee on the Enemy Oil Position. (P.O.G. (L) (41) 11.)

(9) See Graph facing page 28. The urgency of the situation was pointed out in a letter dated 3.9.41 from Ritter, Krauch's executive, to Milch in which he stated that Goering's aviation spirit plan was dependent upon the Caucasus oil being available.

(10) Speer gave the economic reasons for the war against Russia as food and oil (FD. 4548/45). This view is supported by Thomas in the following statement:—

"Goering sent for me at the end of November 1940 and gave me the order to compile all the possible economic material on Russia, as Hitler had decided to defeat Bolshevism once and for all. When I objected that this decision to wage war on two fronts was fraught with tremendous danger . . . . Goering said: "The *Fuehrer* has decided on this war because—

(1) The Bolshevik will attack us one day in any case and his industry must be smashed before he is ready for war.
(2) The war against England is going to last longer than we expected and for food reasons we must therefore break through the British blockade in an Easterly direction; Central Europe can only be fed with the help of the Ukrainian harvest.
(3) We must break through to the Caucasus in order to get possession of the Caucasian oilfields, since without them large-scale aerial warfare against England and America (*sic*) is impossible.'" (FD. 4503/45.)

(11) Interrogation of Speer. 5th Session, 30.5.45.

19. The extent to which this need for Russian oil influenced the strategy of the German forces in 1942 is the subject of different opinions.([12]) The available evidence does not show whether the denial of these oilfields to the Russians was regarded as of more importance than their prospects of relieving the fuel shortage. Nor is it clear whether a major purpose of the Caucasus offensive was to divert Russian forces from the defence of Moscow.

20. It is, however, certain that elaborate preparations were made for the exploitation of the Caucasus oilfields. Following closely behind the spearhead of the advancing German forces was a specially equipped task force called the *Technische Brigade Mineraloel (Kaukasus)*. This force had undergone extensive training in the rehabilitation of "scorched" oilfields and was well provided with equipment. No less than 75 drilling rigs had been manufactured in Germany for the prompt exploitation of Maikop and 40 more were to be made available from German industry.([13])

21. The prospects of considerable production from Maikop in its first year of operation under German control were reflected in the planned crude oil production schedule for the year 1943, in which the output of Maikop for the first twelve months of operation is put at 250,000 tons. In October, however, the plan had already gone awry.([14]) The programme had been to start drilling in the Autumn and to reach 30,000 tons per month by April 1943. But so many difficulties were encountered that it was found that no drilling could start before May 1943.([15]) This was due to the thoroughness of the demolitions and to the continued organised guerilla resistance which did not permit any repair work to be started until December 1942, though Maikop had been entered in August. Thus the amount of oil obtained by the Germans from Maikop was almost nil. Some production was developed at Romni, in the Ukraine, but this was too small to be of any importance.

22. The Germans planned to exploit the Caucasian oilfields on an extensive scale. The production brigade with their hundred or more rigs were to be followed in the Spring of 1943 by refinery construction brigades who would erect topping plants in the fields (4 plants with combined capacity of 255,000 t.p.a.) and probably at Kherson (400,000 t.p.a. planned capacity), these units being

([12]) There is a divergency of views among the military leaders upon the basis for the strategy of the Caucasus campaign in 1942.

Jodl has affirmed that the main object of the campaign was not to secure the oilfields of Maikop and Baku but to deny them to the Russians. (FD. 4472/45.)

Milch is of the opinion that the acquisition of the Caucasian oil was the main aim of the campaign, as it was felt by the High Command that the War could not be carried to a successful conclusion without an addition to the supplies under German control.

On the other hand, Koller considered that the campaign was intended to safeguard the Ukraine from Russian counter-attacks and to cripple Russian resistance by cutting the supply lines along the Volga valley. The acquisition of the Caucasian oil would have been a pleasant additional item on the profit side of the account and the Russians would have lost the benefit of the Persian supplies.

Ruhsert believed that the primary objective of the Caucasus offensive was to facilitate an eventual assault on Moscow by diverting Russian troops to the South. At the same time oil supplies would have had to have been captured in Russia itself in order to bring the War to a satisfactory conclusion and consequently the seizure of the Caucasian oilfields was an essential part of the plan. (A.M.W.I.S., No. 318, 8.10.45.)

A file note by Thomas, recording a discussion with Goering on the 4th September, 1941, indicates that Goering requested Hitler to concentrate upon the capture of Maikop:—

> "The *Reichmarschall* is aware that we will be in a specially difficult position this winter. For these reasons he has demanded from the *Fuehrer* that operations be now concentrated on the southern flank and that as the next target the Maikop district be attacked. He is aware that the (mineral oil) installations will be destroyed but hopes to balance the mineral oil situation within half a year by new drilling." (FD. 4809/45.)

([13]) Minutes of a meeting of the *Zentrale Planung*, October 1942.

([14]) Of interest is a minute signed by Cavalry Major Will, of the Economic Armaments Office, to the Chief of Staff of the Supreme Command, dated 6th December, 1941, pleading for a sudden thrust of the German armies to the Urals [*sic*] mineral oil district. It was pointed out that Russian denial plans in the Caucasus would make it improbable for any substantial quantities of oil to be obtained from these fields in 1942. It was suggested that the occupation of the area between the Volga and the Urals would not only secure the large quantities of oil production equipment that had been evacuated to that area, but would also yield a sizable output of oil suitable for German warfare much quicker than the Caucasus district, at the same time depriving the Russians of their last intact source of oil. (FD. 4809/45.)

([15]) Report to Goering by Bentz.

In a report of the Hartley Committee of 12.12.42 it was estimated that Maikop might contribute 100,000 tons of oil in the first six months of operation, an estimate which is well borne out by the German planned production figure of 250,000 tons in the first year.

made up of equipment dismantled from French refineries. These plants, which were given the priority of a Hitler decree, would supply local requirements of gasoline and diesel oil while the residues were to be sent back to Germany.([16])

23. The end of German hopes for the Caucasus was marked by the Battle of Stalingrad. In Speer's view this particular disaster was not in any way dictated by economic reasons. "It was a purely military measure, which is still to-day quite incomprehensible to us in its tragedy."([17])

The Revision of Plans.

24. Throughout this period German economic planning had lacked positive direction. While this was partly due to uncertainty as to the probable duration of the War and of future needs it was also due to the short-sighted planning of Hitler and his associates.([18])

25. In the Autumn of 1941 it was decided that the maximum industrial effort must be made in the first six months of 1942, so that the arms and equipment necessary for the successful conclusion of the Russian War could be provided. Hence the construction of a number of new synthetic plants was delayed by the withdrawal of labour into the munitions industry.

25A. This policy was justified on the grounds that the capture of Caucasus oil would ease the oil supply difficulties and that, when victory was assured, the labour could be returned; the programme would, in fact, only be delayed by the period of the intense armament drive and would be less urgent on account of the additional supplies which victory would bring.([19]) One consequence of this policy was that, by the 1st June, the chemical and synthetic oil industry was short of 52,000 constructional workers and 52,000 operatives. Furthermore, the labour turnover of the industry, in both construction and operation, was almost 100 per cent. in the first six months of the year.([20])

26. In March 1942, a further demand was made for a restriction of work and for the release of labour. This, however, was partially countermanded by the decision that the aviation spirit production plan could not be modified. The rising demands of the *Luftwaffe* were causing considerable anxiety. Immediately after the attack on Russia Goering had produced a plan for a greatly increased supply of aviation spirit by the construction of extensions to the hydrogenation plants. This plan, known as the "Special Plan for Aviation Spirit and Light Metals of the 23rd June, 1941," was amended as a result of conversations with Milch and Udet and was finally agreed at a planned output in the middle of 1942 of 200,000 tons of aviation spirit a month rising to 300,000 tons a month by the middle of 1943.

27. On the 15th March, 1942, a review was made of the Karinhall (New Four-Year) Plan of 1938 and of the Special Plan for Aviation Spirit for the purpose of ascertaining what construction work could be further reduced.

([16]) Foreseeing the consequences if the Germans acquired the Caucasus oilfields, His Majesty's Ambassador in Moscow (Sir Stafford Cripps), in November 1941, made an offer of assistance to the Russians in making good their oil supplies both during and after the War in so far as they destroyed them in order to deny them to Germany. The Americans, by preconcerted arrangements, made a similar though more qualified offer. At the same time His Majesty's Ambassador conveyed an offer to send demolition experts to help with denial schemes. The Russians accepted both offers and asked for the British offer to be made into a formal agreement. An official (Mr. E. A. Berthoud, C.M.G.) was sent to Kuibyshev to discuss such an agreement and a team of demolition experts was sent to the Caucasus.

Owing to Russian successes in the South in the winter of 1941–42, the Russians dropped the proposal for an agreement. The demolition experts, led by Colonel W. L. Foster, C.B.E., were able to give much practical advice and, upon their return, reported on the thoroughness of Russian denial schemes.

(Notes of the P.O.G. Committee, 4.3.42.)

([17]) FD. 4548/45.

([18]) Thomas has made a caustic comment on the lack of efficient organisation at this time:—

"In the economic sphere Hitler's so-called Fuehrer State was one of complete "Fuehrerlessness," muddle and duplications, because Hitler did not realise the necessity for firm, far-sighted planning, Goering did not understand anything about industry and the technicians in charge had no plenary powers. Not until Speer came on the scene, all industry thus being placed in the hands of one of Hitler's trusted men, did Hitler give in and things began to change." (FD. 4503/45.)

([19]) Krauch files 20 (*g*).

([20]) Letter, Krauch to Speer, 25.6.42.

The conclusions reached were that work was to be retarded on only those synthetic plants which would come into production in 1944. This would entail a loss in 1944 of 73,000 tons of production a month unless the loss could be made good "from some other source, for example, the Caucasus." Although Goering at first refused to allow any modification of his Special Aviation Spirit Programme because it was "vital to ensure technical equality or superiority over the Anglo-Americans," these decisions were put into effect at the end of March and work was retarded on hydrogenation plants at Zeitz, Blechhammer, Heydebreck and at a plant at that time planned at Gladbeck.([21])

28. The plans for crude oil production were also undergoing revision. Goering had ordered that the liquid fuel supplies for the forces on the Eastern front were to be ensured by the ruthless exploitation of the Austrian fields. The extent to which these fields were prodigally over-produced is seen by a comparison of planned and actual production from Austria in 1942 and 1943. Planned production in 1942 was 675,400 tons and actual production was 868,600 tons. In the following year planned production was 1,505,000 tons and actual production 1,106,000 tons.

29. It was not until the 26th February, 1943, that the *Zentrale Planung* decided to write off the prospects of Russian oil, a decision which meant a reduction in the hoped-for increase in aviation fuel.([22]) And, in spite of the modification of the plans for increasing synthetic oil output, the progress of construction was continuing to meet with many difficulties. Krauch was complaining that the demands for armaments were so pressing that the steel allocations for the oil plants were being repeatedly reduced. He needed 60,000 extra workers, production was only being maintained by turning construction workers into operators, and, of the 23,000 labourers supplied to work at the Blechhammer and Heydebreck plants, in the last quarter of 1942, 15,000 had "run away."([23])

Consumption and Stocks in 1942.

30. Oil consumption of the German Armed Forces in 1942, of which the great majority was consumed in Russia, was at very much the same level as in 1941. Consumption had averaged 400,000 tons a month and, with the position barely in balance, the year ended with an aggregate stock decrease in the three main products of a few thousand tons.

Stocks on—	*Aviation Spirit.*	*Motor Gasoline.*	*Diesel Oil.*	*Total.*
1st January, 1941 ...	613,000	499,000	296,000	1,408,000
1st January, 1942 ...	254,000	379,000	164,000	797,000
31st December, 1942	299,000	334,000	156,000	789,000

31. The figures do not, however, clearly show the critical nature of the position. The full import of these statistics only becomes apparent when compared with German estimates of their lowest safety stock margin or "distributional minimum." In 1941 the lowest safety stock margin for the three products was 800,000 tons to ensure efficient distribution throughout the Greater Reich together with occupied Poland and occupied Western Europe. After the invasion of Russia the lowest safety stock margin was estimated at 1,300,000 tons.([24]). Thus, in 1942, the German forces had been supplied from stocks that were at a level of only 60 per cent. of the distributional minimum.

([21]) This explains the slow progress in the construction of the new plants which was duly observed by aerial reconnaissance and which seemed at the time so inconsistent with the stringency in the oil position.

([22]) In this connection the views of Hettlage are of interest:—

"Despite the *débâcle* at Moscow the Nazis continued to regard Russia as being mortally wounded. When von Brauchitsch demanded a radical shortening of the front line up to the Don or even to the German frontiers he was immediately dismissed. The same occurred when Halder and Zeitzler made the same proposals one and a half years later. As late as 1942 Hitler continued to regard the Russian War as being all but won. As a result of all this the aspect of industry did not change to any degree and it was not until late in 1942 that any adaptations were made for a longer war." (U.S.S.B.S. Interview 12A.)

([23]) Krauch recommended the use of "*Ostarbeiter*" and for the reason that "they cannot run away." (Krauch files.)

([24]) Lecture by Krauch to the *Generalrat* on 24.6.41.

32. At a meeting of the *Zentrale Planung* in October 1942 it was reported that all *Wehrmacht* reserves of motor gasoline were exhausted. In the case of aviation spirit the position was hardly any better. Under the revised plans of Goering five of the hydrogenation plants were to increase their output of aviation gasoline to the total extent of 31,500 tons a month. This was to be done at the expense of an equivalent amount of motor fuel which, in turn, was to be replaced by the use of wood-gas generators.([25]) In spite of these arrangements supplies of aviation fuel were not sufficient to meet the peak demand of the Summer. There was a period in September when stocks were down to 202,000 tons, or less than two months' consumption. It was at this time that intelligence sources began to report that there was a serious shortage of supplies, that flying training was being curtailed and that even operational flights were being to some extent affected.

33. Heavy fuel oil for Naval purposes was also in short supply and, for several months, at least one of the hydrogenation plants had to be called upon to produce furnace oil which could only be done at the expense of gasoline output.

34. A breakdown in supplies of fuel for the war-machine had been averted only by two factors, the drastic reduction of civilian consumption and the gradually expanding output of the synthetic oil industry.([26])

35. Whereas, in 1941, the civilian economy of Axis Europe had been allocated 8 million tons of oil products, this was reduced to 5·2 million tons in 1942; of these two quantities Germany's share was reduced from 3·97 million tons in 1941 to 2·47 million tons in 1942. Although these reductions had no measurable effect on industry they, nevertheless, added to the difficulties of increasing output.

36. The increasing output of synthetic oil was largely due to the coming into operation of the plant at Bruex, in Czechoslovakia. At the beginning of the year total synthetic production had been averaging some 200,000 tons a month and by the end of the year this figure was over 260,000 tons and still increasing.

37. Although 1942 had been a year in which disaster had narrowly been averted and stocks had become reduced to the meagre levels of 1939, a critical period had been survived. However, the reversion to the defensive after Alamein and Stalingrad, was to mark a gradual improvement in the German oil position which was to continue until the Spring of 1944.

([25]) A.D.I.(K) Report No. 391A/1945.

([26]) Following a meeting with Goering on 6th July, 1942, Thomas recorded the following note in his files:—

"By this file note I wish to place on record that already prior to the war and continually during the war the Economic Armaments Office has demanded that indirect war requirements, especially basic industries, be more largely built up. These suggestions, however, were always turned down by the *Fuehrer*, the *Reichsmarshal* and the Chief of the Supreme Command because interest was always and is still centred on direct war material. This fact is the reason why the raw material's industry was not built up to an extent as perhaps it might have been. Time and again iron, raw materials and man-power were invested in plain armament plants whereas the requirements of the raw materials' industry were set behind.

The same conditions apply to-day to mineral oils. I state to-day that the day will also come when culprits will be searched for the reason why the mineral oil basis was not enlarged. But furthermore I state that for years the Economic Armaments Office has demanded that the building up of mineral oil plants be brought to the foreground still more than has hitherto been the case. However, at present, leadership and also the Minister of Armament and Munitions, are still much more interested in direct weapons because the *Fuehrer's* wishes in this respect must be fulfilled" (FD. 4809[illegible]).

SECTION VIII.

The Oil Requirements of Italy.

Whatever the strategic advantages of Italy's participation in the Axis, she was a costly liability to Germany in terms of oil. Furthermore, lack of oil contributed to the hastening of Italy's defeat. A short account of the part played by oil in Italy's war is therefore pertinent.

Oil Resources.

2. As in the case of Germany, Italy was not self-sufficient in oil. Her position was, in fact, worse than that of her ally in that the lack of an indigenous supply of coal precluded any possibility of establishing a synthetic industry. With the exception of the negligible production of a small field at Emilia, North of Genoa, the country has no crude oil resources.

3. The conquest of Albania gave access to the small production obtained from the Devoli area. This oil has a low commercial value on account of its small petrol yield and high sulphur content. The output in 1938 amounted to 65,000 tons, increasing to a maximum of about 150,000 tons in 1943. In order to make the best use of this poor quality oil a Government-owned company erected a refinery, with hydrogenation facilities, at Bari and it was designed principally for the production of Naval fuel oil and aviation spirit. In 1943 this plant was processing crude at the rate of 14,000 tons per month. Another plant of similar design was erected by the same company at Leghorn. In addition to these two refineries, the Government had coerced the oil industry into the erection of six modern plants for the processing of imported crudes.

4. For a number of years before the War attempts were being made by the Italians to remedy the lack of domestic crude production by exploration activities in other countries. Some costly but abortive drilling was carried out in a group of islands off the coast of Eritrea. There was also financial participation in some unsuccessful exploration in Hungary.

5. The extent to which the Italian Government coveted the oil resources of other countries is indicated in the diary of Count Ciano in which he writes of Mussolini, in September 1939 ". . . . dreaming of heroic undertakings against Yugoslavia which would bring him Roumanian oil."

Blockade Considerations.

6. In September 1939 the question of limiting Italy's oil supplies was one of the most thorny of the many problems arising from the blockade of Germany. For some time Italy had been gradually building up stocks and by September they had reached a figure of rather more than two million tons which would have been equivalent to about nine months' peacetime consumption. Moreover, it was evident that Italy intended to continue this accumulation of reserves.

7. The Ministry of Economic Warfare held the view that the best way to stop Italy from entering the War on Germany's side was to limit her capacity to fight. On the other hand, the Foreign Office, supported by the Service Ministries, advocated a policy that would incur no risk of provoking Mussolini to abandon neutrality. The War Cabinet agreed with the proposals of the Foreign Office and the adoption of this policy automatically ruled out the various drastic courses that could have been taken to impose a check upon Italy's preparations for War.

8. Any study of the merits of these opposite views would have to take varied circumstances into account. If it had been agreed that a strict limitation of oil imports would have given promise of leaving Italy too weak to fight there might have been difficulty, possibly not fully realised at the time, in preventing Italy from securing substantial imports of Roumanian oil by the overland route through the Balkans On the other hand, a hostile attitude to Italy might well have stampeded Mussolini into a rash declaration of war. It might also be contended that Italy would not have entered the War at all but for Mussolini's view that the collapse of France would imply an early end to hostilities, in which case a weak materials position would not have mattered.

9. As it was, when France was stabbed in the back, the stocks of oil in Italy were fairly substantial and it is probable that no particular anxiety was felt by the Italian High Command as to whether or not these supplies would suffice. The Navy had in reserve the equivalent of more than a year's consumption and the Army had sufficient supplies to maintain full-scale operations for many months. Roumania could still be depended upon to furnish substantial quantities of oil

even though it meant bringing it by a circuitous overland route on account of the blockade of the Eastern Mediterranean. Supplies could also be made to go further by economies in civilian consumption and the savings by these means were substantial, allocations being increasingly curtailed until, in 1943, consumption was less than a quarter of the pre-war level of 2,800,000 tons a year.

10. After the fall of France there was no abnormal increase in consumption and supplies were adequate for the attack upon Greece in October, 1940. The Navy was consuming about 75,000 tons a month of black oils and with stocks on hand of upwards of a million tons.

11 The gasoline requirements of the Army and the Air Force were increasing but the level of consumption was being offset to some extent by restrictions in civilian allocations. Total consumption of all oil products was, however, in excess of production and imports. Before the middle of 1941 the Italian General Staff became aware of difficulties ahead if quick successes were not achieved in Greece and North Africa. The drain on reserve stocks had also been aggravated by a sharp decline in imports from Roumania which was largely caused by transport difficulties. In 1940 imports fell to 342,943 tons as compared with 629,350 tons in the previous year.

12. A new trade agreement with Roumania was concluded in 1941 which resulted in total imports for that year of 761,667 tons. Hungary was also called upon to supply what oil she could and supplies from this source eventually amounted to some 10,000 tons a month. Notwithstanding these imports consumption continued to be greatly in excess of the amount of oil being obtained and it soon became apparent that marine fuel oil was going to be the most critical product. While the consumption of gasoline could be reduced by restricting activity on the home front and by curtailing military training, there could be no such saving in the case of fuel oil with the lines of communication across the Mediterranean and Adriatic to be kept open.

13. The flow of oil from Roumania consequently became of vital importance and a critical situation arose when there occurred a sudden decline in these supplies in the Winter of 1941/42. While this decline was partly due to a large internal demand in Roumania combined with a falling off in production there were other reasons preventing Italy from receiving her share. Germany was likewise suffering from a fuel oil shortage and her aggressive demands upon Roumania not only competed with those of others but also ensured that Germany was given preferential treatment. The freezing of the Danube added to the difficulties in transporting such oil that was made available for Italy.

Naval Fuel Oil Difficulties.

14. Towards the end of 1941 the Naval fuel oil position became critical. Consumption had been at the rate of about 75,000 tons a month and, as imports from Roumania had been averaging less than half this amount and the Bari refinery had been contributing only 7,000 tons a month, the pre-war reserve rapidly approached exhaustion. A curtailment in Naval activity became inevitable. As supplies to North Africa had to be maintained and the convoys escorted, their needs received first priority. Any consumption by the main units of the Italian fleet was precluded and the oil from their bunkers was, in fact, transferred to cruisers and destroyers to enable essential escort duties to be carried out.

15. Early in 1942 it became imperative that Germany should make some contribution if the lines of communication across the Mediterranean were not to be impaired through lack of fuel. Urgent entreaties for supplies were made by the responsible member of the Italian General Staff(1) in person to Field Marshal Keitel and Admiral Raeder. These requests were unwelcome to the Germans as they came at a time when the shortage of fuel oil had reached an acute stage throughout Europe. How unwelcome these representatives were is confirmed by Admiral Raeder's written request to the Chiefs of Staff in Rome pressing them to reduce the operational activities of the Navy. In one of these communications Raeder admitted that the Germans themselves were at that time being obliged to limit operational commitments in the North Sea owing to the shortage of fuel oil.

16. It was, however, necessary for the Italians to be helped out of their predicament and the Germans began to supply Italy with tar oils, the quantities

(1) Admiral Robertelli.

varying from 10,000 to 1[illegible] tons per month. This help was sufficient to ensure the maintenance of shipping movements to and from North Africa, but a major part of the main fleet units still had to be confined to port.(²)

17. It was not until after the fall of Tunisia, and with the consequent saving in marine fuel oil consumption, that a number of capital ships were restored to full complement and permitted to carry out curtailed exercises. The crippling shortage of fuel, nevertheless, prevailed up to the time of Italy's surrender. The lack of oil was such that, during the surrender negotiations, it was questionable whether sufficient supplies were available to enable the Italian fleet to reach Allied ports.

Army and Air Force Supplies.

18. The Italian Army was not a large consumer of gasoline in comparison with the requirements of the other Powers. Some 15,000 tons a month met the needs of military transport and armoured vehicles. Supplies at the time of the attack on France were sufficient for six to eight months and imports of motor gasoline were steadily maintained at the rate of about 17,000 tons a month. These supplies were extended by admixing with alcohol in proportions varying from 25 per cent. to 30 per cent. Civil consumption was progressively restricted from a pre-war level of 38,000 tons a month to less than 10,000 tons a month in the early part of 1943. Although supplies of motor gasoline could therefore have been theoretically adequate to meet military needs the Army suffered from an almost continuous shortage.(³) While this was partly due to bad staff work, defective lines of communications were responsible in large part for shortages in the field.(⁴)

19. In the case of the Air Force the position was still more unsatisfactory. Requirements were about 20,000 tons a month. Production from Italian refineries never exceeded 6,000 tons a month and, when pre-war reserves had been exhausted by the end of 1941, allocations from Germany did not exceed 9,000 tons a month. Evidence is lacking of any restrictions in offensive operations directly attributable to a basic shortage but training was restricted.

Tanker Sinkings and the North African Campaign.

20. The Axis forces in North Africa consumed an average of about 13,000 tons of liquid fuel a month. At no time did circumstances permit of the building up in Tunisia or Libya of a stock sufficient to maintain operations for any considerable length of time and the continuity of supplies was consequently dependent upon the regular arrival of tankers from Italy.

21. Throughout 1941 and in the early part of 1942 tankers were able to make the Mediterranean crossing without incurring very serious risk. The routes taken were generally out of range of Allied aircraft and movement by night from Sicily to the Tunisian coast minimised the danger of submarine attacks. However, as the year progressed the Allied attacks upon enemy shipping became intensified.

22. On the 28th August, 1942, Rommel launched his offensive into Egypt. It began promisingly but the British forces were in a good defensive position which Rommel was unable to break. In four days' time, as Count Ciano recorded in his diary, "Rommel is stationary because of lack of fuel." That this marked the turning point in the African campaign is confirmed by Mussolini in his reminiscences. In referring to the catastrophic fortnight that began with the opening of the Allied offensive at El Alamein on the 23rd October he is reported to have recorded—

> "It (the Allied victory at El Alamein) was of incalculable historical importance. It opened the enemy's march from both East and West. And the strategic initiative passed to the Allies. . . . This is the time when

(²) Any tendency to ascribe Italian Naval inferiority to the shortage of oil would not, however, be in accordance with the facts. There was no shortage in the first twelve months of operations when, on more than one occasion, superior Italian Naval forces avoided action with considerably weaker British forces. It might not be unreasonable to suppose that the Germans did not think that the sending of fuel oil to Italy would be a good investment but the general picture presented shows that, not only the Italians but also the Germans were short of Naval fuel, so that they could probably not have spared supplies for Italy even if the Italian fleet had been a more effective Naval force. However, it should also be recorded that the Italian fuel oil position was critically weak at the time, in the spring of 1942, when their Naval chances were at their best.

(³) General Rossi.

(⁴) The 10 Italian divisions on the Russian front also had their fuel shortages, but these were ascribed to transport difficulties.

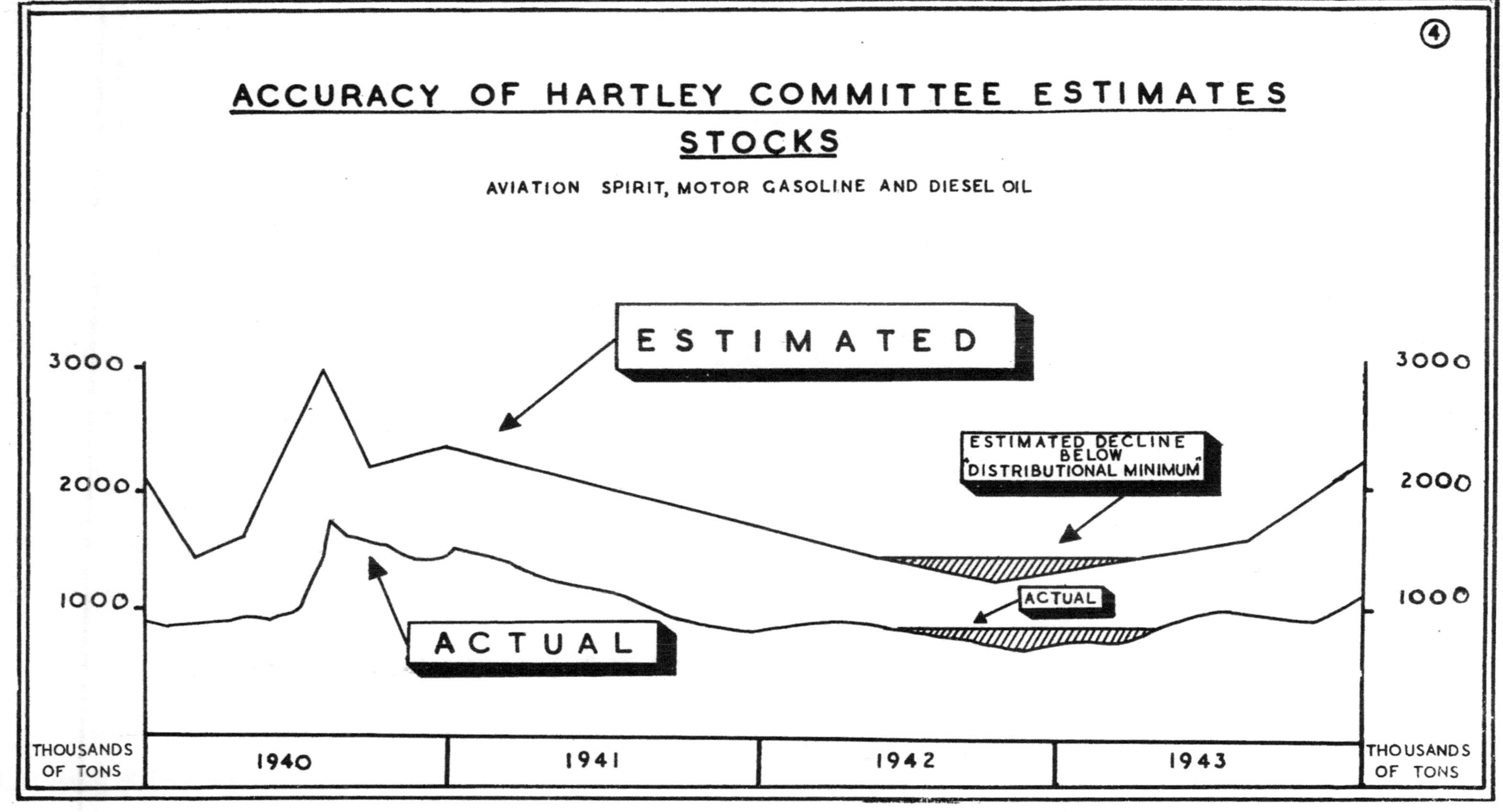

Printed by Photographic Reproduction Branch, Air Ministry, London S.W.1.

Rommel should have abandoned the El Alamein line. He could have withdrawn his Italian troops, who had no motorised transport, to the Sollum–Halfaya defences, since I had some time before told Marshal Bastico to put these in order and fill them with reserves. Then he could have got his German motorised units back in the same way. Thus Rommel would have put 300 miles between himself and the enemy at a single bound. But the German and Italian Command decided to stand at El Alamein and await assault. The battle that broke out so violently on the 23rd October revealed the crushing superiority."[5]

From September 1942 onwards Tripoli was unable to provide with any regularity the fighter escorts necessary for the fuel convoys. Attempts were made to give cover from the mainland, but the units there were themselves short of fuel, and the convoys suffered heavily. Rommel was faced with almost daily crises on account of fuel shortage throughout his retreat.

23. A captured document dated April 1943 gives an indication of the difficulties encountered because of irregular supplies of aviation spirit:—

"Although the intake of B4 fuel was almost twice as great as the quantity received during the previous months it covered only two-thirds of the consumption. As the quantity brought forward on the 1st March was down to about 1,125 tons of B4 fuel, the complete inadequacy of supplies received during the month was accentuated. The fact that stocks of B4 have been allowed to go down from 2,250 tons to 300 tons in the course of two months which have been of decisive importance in the African theatre of war, must be regarded as a dangerous indication of our apparent inability, with the means at our disposal, to overcome problems of supply to Africa. This comes at a time when we must be ready to face most carefully planned and heavy attacks by a numerically greatly superior enemy within an ever decreasing area in Tunisia."

At the end of April 1943, in a desperate attempt to remedy the catastrophic fuel situation, the Army Group in Africa went without *Luftwaffe* ground support so that all available air forces could be freed to escort a fuel convoy.

24. In the closing stages of the campaign the Allied strategic air forces applied the last turn of the screw by attacking the ports through which the fuel supply passed, and the movement of liquid fuel to the German and Italian forces in North Africa was entirely stopped. The number of tankers put out of action by Allied operations is not known. From the 11th January, 1942 to the 19th May, 1943 tankers totalling 82,000 g.r.t. are known to have been sunk in addition to which a considerable number must have been disabled. As a result of these sinkings the Germans resorted to flying supplies of gasoline from Italy to Tunisia in transport aircraft. These were shot down in large numbers during April and early May.

25. In withdrawing to their final positions in the area of Tunis and Cape Bon the enemy had fallen back upon substantial supply dumps of food and ammunition. These dumps had necessarily been well dispersed as a precaution against attacks from the air. The exhaustion of motor fuel and the lack of alternative means of transport left these supplies in isolation. Neither shells could be brought to the artillery nor could the troops be adequately rationed. The rapid advances of the Allies from East and West, combined with the complete breakdown of the enemy's transport system, had rendered 25 enemy divisions incapable of further resistance. From the 1st May to the 11th May the total enemy casualties in killed and prisoners were approximately 110,000. Allied casualties during this period were almost negligible.

---

(5) Cramer has given the following opinion of the battle:—

"Alamein was lost before it was fought. We had not the petrol. Vast stocks of petrol and material were lying around in Italy and the Italians were supposed to bring them over, but they could not do it. Rommel for a long time had known that the campaign in North Africa was hopeless, not because we lacked weapons or reserves, but because of the petrol shortage, and he appealed to Hitler to end the campaign as soon as Alamein was lost and thus save us much greater losses later on—which in fact we suffered at Cape Bon."

(*The Times*, 26.10.45.)

SECTION IX.

GERMANY ON THE DEFENSIVE.

In 1943 all the belligerent Powers were preparing for the major battles to come and, as an essential part of these preparations, the Germans were making haste to consolidate their oil position. Throughout that year, and until the beginning of the Allied air offensive against oil in the Spring of 1944, a determined attempt was made to restore stocks to a more comfortable level. The success of these efforts is shown by the table on the next page. By the end of 1943 stocks of the three main products had increased in twelve months from 775,000 tons to 1,120,000 tons. An account of how this improvement was brought about not only serves to show what Germany was able to achieve during this difficult period, but it also outlines the nature of the target that later became the major objective of the Allied bombing effort.

2. Early in 1943 a number of changes were made in the administrative organisation of the oil industry. The Ministry of Economics was amalgamated with the Speer Ministry and the work of the former was taken over by two departments under Speer: the Raw Materials Department and the Planning Department. Both these Departments were headed by Kehrl. These changes resulted in a closer co-ordination of effort both in the field of production and in ensuring appropriate distribution to consumers.(3)

The Increase in Crude Oil Production.

3. In each successive year since 1940 the decline in production of crude oil in the German *Altreich* had been more than matched by an increase in Austrian production, the total figures for both countries for 1941 and 1942 being 1,465,000 and 1,532,000 tons respectively. This consistently upward production trend had been maintained largely by the policy of directing almost the whole of the drilling effort to producing fields and thereby neglecting exploratory drilling to the inevitable detriment of subsequent production.

4. In the six years 1933–1938, 57 per cent. of German drilling had been done in proven producing fields, the remainder being exploratory drilling; in the next four years, however, 78 per cent. of drilling was in known producing fields and only 22 per cent. for exploratory purposes. Exploratory drilling, by which alone, in the long run, production could be maintained was, therefore, more than halved, and as a result German production consistently declined after 1940.(4)

5. On the other hand, the Austrian fields provided substantial scope for immediate large-scale production. By 1942 Austrian production had outstripped German production, though this would not have occurred until a year later had not Goering ordered that the maximum possible production should be obtained from Austria in 1942 so that sufficient supplies should be available for the German advance that was to envelope the Caucasus. In spite of the opposition of his technical advisers Goering ordered this prodigal expansion in production notwithstanding the consequent jeopardising of the future productivity of the fields.(5)

(3) Speer and Kehrl were faced with the immense task of co-ordinating the control of industry and especially the reorganisation of priorities, the mal-administration of which was stifling a rapid increase in the output of urgently required munitions. Priorities had been by class of industry rather than by individual needs. For example, the tank industry had such a high priority that certificates for raw materials and labour were issued to this industry almost without checking immediate needs. The worst problem, which was the steel industry, was tackled first, and the chemicals industry (which included synthetic oil) was left until later as the industry appeared to be reasonably well integrated under the general influence of *I.G. Farbenindustrie*.

When, early in 1943, Speer and Kehrl turned their attention to the more efficient development of chemicals and oil production, they found *I.G. Farbenindustrie* a reluctant participant in emergency schemes for rapid expansion. The *I.G. Farbenindustrie* were always wanting to see the colour of their money first and they tried to drive hard bargains. Undoubtedly this aloofness, or tendency to put profits before National Socialism, affected, if only indirectly, the speed at which the output of synthetic oil and other products could have been increased. Dealings with the *I.G. Farbenindustrie* were also made more difficult by the fact that they had managed to get their employees into a number of important Government posts and in this way Government policy had hitherto been shaped in their favour. No love was lost between them and Kehrl, whom they regarded as a "small industry" man and an interloper.

(4) Details of the German oilfields are summarised in Appendix 8.

(5) In 1944 the production of Austria reached a peak of about 1,200,000 tons. Since the conclusion of hostilities it has been ascertained that the current economic rate of production should be about 600,000 [illegible]r.

The Annual Stock Position.
(In thousands of tons).

—	Aviation Spirit.					Motor Gasoline.					Diesel Oil.					Total of Three Products.	
	Average Monthly Consumption.			Total Cons.	Total Prod.[1]	Average Monthly Consumption.			Total Cons.	Total Prod.[1]	Average Monthly Consumption.			Total Cons.	Total Prod.[1]	Total Cons.	Total Prod.
	Mil.	Civ.	Exp.			Mil.	Civ.	Exp.			Mil.	Civ.	Exp.				
Stocks at 1.1.40 ...	511					280					138					929	
1940	70	...	...	863	966	77	70	2	1,811	2,130	23	85	1·7	1,335	1,417	4,009	4,513
Stocks at year end ...	613					626					296					1,535	
1941	100	...	...	1,274	910	134	53	21	2,504	2,284	48	78	27	1,856	1,726	5,634	4,920
Stocks at year end ...	254					379					164					797	
1942	101	6	10	1,426	1,472	120	28	24	2,089	2,023	46	54	25	1,519	1,493	5,034	4,988
Stocks at year end ...	324					313					138					775	
1943	134	11	6	1,825	1,917	128	24	21	2,088	2,148	74	46	24	1,307	1,793	5,220	5,858
Stocks at year end ...	440					436					244[2]					1,120	
1944	116	...	...	1,403	1,105	121	23	5	1,805	1,477	76	33	9	1,435	1,260	4,643	3,842
Stocks at year end ...	146					118					121					385	

(1) Production includes imports and loot.
(2) There is no apparent explanation in German Statistics for this discrepancy. The surplus of production over consumption of 486,000 tons is not fully reflected in the figure

6. This uneconomic exploitation, which resulted in a much lower yield from Austria in 1943 than had been anticipated, and this neglect of exploratory drilling, both cardinal sins against the canons of efficiently planned production, nevertheless enabled the total production from Germany and Austria to reach a new high level in 1943 of 1·8 million tons.

Adjustment of Refining Capacity.

7. Until the penultimate stages of the War, when the crude oil refineries were subjected to attack, there was ample capacity for refining the crude oil available. A German report summarises the position as follows :—

	Crude Available. 1943.	Crude Available. 1944.	Annual Refining Capacity.
	(In 1,000's of tons.)		
Hannover Area	515	514	605
Hamburg Area	175	164	2,024
Dortmund Area	...	...	103
Alsace Area	77	65	72
Austrian Area	1,106	1,212	580
Hungarian Area	837	809	1,200
Roumanian Area	5,274	2,142(6)	8,500

The figures exclude limitations in refining capacity due to bomb damage. Poland and Czechoslovakia are not given in this tabulation. The former had a substantial surplus of processing capacity for the production of the Polish fields of about 400,000 tons; the latter had several refineries, all of them designed for processing imported oil. No mention is made of the refineries in Axis hands in Italy, Holland, Belgium and France.

8. These omissions serve to emphasise the overall strength of the position. Taking the figures as listed there was in 1943, and before any attacks were made upon these plants, a total of about 12·5 million tons of more or less conveniently located refining capacity for processing some 8 million tons of crude oil.

9. At the end of 1943 there were some 40 refineries in operation in Greater Germany.(7) Twenty of these were responsible for 85 per cent. of the output, the remainder being principally for lubricants. In addition there were the important refining operations being carried on in Roumania, Hungary, and in other German dominated countries.

10. Notwithstanding this surplus capacity these facilities were not entirely suited to current and future requirements. Additional capacity was required in areas where production was being expanded. There was also a need for cracking capacity for increasing the yield of high octane gasoline, although this need did not become critical until the existing cracking plants and the hydrogenation plants had been seriously damaged by bombing.

11. The promise of Caucasus oil and the increasing output of the Austrian fields resulted in the erection in the Danube Basin of three new plants and the enlargement of an existing plant (the Schwechat refinery of *Deutsche Erdoel*). Only one of these new plants came into operation and then only on a partial scale; this was the Lobau plant of the *Ostmarkische Mineraloelwerke G.m.b.H.*, a joint company of the Rhenania Ossag and Deutsche Vacuum interests. An *I.G. Farbenindustrie* subsidiary, *Donau Chemie A.G.*, put into operation at Moosbierbaum a hydroforming plant for the conversion of white spirit into toluol and aviation fuel base stock, and an oil refining plant was in the course of completion when the war ended. A third plant, with distillation and cracking facilities, which was being erected at Deggendorf by the *W.I.F.O.* organisation, was not ready for operation at the time of capture. All of these plants would have been conveniently placed for the processing of crude from the Black Sea and the Lower Danube.

12. The erection of these plants was assisted by the removal of modern equipment from certain refineries in France. This material was acquired as the result of pressure placed upon the Vichy Government and under promise of its eventual replacement. A large quantity of this equipment was removed in the first instance to provide for the projected refineries in South Russia. Part of it was later used to modernise the refinery at Trzebinia in Poland and the remainder was consigned to the Danube Basin.

(6) Up to 20th August.

(7) Details of the principal German refineries are given in Table 4 on pages 161 and 162.

The Exploitation of Shale Oil Deposits.

13. Shale was a source of oil which, in spite of its comparatively small yield, the Germans did not neglect. Though there are other shale deposits in France and in Germany, the deposits in Estonia were the most profitable to work because of their higher oil content.

14. When the Germans captured the Estonian shale oil area in the summer of 1941 the plants had been thoroughly demolished and no output was possible until 1942 when 60,000 tons of oil were produced as compared with a pre-war production of 250,000 tons a year. However, perhaps because the German Navy had always been a consumer of Estonian diesel and fuel oil, ambitious plans were made for the complete rehabilitation and extension of the industry, this work being entrusted to a subsidiary of the *Kontinentale Oel A.G.* named the *Baltische Oel G.m.b.H.*(8) Planned production for 1943 was 120,000 tons but only 107,000 tons were produced. For 1944 a production of 250,000 tons of oil was planned and for 1945 550,000 tons, the increase being due to a new plant which was to come into operation at the end of 1944. Production in 1944, until the 20th September, was 150,000 tons. Two large plants which had been planned had to be abandoned because of danger from air attack and proximity to the front, and another existing plant at Slantsi had to be dismantled and evacuated. The Germans therefore obtained some 320,000 tons of oil from Estonian shale during their three years occupation of the country and this quantity was obtained at a heavy cost in man-power and equipment for rehabilitation.

15. The proximity of these operations to the Russian front led to a decision in August 1943 to cease the expansion of the shale workings. Later, in May 1944, it was decided that nearly half the 15,000 labourers engaged in the processing of Estonian shale should be brought to Germany. At least 1,000 of these workers were drafted to South Württemberg to help in opening up the new shale workings in that area. In spite of this detraction of effort from Estonia, members of the German High Command have affirmed that Hitler's anxiety to retain this relatively small source of oil in the latter part of 1944 was one reason for the disastrous loss of the divisions on this sector of the front.(9)

16. Estonia was not Germany's only source of oil shale. There is a considerable belt of shale in Württemberg and a certain amount of research and development work had been done there on processes for shale oil extraction in 1942.

17. In the Summer of 1943 the *Planungsamt* directed that the exploitation of these resources should be intensified and work was started in several locations in the vicinity of Frommern and Schoerzingen.(10) At that time only one plant was operating and it was yielding only a small quantity of oil.(11)

The Expansion of the Synthetic Oil Industry.

18. The orderly development and undisturbed operation of the synthetic oil plants remained the sheet anchor of German oil supplies. Since 1940 synthetic production had been Germany's largest single source of supply, and it was, to

(8) See Appendix 4.

(9) Koller has stated:—

"Although in the closing years of the War Hitler was repeatedly advised to withdraw from the northern sector of the German front in the Baltic area, he steadfastly refused to do so in order to maintain his hold on the local shale oil deposits."

(A.D.I.(K.) Report No. 374/1945.)

The same view has been expressed by Doenitz (FD. 4478/45) and Warlimont (FD. 4477/45). These opinions should, however, be noted with caution. Speer has affirmed that there were occasions when Hitler deliberately exaggerated to the military commanders the economic importance of retaining economic conquests. It would appear that this was done to afford an economic pretext to override military logic.

(10) Krauch files (20*j*).

(11) In August 1944 the construction of ten further plants was planned under the Geilenberg programme. These plants, which were known as "*Wuesten*," were to produce 20,000 tons a month of shale oil from the Württemberg deposits by a novel process of carbonisation and electric deposition of oil. Work was pressed forward on these plants, but they were not completed at the time of Germany's collapse. These activities were, however, of an emergency nature and oil from German shale played no more significant rôle in the war effort than that the plans for obtaining it were executed too late and the effort in labour and materials expended on them were largely wasted. Further details are given in Appendices 9 and 21.

all intents and purposes, [illegible] source of aviation spirit. The importance of the hydrogenation plants is shown in the following table.

German Output of Fuels and Lubricants in Terms of Finished Products.
(In 1,000's of tons.)

	1940.	*1941.*	*1942.*	*1943.*	*1944.*
From Hydrogenation ...	1,504	2,107	2,772	3,431	1,875
„ Fischer–Tropsch ...	449	474	446	484	306
„ Crude oil refineries	1,454	1,612	1,729	1,933	1,653
„ Benzol	553	597	585	657 }	1,568
„ Coal tar	612	692	830	985 }	
„ Alcohol	80	60	6	18	10
	4,652	5,542	6,368	7,508	5,412

19. In 1940 there had been nine hydrogenation plants in production with a combined monthly output at the end of that year of 136,000 tons or an annual rate of over 1·5 million tons. In 1943 there were thirteen plants in operation producing oil at the end of the year at a rate exceeding 3·5 million tons a year.

Monthly Production Rate at Year End.
(In 1,000's of tons.)

	1940.	*1941.*	*1942.*	*1943.*	*1944 (April)*
Leuna	42	45	54	54	51
Bohlen	17	19	21	20	23
Magdeburg ...	19	19	18	18	18
Zeitz	15	23	26	20	22
Scholven	21	19	20	18	20
Gelsenberg ...	14	30	36	35	36
Bottrop	7	10	11	12	12
Poelitz	6	11	37	60	62
Lutzkendorf ...	5	2	2	2	...
Wesseling	...	2	13	18	20
Ludwigshafen ...	...	2	4	4	4
Moosbierbaum ...	...	...	5	6	6
Bruex	...	...	3	32	30
Blechhammer ...	...	...	...	...	2
Heydebreck ...	...	...	...	...	1
Auschwitz	...	...	...	...	...
Monthly rate of production at year end ...	146	182	250	299	307 (April)

20. Although the decisions of the General Staff by which the construction was checked of the three plants scheduled to come into operation in 1944 (Blechhammer, Heydebreck and Auschwitz) had reduced the expected output from the hydrogenation plants by over 70,000 tons a month, and although the monthly output of 300,000 tons of all products was the figure which Goering had once fixed as the target for aviation spirit alone, the output that was achieved in 1943 was substantial.

21. During 1943 the plant at Poelitz attained the full output for which it was designed and became the largest producing plant in Germany; Bruex reached half its planned capacity of some three-quarters of a million tons a year, and by the end of the year Blechhammer and Heydebreck were due to start production.

22. Among the plants which had shown the greatest production increases were those which specialised in the production of the components for aviation spirit. This process of specialisation, which had been pursued since Goering's "*Flugbenzin Plan*" of 1941, deserves special notice. These plants were largely devoted to meeting the requirements of the *Luftwaffe*, which, in turn, was the

principal means by which these and other plants could be defended. Thus the circumstances were such that Allied fighter superiority over the *Luftwaffe* units defending these plants would create a situation whereby successful attacks on the comparatively small number of aviation fuel producers would eventually hamstring the *Luftwaffe* and permit the Allied strategic bombers an unrestricted range of action against Germany's industrial resources.

23. In contrast to the rapid expansion of the hydrogenation plants the output of liquid fuel from the Fischer-Tropsch plants showed no increase since **1940**. And, by **1940**, production which had been scheduled under the Four-Year Plan of **1936** to reach **790,000** tons a year had by that year only reached **449,000** tons. The process was, in fact, a disappointment. With the exception of Kamen-Dortmund, not one of the plants ever attained the designed output capacity. In the case of the older plants this was partly due to the poor design of the ovens. The principal difficulties were, however, due to impurities in the synthesis gases and these led to catalyst deterioration.([13])

24. For increased synthetic output all reliance was therefore placed on the coal hydrogenation plants and the only deviation from this policy was a plan to extend the capacity of the Deschowitz (Odertal) Fischer-Tropsch plant, a project which was apparently never realised.([14])

Substitute Fuels.

25. A valuable contribution towards the maintenance of road transport in the face of declining supplies of fuel was afforded by the use of substitute fuels and especially producer gas. The effort to convert to these fuels was, however, too small and too late to have achieved the full benefits that might have been obtained. A German commentator has summed up the position as follows :—

> "All countries without mineral oil resources have long recognised the necessity to convert motor vehicles, particularly lorries, to producer-gas propulsion. Germany alone lagged behind, relying on the production of synthetic petrol and deliveries from Roumania. Also she was certain she would conquer the Caucasian oilfields. When the Red Army dashed this crazy hope and Roumania ceased to be a source of supply, when the bombing continually reduced the production of Germany's synthetic petrol plants, a quick conversion to producer gas was attempted under a scheme of total planning with high priority. The scheme, however, failed. The armaments industry was too fully extended and steel was short. Thus Germany was once again too late. . . . ."([15])

26. Before the War the Germans were already actively developing the use, as motor fuel, of bottled propane-butane, being surplus gases from the hydrogenation plants. But it was not until **1941** that the extensive introduction of the more cumbersome producer-gas form of propulsion was undertaken. From then on the wholesale conversion of civilian road transport to gas generated by the burning of wood, charcoal and anthracite became a question of major importance. It was, in fact, a question that received Hitler's personal attention and the record of his conferences with Speer contains periodical references revealing the anxiety of the *Fuehrer* over the slow progress in the completion of the programme for conversion.

27. The consumption of motor gasoline by civilian and industrial users in Greater Germany in 1940 amounted to 850,000 and in the three following years the figures were 640,000, 345,000 and 298,000 tons respectively. If this consumption had been largely eliminated by conversion to producer gas, which would have been an entirely possible achievement, anxieties over motor-gasoline supplies would have been alleviated and the ground forces would have been more adequately supplied.

28. The plan for converting road transport to substitute fuels was, like other plans, both disrupted and also relegated from first to second priority by the

([13]) Interrogation of Martin. The Japanese also placed considerable hopes upon the Fischer-Tropsch process. They erected a number of plants and their disappointment must have been greater than that of the Germans in that they lacked the skill of the Germans in making the best of the process.

([14]) Meeting of the *Planungsamt*, 25.8.43. Krauch files (20*j*).

([15]) Radio speech by Head of the Economic Department of Berlin Municipality, 8.8.45.

promise of Caucasus oil; [illegible] before this promise appeared unlikely to be fulfilled the conversion programme was lagging considerably behind schedule. In July 1942 Hitler ordered the dismissal of the responsible official([16]) and revised proposals were put into effect. The aim of these plans is shown in the following extract from the record of one of Hitler's conferences with Speer in the summer of that year :—

> "After perusal of documents, the *Fuehrer* approved of (Schieber's) producer-gas programme. A saving of 100,000 tons a month is expected by the middle of 1943. The *Fuehrer* will arrange for it to be ascertained to what extent the heavy supply columns of the *Wehrmacht* can be converted."
>
> "On the 1st June, 1942, 145,000 motor vehicles were converted to substitute fuels. By the 1st October 182,000 will be converted—thereby making an additional saving of 20,000 tons a month. For the period October 1942–June 1943, 300,000 conversions are planned. The fuel supply is assured. The liquid fuel saving will then be over 1 million tons a year."

29. These expected savings were not achieved.([17]) Numerous difficulties conspired against the programme being carried through as planned. In addition to the lack of labour, the inflexible system of allocating raw material priorities that was in force at that time made it impossible for steel and other materials to be supplied in the required quantities. The extent to which the output of producer gas units fell short of the savings in liquid fuel that had been anticipated is shown by the following estimates of the quantities of gasoline made available for other uses as a result of these conversions([18]) :—

	Tons.
1942	125,500
1943	320,600
1944	497,500
1945 (to end of April)	184,700

These figures represent the savings from producer gas only, and in the absence of statistics covering the use of other substitute fuels, an addition of about 50 per cent. might cover the contribution to these savings by the use of bottled gas, town gas, methane and electricity.

30. Even by the middle of 1944, by which time the conversion programme should have reached a peak, the number of vehicles being converted to producer gas and other substitutes was only averaging from three to four thousand a a month.([19])

31. In addition to the inefficiency of these substitutes in comparison with gasoline, the provision of the necessary solid fuel was a costly liability for the German economy. In order to achieve a saving of nearly half a million tons of gasoline in 1944 the following fuel was required :

Wood	4,400,000 cubic metres.
Peat	58,000 tons.
Charcoal and Peat Coke	7,000 tons.
Anthracite	120,000 tons.
Low Temperature Coke	40,000 tons.
Brown Coal Briquettes	200,000 tons.

32. The responsibility for supplying these fuels was largely allocated to a company entitled *Generator A.G.* which was formed in 1940. The capital of this company, which was originally Rm. 1,800,000, was increased to Rm. 4,500,000 in 1942. Later, the Government took a 50 per cent. interest in the company and its capital was increased to Rm. 9,000,000. By 1944 the company's sales had reached a figure of Rm. 100,000,000.

33. One outcome of these attempts to diminish the consumption of gasoline by non-military vehicles was that the production of aviation fuel in the latter part of the war was indirectly impeded by the very plans that were intended to have the opposite effect. This unexpected situation was due to the widespread use of bottled propane and butane gases from the synthetic plants and

([16]) Von Schell.
([17]) Statistics are given in Table 17 on page 170.
([18]) Records of the Mineral Oil Section, *Rohstoffamt.*
([19]) Nagel. (U.S.S.B.S. Interview No. 15.)

UNDERGROUND OIL STORAGE.

The underground *WIFO* storage installation at Nienburg, on the River Weser, photographed by a reconnaissance aircraft on 6th August, 1944, the day after it was attacked by the United States 8th Air Force with 665 tons of bombs. This main Army and Air Force supply depot was constructed in the latter part of 1934. Some 30 tanks, with a total capacity of almost 100,000 tons, are deeply buried beneath a cover of evergreen trees and scrub. The arrows show where the bomb craters have penetrated the tanks in some seven places, approximately one-quarter of the storage capacity being damaged. The attack fractured the buried pipe connections between the tanks and, on account of the difficulties in making repairs, the installation was totally out of action for several months.

refineries as a fuel for motor vehicles. When, in 1942, it was decided that a much greater proportion of aviation spirit must comprise high octane fuel, the logical source of this fuel would have been those gases. However, the withdrawal of bottled "*Treibgas*" from public use would have created such dislocation that it was decided that the two immense plants at Heydebreck (Blechhammer South) and Auschwitz would have to produce those high octane fuels by the cumbersome, low-yield process of converting iso-butylene to iso-octane. The inability to make extensive use of butane and propane for the manufacture of high octane spirit represented a theoretical loss of between a quarter and a half a million tons of aviation fuel a year.

Consumption Requirements.

34. While production was steadily increasing a close control was being maintained of consumption. The defensive rôle of the German armies was helping to limit consumption. In addition the collapse of Italy conferred an indirect benefit by making available to Germany the share of Roumanian oil that had been going to the Italians. In 1942 Italy had imported 862,179 tons of oil from Roumania and in 1943 these imports amounted to only 391,354 tons. Total German acquisitions from Roumania in these two years were 2,191,659 tons and 2,511,304 tons respectively.

35. The aviation spirit position was showing a steady improvement. The average monthly consumption of aviation spirit through 1943 and for the first five months of 1944 was approximately 150,000 tons, of which 135,000 tons were consumed by the Luftwaffe.[20] Service consumption necessarily varied with operational activity, peak figures of 176,000 tons having been reached in July 1943 and 195,000 tons in May 1944. Industrial consumption, which was mostly for the testing of engines, was almost totally eliminated by August by the use of alternative fuels, and thereafter any quantities used for these purposes were too small for inclusion in consumption statistics. In the same month the export of aviation spirit to the Italians was discontinued. The Italian Air Force had for long been completely dependent upon German supplies of aviation spirit. Although these allocations were small they were a steady drain on German supplies and, in combination with the allocations for industrial uses, effectively prevented any German stockpiling until both were discontinued. After August 1943, since it was no longer necessary to supply the Italian Air Force and at the risk of mechanical inefficiency through limitations in engine-testing, aviation spirit production was, for the first time since the start of the War, well ahead of consumption. An average of 160,000 tons a month was available to meet an average monthly *Luftwaffe* consumption of 135,000 tons. This situation continued until May 1944, when aviation spirit stocks at 574,000 tons were higher than at any other time during the War with the exception of the period of eight months immediately following the fall of France when the capture of 245,000 tons raised the stock level to a height out of all proportion to the production/consumption balance.

36. Army requirements of motor gasoline in 1943 and early 1944 averaged about 135,000 tons a month, and civilian and export requirements at 25,000 and 21,000 tons respectively brought the total average monthly consumption to about 180,000 tons a month. To set against this, production was averaging 110,000 tons monthly and imports were providing a further 88,000 tons, leaving an average surplus of production over consumption of slightly more than 10,000 tons a month. There were fluctuations in consumption requirements and the stock increase did not take place with the regularity which consideration on the basis of averages would imply. The position was no more than in uneasy balance Nevertheless, stocks had grown by April 1944 to 510,000 tons, which, as with aviation spirit, was the highest figure recorded except immediately after the capture of 300,000 tons in France.

37. The situation with regard to diesel oil during this period was comparable to that of the other fuels. Average monthly consumption of the sea and land forces was 75,000 tons, civilian requirements averaged 45,000 tons

[20] An indication of Allied air superiority at this time is given by comparing this figure with the Allied consumption of aviation fuel in the European theatre of war which was then averaging between 500,000 and 600,000 tons per month.

and exports 25,000 tons. Production and imports averaged 150,000 tons a month, thereby allowing a small monthly addition to stock which continued until April 1944 when stocks stood at almost 300,000 tons, very little short of the total which had been reached two years previously when diesel oil had production priority and stocks were swollen with French loot.

38. Lubricating oil supplies had dropped to dangerously low levels in the early months of the year and in August it was necessary to impose a rationing scheme. Production was, however, steadily increased and especially by means of securing a higher yield from domestic crude oils.

Stocks and the Preparations to Resist Invasion.

39. On the 1st May, 1944, German fuel stocks were as follows :—

	Tons.
Aviation Spirit	574,000
Motor Gasoline	506,000
Diesel Oil	256,000
	1,336,000

This total was considerably more than the comparable figure of September 1939. The aviation gasoline was sufficient to meet up to three months' requirements on the basis of previous rates of consumption. The motor gasoline and diesel oil stocks were each equivalent to over two months' consumption.

40. Territorial losses had to some extent reduced the distributional minimum but, nevertheless, the margin of free stocks to meet future emergencies was hardly large enough for the position to be reviewed with equanimity. Moreover, with invasion threatened upon a coastline of great length there was no prospect of laying down a comfortable sufficiency of local reserves to meet all contingencies.

41. The actual stocks established on the threatened fronts are not known but the total quantity more or less immobilised in this manner was not large. In France, where the threat was greatest, the total strategic reserves, exclusive of tactical reserves and current supplies, were to the order of only 10,000 tons. In the opinion of one authority([21]) the strategic reserves in France were as large as the Germans could risk sending there. With production increasing, the Germans were hoping to meet without undue difficulty the increased consumption requirements of the battles to come. But this was dependent upon the continuous operation of the producing plants.

SECTION X.

THE PREPARATIONS FOR THE OIL OFFENSIVE.

A study of the German oil position was made by the Air Ministry in the summer of 1939, when it was concluded that the situation was exceptionally vulnerable. Oil was consequently earmarked as a priority bombing target ranking second only to certain military objectives.(¹)

2. In the list of targets, priority was given firstly to the stocks of oil at refineries and above-ground storage depots, secondly to the hydrogenation plants, thirdly to refineries operating on domestic crude, and fourthly, to the Fischer–Tropsch plants.

The Early Attacks.

3. On the basis of this appraisal and upon the reports of the Industrial Intelligence Centre, systematic attack of oil production and oil storage as a primary objective was attempted, with more enthusiasm than success, by the inadequate forces available to R.A.F. Bomber Command in the summer of 1940. In view of the small amount of damage that was inflicted, the description of these attacks need only be confined to a brief record of what occurred.

4. Between May 1940 and the end of March 1941 twenty-nine oil plants and storage installations were attacked on a total of 344 occasions. The total bomb load carried, both high explosive and incendiary, amounted to about 1,500 tons. The attacks were carried out by small numbers of Blenheims, Wellingtons, Whitleys and Hampdens, the weight of bombs carried necessarily being small in relation to later standards. Some of the attacks consisted of single sorties, and others averaged from seven to eight aircraft. The two biggest attacks comprised one of thirty-eight aircraft on Leuna on the 16th August, 1940, and one of fifty-six aircraft on Gelsenkirchen on the 9th January, 1941.

5. Reports on the results of these attacks are few and far between, and this absence of any detailed record of damage done may be taken as significant of their ineffectiveness.(²) Although on many occasions pilots reported bursts on the targets or fierce fires, the fact remains that notwithstanding the best efforts of the air crews the technique of bombing had not, at that time, been sufficiently advanced to permit of either the accurate pin-pointing of the target or the means for observing results. In general it would seem that these efforts by Bomber Command had no effect on German morale except possibly to stimulate it. Perhaps the best that can be said for these attacks is that they provided a school of experience the benefits of which were reaped to the full at a later date. It could also be recorded that the optimistic reports that were being given to the British public at that time on the extensive damage being done to Germany's war potential were proving an anodyne to the gloomy despatches of the reverses on the fighting fronts.

6. The one reported exception to these otherwise abortive attacks was the mission against the hydrogenation plant at Gelsenkirchen Nordstern, in the Ruhr, on the 10th October, 1940, when five Wellingtons, carrying a total of 4·7 tons of bombs, secured hits on certain vulnerable parts of this plant and caused a stoppage of production for three weeks. Although the air crews reported large explosions in the target area a confirmatory report on the damage done was not received until two months later. This plant, one of the largest in the Ruhr, was not again made the subject of a specific attack until the 13th June, 1944.

7. In the Spring of 1941 it became apparent that the weight and accuracy of the attacks which could then be made were inadequate for the purpose and, as counter-invasion measures developed an increasing urgency, the attack upon oil was soon suspended. No further action against the enemy's oil plants was taken until two years later.

(¹) An account of the Intelligence appreciation of the German oil position is given in Annex A.

(²) There is, however, a report in the Krauch files (20 *f*) which refers to losses due to air attacks from the beginning of the War to 22.6.41 as being 35,000 tons of aviation spirit and 35,000 tons of other products. It is presumed that these are losses of products in storage and that the figures do not include loss of output due to inability to manufacture.

In the Annual Report of the *Planungsamt* of 29.6.44 losses from enemy action to the end of 1943 are put at "approximately 150.000 tons." (F.I.A.T. Final Report, No. 403.)

The Casablanca Directive.

8. On the 21st January, 1943, the Combined Chiefs of Staff issued a Directive (subsequently known as the "Casablanca Directive") to the Commanding General, Eighth Air Force, and the A.O.C.-in-C., R.A.F., Bomber Command, as follows:—

"Your primary object will be the progressive destruction and dislocation of the German military, industrial and economic system and the undermining of the morale of the German people to a point where their capacity for armed resistance is fatally weakened.

2. Within that general concept, your primary objectives, subject to the exigencies of weather and of tactical feasibility, will for the present be in the following order of priority:—

(*a*) German submarine construction.
(*b*) German aircraft industry.
(*c*) Transportation.
(*d*) Oil plants.
(*e*) Other targets in enemy war industry.

The above order of priority may be varied from time to time according to developments in the strategical situation."

9. To implement the policy laid down in the Casablanca Directive, the Air Staffs of the British and United States Air Forces examined in detail the specified target systems and produced a comprehensive plan for its execution, nominating the targets which were to be attacked and indicating the effort which would be required to achieve the desired result. In dealing with oil as one of the primary objectives, the Combined Bomber Offensive plan stated that—

"The quantity of petroleum and synthetic oil products now available to the Germans is barely adequate to supply the life blood which is vital to the German war machine. The oil situation is made more critical by failure of the Germans to secure and retain the Russian supplies. If the Ploesti refineries, which process thirty-five per cent. (35%) of current refined oil products available to the Axis, are destroyed, and if the synthetic oil plants in Germany, which process an additional thirteen per cent. (13%), are also destroyed, the resulting disruption will have a disastrous effect upon the supply of finished oil products available to the Axis."

In addition, the plan asserted that a successful initial attack on the key elements of the system would demand the concentration of effort on the remaining elements of the system to exploit the initial success.

10. The plan envisaged the integration of the attacks of the Royal Air Force and the United States Air Forces, and stated that—

"All-out attacks imply precision bombing of related targets by day and night, where tactical conditions permit, and area bombing by night against the cities associated with these targets."

The Combined Bomber Offensive Plan.

11. The fulfilment of these plans, however, had to await the time when the operational capabilities of the Air Forces had been built up to the level judged to be equal to the task. Another factor also delayed the execution of the task. In the weeks following the Casablanca Directive the strategic air situation had altered. As a result of the increasing scale of destruction which was being inflicted by the night bomber forces and the development of the day bombing offensive, the Germans were deploying day and night fighters in increasing numbers on the Western Front. It became evident that, if the aims of the Combined Bomber Offensive were to be achieved, it would be necessary to check this increase in enemy fighter strength. Accordingly, a Directive was issued to the Air Commanders on the 10th June, 1943, giving effect to the conclusions reached in the Combined Bomber Offensive Plan and placing the German fighter strength on the highest priority for attack by the Eighth Air Force The list of priority objectives was thus stated:—

Intermediate Objective—

German fighter strength.

Primary Objectives—

German submarine yards and bases.
The remainder of the German aircraft industry.
Ball-bearings.
Oil (contingent upon attacks against Ploesti from the Mediterranean).

Secondary Objectives—

Synthetic rubber and tyres.
Military road transport vehicles.

The Directive also stated that—

" While the forces of the British Bomber Command will be employed in accordance with their main aim in the general disorganisation of German industry, their action will be designed, as far as practicable, to be complementary to the operations of the Eighth Air Force."

The Low-Level Attack upon Ploesti.

12. Although it was not possible to begin the all-out attack on oil until May 1944, one special preparatory operation was carried out on the 1st August, 1943. This was an attack on the major oil refineries in the Ploesti area and it was executed by Mediterranean-based heavy bombers of the U.S.A.A.F., reinforced for the purpose by formations transferred from Great Britain.

13. The attack was delivered from low-level by 175 Liberators and pressed home with extreme gallantry. As the operation involved deep penetration into enemy territory it could not hope to achieve tactical surprise. In addition to the lack of fighter protection the aircraft were exposed to heavy fire from the defences at the low height at which they were flying. The operation consequently involved a heavy loss in casualties.

14. The damage inflicted was not as thorough as had been intended. This was due to a navigational error on the part of one section of the attacking force causing some of the objectives to be hit twice, thus minimising the damage to certain other refineries. Furthermore, an appreciable number of the bombs dropped failed to explode. Nevertheless, although this operation did not succeed in causing decisive damage to all the principal units in the Roumanian refining industry, considerable destruction was done and a temporary loss of oil output occurred. The main achievement of the attack was to eliminate permanently the surplus of effective refinery capacity,(3) in relation to crude production, which had formerly existed in Roumania, and thereby to render more immediate the effects of damage subsequently inflicted.(4)

15. However, the beneficial results of the attack on the refineries were to a certain degree offset by the warning which was provided to the enemy that the Roumanian oil industry was at risk. As the attack was not immediately followed up, the enemy was able to employ the following months in devising additional protection for the equipment of the refineries, in reinforcing the fire-fighting services, in installing smokescreens and in strengthening the defences. All these measures increased the difficulties of the United States 15th Air Force in inflicting decisive damage in 1944.

(3) See page 44.

(4) If the operation had been wholly successful the output in Roumania of refined products, and especially gasoline, would have been reduced to a trickle for a period of at least a number of weeks and for some time thereafter output would have been much below essential needs. The two most important consequences would have been the blow to Roumania's internal economy and the cessation of fuel exports to the German army.

In the case of the former, rail communications, which were almost wholly dependent upon fuel oil for their operation, would have been severely curtailed for at least several weeks and Roumánian economy as a whole would have been abruptly shaken. On the other hand, Roumania was at that time inaccessible to any Allied land attacks that might otherwise have been made to take advantage of her weakened condition.

In regard to the possible stoppage of supplies to the German Army, Roumania was at that time supplying directly to the *Wehrmacht* an average of about 60,000 tons of liquid fuels a month, a large proportion of which was gasoline. Practically all this oil was being sent to the forces on the Rusian front. On the 5th August the Russians took the offensive, capturing Orel and Byelgorod, and by the 7th November Kiev had been retaken. German stocks of gasoline in August and September were about 400,000 tons and monthly consumption was slightly more than half this quantity. While these stocks would probably have been sufficient to cushion a sudden stoppage of supplies from Roumania, an already weak gasoline supply position would have been further weakened.

16. The operation also delayed to some extent the execution of the other obligations of the Combined Bomber Offensive Plan. The transfer of bomber formations from Great Britain and the time expended in low-level attack training in North Africa represented some diversion of offensive effort.

Inability to Begin the Oil Offensive before 1944.

17. In the course of interrogating the German leaders surprise has been expressed at the apparent failure of the Allied Air Forces to begin the systematic destruction of liquid fuel supplies before the spring of 1944. However, there were a number of reasons why this task was not attempted earlier.

18. It was considered first necessary to reduce the surplus refining capacity in Roumania. This was achieved by the costly *tour de force* of the 1st August, 1943. However, the lessons of this attack were not necessary to make the planning staffs realise that oil could not be successfully attacked until air supremacy had been won. In the spring of 1943 the United States 8th Air Force in Great Britain had only six groups of bombers as compared with forty groups a year later. They had no long-range escort fighters and as their first deep penetration into Germany(5) cost a loss of 20 per cent. of the aircraft engaged, a decision was taken to attempt no further long-range sorties until adequate escort forces were available.

19. It was not until January 1944 that the necessary long-range escort of P-51 fighters became available and deep penetrations could be made into Germany without occurring expensive losses. With the whole of the *Reich* thus made available for daylight attacks, the first task to be performed was the suppression of the *Luftwaffe.* This was necessary both to ensure the effectiveness of the strategic bombing offensive and also to reduce the striking power of the *Luftwaffe* in resisting the projected Normandy landing.

20. It was decided that this objective would best be attained by a concerted attack upon aircraft production and a week of perfect weather at the end of February favoured the execution of this task. Following the February attacks upon the aircraft plants the preparations for the landing in Normandy were imminent and the Strategic Air Forces were put under the control of the Supreme Commander. It was only when S.H.A.E.F. was satisfied that the conditions requisite for the landing had been created that effort could be diverted to the attack of oil.

21. As regards operations in the Mediterranean theatre no sustained effort against the Roumanian refineries could be made by the United States 15th Air Force until the airfields at Foggia were in Allied possession. These were occupied in November 1943, but the bombers were bogged down by winter conditions until the following spring. However, when operations became possible it was decided that other tasks than oil would first have to be undertaken. The penetration of the Russian armies to the frontiers of Roumania had cut the Lwow–Cernauti railway, by which the southern group of German armies had hitherto been largely maintained, and the main weight of the German supply traffic was suddenly thrown upon the Roumanian and Hungarian railways which were ill-prepared to receive this additional burden. In the operations of the United States 15th Air Force, first priority was for a time given to the attack of communications in Eastern Europe, including the mining of the Danube. However, in the course of these attacks the proximity of several Roumanian refineries to rail centres resulted in damage being inflicted upon them. In addition, the damage done to communications caused some interruption in the distribution of oil products.

---

(5) An attack upon the ball-bearing industry at Schweinfurt.

SECTION XI.

The Beginning of the Strategic Bombing Offensive against Oil.

Early in 1944 those directing the Combined Bomber Offensive were chiefly concerned with plans for the best use of the Strategic Bomber Forces in connexion with the coming invasion of the Continent. The main plans under review at this time were :—

(*a*) Reduction of the German Air Force, by attacks on aircraft factories, airfields and associated installations, and on the ball-bearing industry.

(*b*) Support of the landings and subsequent land operations designed to establish a lodgment area in Normandy.

(*c*) Attack of oil resources.

2. A directive issued by the Combined Chiefs of Staff on the 17th February named the German Air Force as the primary objective of the Strategic Bomber Force, as follows :—"Overall reduction of German air combat strength in its factories, on the ground and in the air . . . . in order to create the air situation most propitious to 'Overlord '(1) is the immediate purpose of the bomber offensive." In a later paragraph the directive stated "preparation and readiness for the direct support of 'Overlord' and 'Rankin '(2) should be maintained without detriment to the Combined Bomber Offensive."

Selection of Plans to Support "Overlord."

3. On the 25th March a meeting was held, attended by the Chief of the Air Staff, the Supreme Commander, the Deputy Supreme Commander, Commanders of the Bomber Forces and representatives of the War Office, at which the relative merits of various plans were discussed. As a result of these discussions the Chief of the Air Staff, with the agreement of the Supreme Commander who was shortly to be made responsible for the control of the Strategic Bomber Force, made the following decisions :—

(*a*) The attack of oil would not have a significant effect on operations in Europe during the vital period of the landings and the first five or six weeks after D-Day.

(*b*) The attacks on the German Air Force must continue.

(*c*) The plan to disrupt enemy rail transportation, recommended by the Air Officer Commanding-in-Chief, Allied Expeditionary Air Force, should be adopted.

4. On the 14th April, 1944, in accordance with the decision of the Combined Chiefs of Staff, control of the Strategic Air Forces passed to the Supreme Commander, Allied Expeditionary Force.

5. In a directive issued on the 17th April, 1944, to the United States Strategic Air Forces and to Bomber Command "for support of "Overlord" during the preparatory period," the Supreme Commander laid down as the particular mission of the Strategical Air Forces :—

(*a*) To deplete the German Air Force and particularly the German fighter forces.

(*b*) To destroy and disrupt the enemy's rail communications.

6. Although the Transportation Plan received first priority, a certain degree of bombing effort was directed against oil targets, and the economic intelligence in favour of attacking oil production was frequently represented to and fully appreciated by the Supreme Commander. In April, the 15th United States Air Force operating from Italy attacked transportation objectives at Ploesti and for tactical reasons directed the surplus effort, necessary to flood the defences, on to the neighbouring refineries. These attacks were followed by a 1,000-ton raid on the railway centre and refineries in Ploesti on the 5th May.

7. By mid-May, after meeting the needs of the Transportation Plan, a surplus of bombing effort was still available from the United States Strategic Air Forces. This effort was directed against oil targets. Two heavy attacks were accordingly made by the 8th Air Force on the 12th and 28th–29th May in which over 2,500 tons of bombs were dropped on the major producing plants at Leuna, Breux, Poelitz, Bohlen, Luetzkendorf, Magdeburg, Zeitz and Ruhland, which were responsible for about 40 per cent. of the total output of synthetic oil.

(1) "Overlord" was the code name given to the plan for the invasion of the Continent.

(2) "Rankin" was the code name for a plan similar to "Overlord" but was designed to deal with conditions which might be met if German resistance collapsed suddenly.

First Phase of the Oil Offensive.

8. The Germans had been in constant fear of the bombing of their oil plants and had been expecting the attacks for some time. Early in **1943** they had concluded that submarine plants and liquid fuel plants were likely to be the primary targets(3), and Speer has confirmed that oil and high grade steel were the two target systems upon which attacks were most feared(4).

9. In spite of these apprehensions little progress had been made with passive air defence preparations and the plants had little or no protection from either direct hits or blast and splinter effects(5). One reason for this unpreparedness was the cost in materials and labour of providing even partial protection; one scheme for the protection of all plant buildings had been calculated to require an expenditure of RM.6,000 million and to involve the employment of **1,200,000** workmen for a year(6).

10. The fact that a blow had been struck at a vulnerable point soon became evident from numerous intelligence reports, and the importance of following up the attacks of the 12th and 28th May was obvious. During the first two weeks of June appreciations of the situation were prepared by the United States Economic Objectives Unit, the staff of the United States Strategic Air Forces in Europe and the General Staff of Supreme Headquarters, Allied Expeditionary Force. The conclusions reached in these appreciations are of interest in the light of later events.

11. The first of these reports expressed the view that "If engagements continued on three fronts, the elimination of two-thirds to four-fifths of German oil output would force collapse on one or more fronts once small military reserve stocks were used up." The collapse would not come on "strategic" grounds but directly through the armies in the field. It was recommended that to cause the main effect from September onwards the elimination of output was advisable in June and July with the elimination of the military reserve stocks the following month.

12. The second of these reports was entitled "Use of Air Power against Enemy Military Transport and Supplies." Apart from introducing a communications interdiction plan this report considered attacks on industry in Germany which could directly affect the enemy's military potential in the West. In considering oil this document repeated that "it is estimated that continued attacks on the enemy's oil industry could vitally affect his military capabilities within a matter of months." It concluded that "the most direct and quick results will be obtained by concentrating air effort on the interdiction of communications in France and the Low Countries, and oil and tank engine production in Germany and the rest of Europe.

13. The third report, entitled "Plan for the Employment of the Strategic Air Forces," took the same view and added that "Germany is facing an oil crisis which can probably be turned into military collapse if the efforts of all available Air Forces are simultaneously directed ruthlessly against this one system of targets." It was emphasised that the primary objective of the suggested strategic bombing programme was the denial of the enemy's armed forces of oil. Other targets such as armoured fighting vehicles were added only because of their vulnerability and importance to the enemy's ground forces.

14. As a result of these appreciations and the accumulating intelligence upon the effectiveness of the initial disruption caused to oil production the Supreme Allied Commander directed that any effort surplus to the support of the Normandy landing should be applied to oil targets. After "D" day on the

(3) Meeting of the *Reichsanstalt Luftwaffe*, February 1943.

(4) FD. 2960/45. Some German opinions upon the bombing of oil are summarised in Appendix 12.

(5) Speer has made an interesting observation upon the lack of passive air defence preparations:—

> "In 1916, when the Leuna plant was built in Central Germany, Ludendorf insisted that the works be put underground, which shows the real greatness of that man. Our chemical industry learned nothing but built their whole industry very carelessly in regard to their attack. One could have given them much more protection against air attacks if just some parts could have been protected by concrete." (U.S.S.B.S. Interrogation, 19.5.45.)

(6) Details of the arrangements made for the protection of the plants are given in Appendix 19.

TOTAL SYNTHETIC FUEL PRODUCTION BY PROCESS

AVIATION GASOLINE PRODUCTION AND TONNAGE OF BOMBS DROPPED ON SYNTHETIC FACILITIES

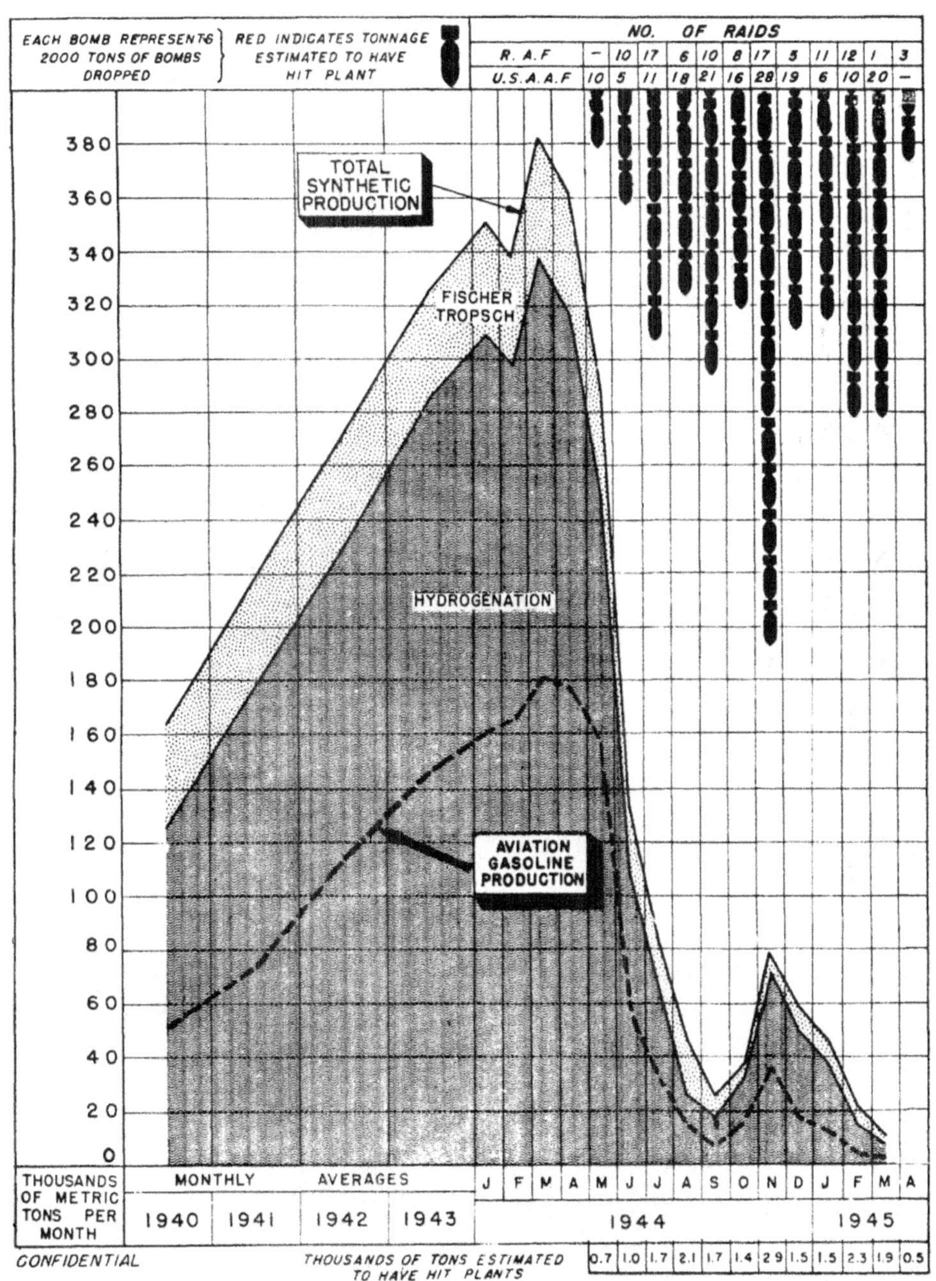

DIAGRAM BY U.S.S.B.S.

Figure 5

6th June, attacks on transportation targets were determined on the basis of current intelligence and not by previously determined targets lists as in the pre-" D " day phase. These changing conditions permitted a total of 17,033 tons of bombs to be dropped on oil targets during June, this representing 12 per cent. of the tonnage dropped in that month.(7)

15. Realising the inter-relation of communication and oil attacks the Supreme Allied Commander directed that R.A.F. Bomber Command should turn its attention to the synthetic oil plants in the Ruhr and Western Germany and heavy attacks were made on Gelsenkirchen Nordstern, Scholven, Wesseling, Sterkrade-Holten and Homberg. At the same time the U.S. 8th and 15th Air Forces were following up their attacks of the previous month by dropping a total of 7,782 tons in the course of 26 missions against oil targets.

16. A large proportion of the effort was devoted initially to the attack of the hydrogenation plants which were the logical choice in view of their importance as gasoline producers and of the importance of gasoline supplies to the enemy's land and air operations in forthcoming battles. For the same reason, and because of their immediate importance in the supply of the Eastern front, the U.S. 15th Air Force concentrated primarily on the attack of the Ploesti refineries.

German Reactions.

17. The initial round of attacks benefited from the relative unpreparedness of the enemy and inflicted serious damage. The output of the Roumanian refineries was reduced at the end of June to 21 per cent. of the April output. But the position in Germany was even more serious. The results of the damage are clearly set out in a personal letter(8) from Speer to Hitler dated the 30th June and the following extracts suffice to show the realistic view that was taken of the probable future consequences of these attacks. The underlining of certain phrases is by Speer himself.

> " My *Fuehrer*,
>
> In the course of June the enemy's attacks on the synthetic oil plants and refineries were carried out with increased strength. By means of his continuous air reconnaissance and espionage, he was again able to inflict heavy damage on those plants which had been hit in the previous month—plants which were mostly just coming into production again. . . . .
>
> In May and June the attacks were concentrated on the German aviation petrol production. The enemy thereby succeeded, on the 22nd June, in bringing the loss of aviation gasoline up to 90 per cent. Only through the most rapid repair of the damaged plants, whose return to production was in every case far in advance of the originally laid down date, will it be possible to restore a part of the catastrophic loss of the 22nd June. Nevertheless, the output of aviation spirit is wholly insufficient at present. . . . .
>
> The drop in production for June and the small amounts which can be expected in July and August in the present state of air attack will doubtless use up the greater part of the reserves of aviation spirit and other fuels.
>
> Thus the repair of these plants, as the experience of June demonstrates, will be useless unless we succeed in taking all possible measures to protect the synthetic plants and refineries. Otherwise it will be absolutely impossible to cover the most urgent of the necessary supplies for the Wehrmacht by September, *in other words, from that time onwards there will be an unbridgeable gap which must lead to tragic results.* . . . .
>
> (Signed) SPEER."

(7) All bomb tonnages are given in short tons of 2,000 lbs. The allocation of bombing effort in June was as follows:—

	Tons.
Tactical Targets	53,772
Transportation	36,431
Cities and Areas	32,080
Oil	17,033
Other Industrial Targets	3,040
Total	142,356

Bomb tonnage statistics are given on pages 171 to 184.

(8) See Appendix 12.

18. Speer lost no time in organising counter measures. Repairs were put in hand with the maximum expenditure of labour and material. Active steps were also taken to provide additional protection for the most vital parts of the plants and for the personnel; this called for a priority programme for the supply of no less than 800,000 cubic metres of concrete.

19. Against the wishes of Krauch, Speer put a stop to all construction work on all new plants that could not be completed within three months. Although this meant abandoning the plan whereby aviation spirit production was to have been increased from 180,000 tons to 210,000 tons by December, there was made available a large quantity of new material which could be used for patching up the damaged plants.

20. Liquid fuel consumption on the home front was cut by 35 per cent. and this cut was to be increased in July to 42 per cent. of the May allocation. The agreement of the head of the army motor transport organisation was secured to a programme for fitting military vehicles with producer gas; supplies of wood were to be obtained by the ruthless felling of trees.

21. Speer also made strong recommendations for various measures to be taken that were outside his province. He urged that flying should be reduced drastically as " in a couple of months we may grievously regret every ton of fuel which is now used up unnecessarily." At the same time, he insisted that the fighter protection for the plants must be considerably increased, " for the *Luftwaffe* must clearly understand that, should the attacks continue successfully, they will, anyhow, only be able to operate with part of their fighters by September, owing to a lack of fuel."

22. Speer recommended considerable increases in smoke and *Flak* protection. " Despite the already strengthened *Flak*, a further *Flak* defence should be provided, even at the expense of the protection of German towns." He also asked for *Flak* protection for the new plants at Blechhammer and Heydebreck, in Upper Silesia, which would not be ready to come into production for one or two months. At the same time an intelligent view was taken of the importance of Allied air reconnaissance :—

> " The importance of shooting down reconnaissance planes was pointed out to the *Fuehrer*. These reconnaissance planes will always enable the enemy to discover the resumption of activity of our hydrogenation plants. It may be assumed that the enemy considerably over-estimates the extent of destruction and time taken for reconstruction, just as we did in our earlier attacks on England, and would considerably postpone his attacks if the reconnaissance planes did not always enable him to discover the early restarting of the hydrogenation plants." (9)

23. Finally, Speer suggested that the *Wehrmacht* must calculate how the War should be carried on if only part of the present quantity of fuel was available.

24. No time was lost in giving effect to these recommendations in so far as conditions permitted. From the latter part of June onwards there was a staccato fire of decrees and directives from the *Fuehrerhauptquartier*. No manpower of any kind was to be withdrawn from the synthetic oil industry. A total of 7,000 technicians were to be released immediately from the Army and used as a central repair reserve. Mine sweeping on the Danube must be improved. Fuel cuts must be made in Hungary. The construction work in the Estonian shale plants was to continue. An Italian refinery must be kept working. To bridge inroads into stocks further severe cuts in consumption must be made in Germany " but the *Fuehrer* asks that small allocations to the Party are not reduced."

25. Speer was exercising to the full his remarkable ability for organisation. And he missed no opportunity for making the most of the powers wielded by Hitler in order to get things done. The records of his conferences at the *Fuehrerhauptquartier* show that not all the talking was done by Hitler and that Speer was able to request action where action was needed.

The Second Phase: July–September 1944.

26. It soon became obvious that every effort was being made to put the damaged plants back into operation. Intelligence sources reported that Hitler

(9) Hitler Conferences. FD. 3353/45. A summary of the technique employed in the interpretation of reconnaissance photographs of oil plants is given in Appendix 11.

had appointed a Special Commissioner for the repair operations. Air reconnaissance was also reporting the construction of large labour camps adjacent to the plants and the photographs were showing ant-like columns of marching labourers entering the plants within a few hours of their attack.

27. In these circumstances it became necessary for the Strategic Bombing Forces to be guided by a planning committee engaged solely upon a close and continuing study of the condition and capabilities of all producers or potential producers in order to regulate the priorities for further attacks. To this end a Joint Oil Targets Committee, representing the Bomber Operations and Intelligence Directorates of the Air Ministry, United States Strategic Air Force and Ministry of Economic Warfare, was set up at the beginning of July—

(i) to keep the Axis oil position under continuous review;
(ii) to assess the effectiveness of attacks;
(iii) to determine priorities.

28. This Committee, at its first meeting on the 6th July, endorsed the policy of giving first priority to the attack of gasoline production. This necessitated concentration of attacks whenever possible on hydrogenation plants, and on refineries in Roumania, Hungary and Poland, which were operating on local crudes giving a high gasoline yield. The first priority list was drawn up on this basis and the same principle was observed throughout the offensive.

29. Priority was not immediately given to the attack of the oil distribution system since there was uncertainty regarding the vulnerability of underground storage and as to the extent to which it might be in use owing to the depletion of strategic oil reserves. However, it became evident that these depots formed an essential part of the distribution system to the armed forces. It also seemed likely that, even if underground storage was not easily vulnerable to attack, the effective working of the depots could be severely disrupted by damage to the loading facilities. The principle depots in Germany were accordingly added to the bottom of the priority list with a view to experimental attacks.

30. During July and August, and notwithstanding unusually bad weather for that time of year, good progress was made by both Air Forces in the Western theatre, the more distant targets being dealt with by means of shuttle missions, using bases in Russia. The success that was being achieved can be gauged by the note of exasperation and despair in Speer's monthly reports to Hitler.([10]) On the 28th July he wrote :—

"The attacks on the synthetic oil plants and refineries in July had the most dire consequences.

It was possible for the enemy, in most cases, to destroy the plants so effectively, shortly after work in them had been resumed, that instead of the expected increase there was a decrease in production, although the reconstruction measures taken lead to the anticipation of a substantial increase. . . ."

31. Speer's report of the 30th August was even more gloomy :—

"The last air attacks have again hit the most important chemical (oil) works heavily. Thereby the three hydrogenation plants, Leuna, Bruex and Poelitz, although only recently in commission again, have been brought to a complete standstill for some weeks. . . .

The effect of these new raids on the entire chemical industry are extraordinary as severe shortages will occur not only in liquid fuels but also in various other important fields of chemistry. . . .

With these results the enemy has hit the chemical industry so heavily that only by abnormal changes in the conditions is there any hope for the retention of the bases for powder and explosives (Methanol), Buna (Methanol), and nitrogen for explosives and agriculture. At the same time the loss in carburettor and diesel fuels is so widespread, that even the severest measures will not be able to hinder encroachments on the mobility of the troops at the front.

The possibility of moving troops at the front will therefore be so restricted that planned operations in October will no longer be able to take place. With this fuel situation, offensive moves will be impossible.

([10]) Appendix 12.

The flow necessary for the supply of troops and the home country will, therefore, be paralysed in the late autumn of this year, since substitute fuels, such as producer gas, are also inadequate to provide the essential help in all sectors. . . ."

32. The United States 15th Air Force was making successful attacks during this time on the principal refineries in Austria and Hungary as well as upon a number of the smaller refineries in South-Eastern Europe. But the maintenance of the initial advantage achieved in the attack of Ploesti proved a more difficult matter. The early warning obtained by the enemy of the approach of the bombers enabled him to have in operation, on most occasions, effective smoke-screens which prevented visual aiming by the bulk of the forces employed. For many weeks the attack on Ploesti appeared to achieve indifferent results and refinery production staged a substantial recovery in July and the early part of August.

33. More disturbing to the Germans at this time, than the attacks upon Ploesti, were the effects of the mining of the Danube.([11]) Shipping was being dislocated to the extent that transport was proving a greater bottleneck than the damage to the refineries.([12])

34. Much thought was given during these weeks to the problem presented by the Ploesti smoke screens and two alternative plans were examined in case it should eventually prove necessary to abandon the direct attack of refineries. The first envisaged the attack of tank farms and collecting stations on the oilfields, the destruction of which would have hampered the handling of crude oil from the fields to the refineries. The second envisaged interdiction of the rail routes by which finished products were despatched from Ploesti together with attacks on the pipeline terminals and other key-points in the handling of waterborne traffic by the Danube. Both plans exhibited serious disadvantages and it was decided that their examination should not in any way detract for the time being from the continuance by all possible means, of direct attacks on the refineries themselves. Perserverance with the original aim proved in the event to be an effective policy. In the middle of August the United States 15th Air Force succeeded in delivering a number of successful attacks against the leading active plants. These successes were rapidly followed by the Russian occupation of the oilfields and of Ploesti itself and by the surrender of Roumania.([13])

35. The capture of the Roumanian oil industry exercised a far-reaching and immediate influence on the situation. Not only did it remove from Axis control productive capacity which had accounted for more than a quarter of the total production in the previous month and deprived part of the Central European refining industry of crude feed-stocks, but it also released the strength of the United States 15th Air Force from its most onerous commitment and led to the immediate intensification of attacks against other targets in the Mediterranean theatre which had hitherto received only the surplus effort. By reason of this relief, the United States 15th Air Force was also able to take an increased share of the task of attacking the synthetic plants in Central and Eastern Germany and made the elimination of the Silesian synthetic industry its particular and primary concern.

36. With the loss of Roumania a further fall in production in September was inevitable. However, the sharpness of the fall which actually occurred was due to the speed with which the United States 15th Air Force exploited the lightening of its burdens among the Central European refineries, and to a long overdue spell of good weather in Western Europe in the second week of September. This opportunity was seized by the Western Air Forces for a series of intensive attacks on the major active synthetic plants and refineries in Germany.

([11]) During the summer of 1944 Wellington and Liberator aircraft of 205 Group, based on Foggia, laid over 1,500 mines in the Danube between Giurgiu and Bratislava. Very few tankers or any other vessels succeeded in reaching the upper river between May and October. Not less than 250 craft of all types, including 29 tankers, were sunk and at least 200 vessels were damaged. The results therefore suggest one ship sunk or damaged for every 3·4 mines laid. This cannot be far from an all-time record for mining. The operations were assisted by a small intelligence group of the Royal Navy which was operating clandestinely in the Danube area during most of this period. One consequence of these operations was that the bombing offensive against rail communications in south-east Europe was made more effective. (A.W.I.R., No. 294, 2.11.45.)

([12]) Speer. (U.S.S.B.S. Interrogation, 19.5.45.)

([13]) The results of the bombing of the Roumanian oil refineries are summarised in Appendix 13.

37. These attacks succeeded in temporarily putting out of action all the remaining plants that had been capable of operation. Between the 11th and the 19th September no plants of any kind were operating. Production in the month of September consequently dropped to the lowest figures recorded in 1944.

Additional Oil Targets.

38. This situation prompted an attempt to find means of broadening the front of the attack on liquid fuel supplies in order to intensify the results and exploit the critical position in which the enemy was now placed. After careful consideration it was decided to extend the oil target system to include the principal sources of benzol production and also selected targets in the oil distribution system.

39. The supply of motor gasoline was being implemented to the extent of about 30,000 tons a month by the production of benzol from coke ovens and gas works. Approximately two-thirds of this production was derived from twenty of the largest coking plants, the majority of which were situated in the Ruhr.(14) On the 13th September these twenty plants were included in the priority target list.

40. Although the principal depots in the oil distribution system had been carried on the attack priority list, in a low position, for several weeks, little effort had been devoted to their attack. While fairly good results had been obtained from attacks on an underground depot at Montbartier in France, the observed results of a heavy, experimental attack on the 5th August by the United States 8th Air Force on a main W.I.F.O. depot at Nienburg, near Osnabrueck, were inconclusive. Reconnaissance photographs showed a large concentration of bomb craters on the target but there was no visible damage to the deeply buried storage tanks. It was not until the 23rd December that aerial reconnaissance perceived extensive excavation work in progress denoting much more extensive damage than the first pictures had led the interpreters to expect. This attack consequently confirmed the vulnerability of these installations, although this vulnerability was dependent upon the attackers being able to locate a cleverly concealed target that was devoid of any easily observable features.

41. There was also the question of the quantity of products that any of these depots contained. Intelligence sources were generally able to confirm which of the principal storage centres were in use, but in view of the heavy decline in production and the large surplus of storage capacity available, there was reluctance at this time to believe that any of these places held stocks sufficiently large to justify the weight of attack necessary to neutralise them. At the same time it was appreciated that the attack of the storage system, in order to be effective, should be comprehensive and carefully timed so that the maximum efforts might coincide in the optimum manner with any Allied ground offensive. For these reasons the occasion did not seem propitious for according these targets a higher priority for attack and precedence was given to the attempt to achieve a deeper cut in production by extending the offensive to the attack of benzol plants.

42. The second phase of the oil offensive ended with the Air Forces being temporarily so much on top of their task of immobilising the synthetic plants and the refining industry that they were able to take on the additional commitments of attacking the benzol industry and, when opportunity offered, the oil distribution system. This had been achieved although oil as a target system was not, during these months, first priority in the bombing directive.

---

(14) The calculation of the actual amount of benzol being produced and allocated to motor fuel at this stage of the War was a complex study involving a number of difficult factors. The Committee's estimate of 32,000 tons a month proved satisfactorily close to the mark.

Although questions were raised at the time as to whether benzol plants so near the front line could still be operating, the importance of these producers was correctly appreciated. On the 15th September Speer reported to Hitler:—

> "The coking plants in the Saar are currently producing benzol, which is distributed to the various divisions to stretch the fuel supply. . . . . It has been proved that a relatively good production is still possible just behind the front line because the enemy concentrates his air attacks on the big shunting stations far behind the front . . . ." (FD. 4734/45.)

SECTION XII.

The Concluding Phases of the Offensive.

As a result of the progress made in the land operations the control of the Strategic Air Forces reverted on the 16th September, 1944, to the Commanding General, United States Army Air Forces and the Chief of the Air Staff, the following directive being issued:—

"1. The Combined Chiefs of Staff have decided that executive responsibility for the control of the Strategic Bomber Forces in Europe shall be vested in the Chief of the Air Staff, Royal Air Force and the Commanding-General, United States Army Air Forces, jointly.

2. The Deputy Chief of the Air Staff, Royal Air Force and the Commanding General, United States Strategic Air Forces in Europe are designated as representatives of the Chief of the Air Staff, Royal Air Force and the Commanding-General, United States Army Air Forces respectively, for the purpose of providing control and local co-ordination through consultation."

2. On the 25th September a new directive was issued in which oil was the first priority:—

"*Directive for the Control of Strategic Bomber Forces in Europe.*

1. In accordance with instructions received from the Combined Chiefs of Staff the overall mission of the Strategic Air Forces remains the progressive destruction and dislocation of the German military, industrial and economic systems and the direct support of land and naval forces.

2. Under this general mission you are to direct your strategic attacks subject to the exigencies of weather and tactical feasibility, against the following systems of objectives:—

First Priority.

(i) Petroleum industry, with special emphasis on petrol (gasoline) including storage.

Second Priority.

(ii) The German rail and waterborne transportation systems.
(iii) Tank production plants and depots, ordnance depots.
(iv) M.T. production plants and depots."

3. With this development the Joint Oil Targets Committee handed over its responsibilities to the Combined Strategic Targets Committee which was entrusted with the duty of furnishing the Deputy Chief of Air Staff and the Commanding General, United States Strategic Air Forces, continuously with advice as to priorities between the different systems of strategic objectives and the priorities of targets within these systems. The Combined Strategic Targets Committee was also in turn to be advised by the combined working committees for each of its target systems.(1) The Working Committee of the Joint Oil Targets Committee

(1) The full terms of reference of the C.S.T.C. were defined as follows (CMS. 606/ACAS (Ops.) of 13th October, 1944):—

"The Terms of Reference of the Combined Strategic Target Committee are as follows:—

(i) To advise jointly the D.C.A.S. and C.G./U.S.S.T.A.F. and to make recommendations in regard to—
  (*a*) the priority of targets within the various systems of strategic objectives selected for attack under the current directive;
  (*b*) the priorities which should be established between the different target systems;
  (*c*) the need which may arise at any time for a major change in the current directive;
  (*d*) any proposals submitted by S.H.A.E.F., the Admiralty and War Office involving the employment of strategic bomber forces.
(ii) To issue on behalf of the D.C.A.S. and C.G./U.S.S.T.A.F. weekly priority lists of strategic targets for attack under current directive.
(iii) To formulate and to submit to the D.C.A.S. and C.G./U.S.S.T.A.F. joint proposals to meet specific situations as and when these may arise.

The Combined Strategic Targets Committee will be responsible for issuing all target priority lists and will be advised by combined Working Committees on—

(i) Oil.
(ii) Army Support Objectives.
(iii) G.A.F.
(iv) Any other target system which may, from time to time, require examination."

Details of the work of this Committee are given in Appendix 11.

accordingly became the Working Committee of the Combined Strategic Targets Committee without any change in its constitution or functions.

The Third Phase: October–November, 1944.

4. The success of the land battle in Western Europe had a considerable effect on Allied bombing operations. The collapse of the enemy's early warning system with the loss of his chain of coastal radar stations enabled Bomber Command to penetrate the Ruhr in daylight with fighter cover. Even more important, "OBOE" stations were set up on the Continent extending the range of this blind bombing and navigational aid to cover targets in the Ruhr and Western Germany, and enabling R.A.F. Bomber Command, in particular, to increase the density of individual attacks in these areas.

5. It was, however, recognised that although oil production had been brought to a low level in September there would be considerable difficulty in preventing output from rising. It was appreciated that the normal progress of repair would bring the majority of the major synthetic plants and refineries back into partial operation during the month unless their prompt re-attack could be ensured.(2) At the same time expectations of deteriorating weather made it doubtful whether successful re-attack would be possible with the regularity previously achieved. Moreover, the effect of deteriorating weather on reconnaissance possibilities would make it much more difficult to check the exact dates of resumption and hence would seriously hamper the effectiveness of the previous policy of allowing major plants to resume production before moving them into high priority for re-attack.

6. It was accordingly decided that if production was to be prevented from increasing, maximum use would have to be made of blind-bombing possibilities, even at the expense of discharging a greatly increased tonnage of bombs in order to achieve the former results. At the same time an important change was made in the policy regulating re-attack. Whereas it had formerly been the policy to spread attacks in order to achieve the maximum immediate effect on production, it was now decided to use every opportunity to inflict major long term damage on the principal plants. It was also agreed to pursue this aim, if opportunity occurred, even in the absence of accurate and up-to-date information from reconnaisance on the status of the target, provided that there was a reasonable presumption that the plant was at least in a "near-active" condition. For it was certain that there would be long intervals during which new photographic cover would be unobtainable owing to the weather.

7. Owing to the weather factor, the task facing the air forces in even maintaining the advantage previously gained was therefore a formidable one, and despite the permanent reduction in commitments resulting from the fall of Roumania.

8. Allied apprehensions as to the possibilities of preventing production from increasing were justified, and October proved to be the most critical month in the whole offensive. On the 6th October Speer sent a courier to Karinhall to advise Goering that the following aviation fuel producers had resumed, or would resume, operations on the dates given:—

		Dates of Subsequent Attacks up to end of 1944.
Poelitz (high octane production)	20th September	7th October; 21st December.
Poelitz (additional production)	6th October	
Moosbierbaum	2nd October	3rd, 6th, 7th November; 11th December.
Leuna	2nd October	7th October; 2nd, 8th, 21st, 25th, 30th November; 6th, 7th, 11th December.
Blechhammer	8th October	14th October; 2nd, 17th, 18th, 19th December.
Bruex	1st November	16th, 20th October; 16th, 20th, 25th December.

(2) The proportion of attacks that resulted in a plant being rendered incapable of further repair was small. A note on the complex subject of aiming error and weapon effectiveness is given in Appendix 20.

He advised that it would be necessary to ensure that the fighter protection would be available and ready a few days before and a few days after the start of production so that the enemy attacks could be "disturbed considerably." A copy of this message was at the same time sent to Galland.(3) Meanwhile the Targets Committee had predicted that these plants would be resuming operations on about the dates given by Speer.

9. Throughout October and November the weather lived up to the worst expectations. During this time R.A.F. Bomber Command was having to resort to the aid of "FIDO"(4) for night operations. Opportunities for attacks in conditions of satisfactory visibility were consequently rare. In addition, on account of the difficult conditions the results of a large proportion of the attacks made were inconclusive.

10. The importance of inflicting long-term damage on these targets was fully understood, but suitable opportunities for attack were limited by the adverse weather conditions. Although the targets remaining to be dealt with were few in number they were mostly situated near the extreme limits of operational range. They were also difficult to pin-point on account of smoke screens by day and the lack of prominent topographical features by night. The plants were elaborately defended, and the two that were the most important at this time, Leuna and Poelitz, were the most heavily protected objectives in Germany. Furthermore, as reconnaissance photographs were only being obtained at rare intervals there was always the possibility that a costly attack, with restricted chances of success, might have been delivered against a plant that would not be ready to resume operations for some time. Finally, on a number of occasions when conditions would probably have been suitable, the needs of the ground forces for direct support had to take priority; these needs included the breaking of the dykes at Walcheren in October and combating the Ardennes counter-offensive in December.(5)

The Achievements of the Plant Repair Organisation.

11. At the same time the tremendous repair organisation for the restoration of oil output was beginning to operate with maximum effect. This organisation was sponsored by Speer who had realised in the early stages of the oil offensive that a prodigious and continuing task of rehabilitation would have to be carried out.

12. Before the end of May, upon the advice of Speer, Hitler had summoned Edmund Geilenberg to his Headquarters and appointed him special commissioner for the repair and dispersal of oil plants. Geilenberg had previously made a name for himself in the organisation of munitions production and he was known as an extremely able executive.

13. To accomplish his task he was given powers which, in the words of Speer, "would make everybody's hair stand on end." In a decree of the 30th May, signed by Hitler, Keitel and Lammers, his operations were given unrestricted priority over all other measures. These operations were to be carried out with the most generous application of labour and materials, and with reckless energy. He was authorised to use Hitler's name in demanding the assistance of military formations in expediting repairs.(6)

14. The importance attached to the whole project was such that manufacturers had only to prefix the word "Geilenberg" to their requirements for any material even remotely associated with the reconstruction programme for

(3) Hitler Conferences, Vol. 99. FD. 3353/45.

(4) "FIDO" was the means of dispersing fog on airfield runways.

(5) In the last three months of the year R.A.F. Bomber Command carried out 38 attacks on oil targets (20 by day and 18 by night). A study shows that there were 7 other nights and 3 days when weather conditions might possibly have permitted attacks but other strategic targets were taken. Operations against other strategic target systems during this period were carried out on 35 days and 46 nights.

In the case of the United States 8th Air Force oil targets were persistently attacked in October. In the following month all flying days were allotted to oil targets with the exception of three days when other strategic targets were taken. Consecutive daily attacks against certain targets was considered inadvisable for operational reasons although at one point Leuna was attacked on no less than four consecutive days. Owing to the bad weather there were 37 days when there were no operations or scheduled attacks had to be abandoned. Attacks had to be limited to conditions not worse than 5/10ths cloud.

The United States 15th Air Force, based on Foggia, maintained, whenever conditions permitted, a persistent attack on all active oil plants throughout this period.

(6) Hitler Conferences, Vol. 90. FD. 3353/45.

PHOTOGRAPH BY U.S.S.B.S.

THE *RHENANIA-OSSAG* REFINERY, HAMBURG.

The refinery of the *Rhenania-Ossag Mineraloelwerke A.G.*, a subsidiary of the Shell Group of companies, had an output capacity of 630,000 tons a year and its equipment included the largest lubricating oil plant in Germany.

During the oil offensive the refinery was bombed on sixteen occasions resulting in the destruction of approximately 75 per cent. of all the buildings and equipment in the plant area.

Until the refinery was put totally out of commission in October 1944, rapid repairs after each attack enabled an output to be achieved that exceeded the estimates of the Targets Committee. However, the timing of each attack was well judged, the attacks invariably occurring just before or just after the resumption of operations.

During the course of the attacks the personnel employed increased from 660 to 1,256, yet the total casualties sustained amounted to only three killed and three injured.

such requirements to be given immediate production and transport priority. Although skilled technicians were not available in the numbers required, and notwithstanding withdrawals of technical personnel from the armed forces, there was an abundant supply of unskilled labour, the great majority of which were foreign workers together with political prisoners. Figures are not available of the numbers employed, but Speer has estimated that the number of workers actually engaged in the repair of the plants eventually reached a total of between two hundred and three hundred thousand, a figure that must have represented a costly demand upon Germany's resources of man-power and especially of skilled workers.(7)

15. Geilenberg, in his preliminary conferences with Hitler and Speer, was faced with the problem of either continuing the uneven contest of repeatedly building up plants to have the bombing forces knock them down again, or to cut the losses, disperse the industry, and "go underground." A momentous decision had to be made. The right policy to pursue was dependent upon the correct answers to such difficult questions as the probable duration of the War, the future fuel requirements of the Army and the Luftwaffe, the extent of success that might be achieved by the Allied bombers under winter conditions, and the possibility of improved defences thwarting these attacks. Furthermore, putting the industry underground was a task of such magnitude that, at first, the possibility of such an undertaking did not receive serious consideration. The larger hydrogenation plants each covered over a square mile in area, and the dangerous nature of the process necessarily demanded wide spacing of the processing equipment and open air conditions for the dispersal of released gases. Each of these plants had taken three or more years to complete, and no less than ten of these plants were being relied upon for the essential and minimum requirements for the defence of Germany. Added to these problems were the difficulties that would be involved in moving these immense plants away from their sources of raw material and also the difficulties of ensuring that access to the new sites and the domiciles of the workers would be invulnerable to attack.

16. Speer himself was originally of the opinion that the industry could not be dispersed, let alone put underground. Geilenberg took the same view, and the first policy adopted was to repair and protect at all costs ten hydrogenation plants and one Fischer-Tropsch plant.(8) Before the attacks on oil began nine hydrogenation plants were producing 90 per cent. of 318,000 tons a month from hydrogenation, the remaining 10 per cent. of the 318,000 tons came from the nine other hydrogenation plants. Total production of the hydrogenation plants at that time contributed 48 per cent. of Germany's available supplies of 662,000 tons a month. It is significant that the ten plants selected to withstand the Allied attacks were referred to as "Hydrierfestungen" or "hydrogenation fortresses."(9)

17. The contest of reconstruction versus destruction achieved an important degree of success during this period of adverse weather. This temporary recovery in output, which is shown in the graph of synthetic fuel production facing page 56, resulted in an output of one-third of a million tons of aviation and motor gasoline, a quantity that enabled the Army and the Luftwaffe to continue the struggle until May.(10)

(7) Exclusive of construction and repair personnel, the total number of persons employed in the oil industry (synthetic and crude oil) in Greater Germany in September 1944, were 139,531, of which 22,846 were women. These comprised 69,525 German civilians, 25,974 Germans released from the armed forces, 34,034 foreign prisoners and Jews, and 9,000 prisoners of war.

(Source: *Wirtschaftsgruppe Kraftstoffindustrie.*)

Fischer has estimated the total employed on oil plant reconstruction as 350,000.

(8) Sauer also confirms (U.S.S.B.S. Interview No. 48) that in July it was maintained that underground plants could not be built and that this decision was not reversed until September.

(9) Krauch files (20*k*).

(10) By the end of September army supplies of gasoline were on a hand-to-mouth basis and the position was strained to the utmost. If adverse weather conditions had not prevented the effective bombing of the active and near-active plants, army supplies would have been rapidly reaching total exhaustion by the end of October. Moreover, the length of the "pipeline" at that time, in terms of delivery time from producing plant to railhead, was not more than about three weeks. This conclusion is confirmed by Speer, who reported on the 30th August, 1944 (see Appendix 12) that under these conditions "the possibility of moving troops at the front will therefore be so restricted that planned operations in October will no longer be able to take place." By November the operations of armoured fighting vehicles, self-propelled guns and a large part of the army's transport organisation would have been crippled.

18. Speer summarised the results of these attempts at reconstruction in a statement he made at that time:—

> " We have sometimes managed to resume full production, even if it was only for three, four or five days, and that after a reconstruction period of six or eight weeks. Then the plant is smashed up again. We start again and reconstruct it, then again everything is smashed up and again everything has to be rebuilt.
>
> This is a tiresome job which absorbs all the energy of the people concerned. It is a terrible thing to get everything smashed up again and again. By this method, in October, November and December, we managed to produce a little more fuel than we consumed so that the aircraft fuel reserve could be increased by 8,000 or 10,000 tons during this period. It must, however, be added that the quantity which we produced per month, and which was about equal to requirements, is only about one-fifth of the aviation (fuel) production which had previously been considered by the *Luftwaffe* to be absolutely essential."(11)

19. Speer was exaggerating in his statement that for three months the production of aviation fuel was in excess of consumption although it was true that a recovery in output during November resulted in a production of 46,000 tons while consumption in that month was reduced to 41,000 tons. Speer had been hoping for a substantial recovery in production during these months. In a telegram sent on the 12th September to prominent Government officials he stated:—

> " After a fair weather period of extraordinary length we may from experience expect a season of predominating bad weather and fog. The bombing of synthetic plants cannot then be carried out with the same precision, our own air force will need less fuel in the same period, and may without disturbance strengthen and reorganise itself, aided by the ever-increasing production of fighter aircraft . . . .
>
> In spite of the really considerable damage done we can, in a period of five to six weeks, restore production to about two-thirds of the level attained by synthetic plants and refineries before the attacks. This production would suffice to cover the fuel requirements of our entire air force, considering the present reduced areas of activity and operational possibilities."

20. At about this time Geilenberg had made his big decision to disperse and go underground. With September as the starting month it was hoped that the first small production from the new plants might start in February. These hopes were not realised.(12)

The December Attacks.

21. Despite the adverse weather a heavy scale of attack had been maintained in November. No less than 35,558 tons of bombs, which was 31 per cent. of the total tonnage dropped, were directed against oil targets. This compared with 13,950 tons in the previous month. Although the accuracy of these attacks was hampered by bad visibility the recovery in production was checked. Decisive damage was inflicted upon a number of important producers. One by one the Ruhr synthetic plants were crippled and fell out of the battle, destined to remain inactive for the rest of the War.

22. There remained, however, a number of plants that were substantially recovering from earlier attacks. Several of them, notably Poelitz and Leuna, were again producing important quantities of gasoline. Their distance from our bases, and the adverse factors created by the season of the year, rendered their successful attack hazardous and difficult. There were also other urgent tasks to be done. The Ardennes counter-offensive demanded the services of the Strategic Air Forces from mid-December until early January, attacks being necessary upon enemy communications and tactical targets.

23. The weather conditions, combined with the need to lend support to the land forces, reduced the bomb tonnage used against oil targets in December

(11) Speech at a Staff Officers' Course, 13.1.45.

(12) A description of the plan, and the progress made with it, is given in Appendix 21. The actual output of oil obtained from the Geilenberg dispersal plants is detailed in Table 9 on page 164.

to 15,779 tons. However, a turning-point in the oil offensive arrived in the middle of the month when the weather at last relaxed its severities. Operations at this period were noteworthy for two developments.

24. Firstly, the United States 15th Air Force carried out, over a period of ten days, what was in many ways the most remarkable series of sustained operations of the whole offensive. This achieved in particular the immobilisation of the Silesian synthetic plants, which was clinched four weeks later by their capture by the Russian Army, and the stoppage of production by the synthetic plant at Bruex which was working up to a substantial output after it had been heavily damaged early in the previous summer.

25. Secondly, R.A.F. Bomber Command carried out successful night attacks against the synthetic plants at Poelitz and Leuna, thus demonstrating that in suitable weather conditions the great weight of attack wielded by the night bombers could now be brought to bear effectively against any major oil target in Germany.

26. With these two developments, the final outcome of the oil offensive was never again in doubt.

The Final Phase : January–April 1945.

27. The opening of the final phase of the oil offensive was marked by the realisation, born of the successful operations of the previous month, that the capabilities of the air forces had become more than equal to the task of finishing the destruction of the enemy's oil industry. However, the setback to Allied offensive plans caused by the German counter-offensive in the Ardennes provoked some doubts as to whether the continued attack of oil would constitute a sufficiently effective contribution by the strategic air forces to the shortening of the War. It was appreciated that if the enemy should succeed in prolonging the War into the summer, action might be necessary on high priority to hold in check the enemy's development of jet aircraft, submarine warfare and long-range weapons, and that this would detract from the effort available for the attack of oil. At the same time prolongation of the War might allow the enemy to complete the Geilenberg programme of dispersed oil production facilities and bring into action producing plant adequate for his minimum needs which would be immune from effective attack.

28. On the question of the further strategic effects of the continuing attack of oil, the Targets Committee sought the opinion of the British Joint Intelligence Committee, who were asked to evaluate the strategic effects of the further destruction of oil production plants which the air forces expected to be able to achieve by the 1st March. The Joint Intelligence Committee expressed the opinion that successful completion of the task specified would result in the almost complete immobilisation of the German Army and Air Force within a period of six weeks.

29. The possible results of the Geilenberg programme had been under detailed consideration for many weeks with the assistance of a number of co-opted industrial experts. In the light of later knowledge the conclusions reached by this investigation were remarkably close to the actual extent of the achievement of the Geilenberg programme. While it was considered probable that a small but increasing production of motor fuel would be forthcoming from a number of dispersed and concealed distillation units from January onwards, it was regarded as unlikely that there would be any production of aviation fuel from new plants before the summer of 1945. These predictions proved to be correct.

30. Consideration of the possibilities of further attacks on oil production, as revealed by these appreciations and by current intelligence as to the effects of previous attacks on the enemy's military capabilities, led to the confirmation of the policy of allotting first priority to this task. This decision was endorsed by the Chiefs of Staff at their meeting on the 24th January, 1945, and who requested that the following note be forwarded to the Commanders of the Strategic Air Forces and with a copy to the Supreme Allied Commander :—

> " In view of the valuable results achieved by the relatively small percentage of the Allied bombing effort which has so far been directed against oil targets, the Chiefs of Staff are of the opinion that, despite the difficulties involved which are fully realised, the utmost efforts should be made to achieve

total success in our attacks on the major enemy oil plants. They believe such a success to be possible and hope that all concerned with the execution of the operations against oil and with the regulation of calls on bomber aircraft for other necessary duties will do all they can to achieve it."(13)

31. On the 31st January it was agreed that, while the main synthetic oil plants should continue to hold first priority for all the strategic air forces, the lesser oil targets should be suspended in favour of attacks upon communications and other targets. Effect was given to this policy by the omission from the priority list distributed to Commands of the bulk of the oil storage targets and of several minor production plants. The possibility of treating benzol plants in the same manner was considered, but in view of the importance which benzol was assuming as a means of eking out the inadequate motor fuel supplies on the Western Front and the convenient situation of the Ruhr coke-oven plants for that purpose, it was decided to continue the attack of these plants on first priority.

32. The suspension at this time of the priority accorded to the attack of the strategic oil storage installations was not considered by the Targets Committee to detract seriously from the overall effectiveness of the attack of the enemy's oil supply since the conclusion had been reached that the attack of the distribution system, in order to achieve its maximum effectiveness, should be comprehensive and carefully timed to coincide with an Allied ground offensive. It would be more effective if, at the same time when it was undertaken, all significant producing plants had been put out of action, thereby producing a situation in which the stocks in the distribution system were the sole effective source of fuel supply for the enemy's immediate military needs. The Committee therefore prepared, at the end of January, a skeleton plan for the attack of storage depots and subsequently kept this up to date against the occasion when the progress of the attack on production and the battle situation might warrant is implementation.(14)

33. During January and February all air forces made good use of improving weather and their own increasing operational capabilities in the methodical exploitation of the oil target priority list. At the beginning of January this still comprised 60 major targets. By the third week in February the "active list" had been reduced to 40 targets. In the West, in particular, such rapid progress was being made in the immobilisation of the synthetic oil plants and the major benzol plants in the Ruhr that it was possible, at the end of February, to add a further 13 benzol plants of secondary importance to the priority list. Despite these and other further minor additions, operations made such successful progress in the next four weeks that by the end of the third week in March the priority list had been further reduced to 22 targets.

34. At this stage, having regard to the progress made in the immobilisation of production, and to the advance of the Allied ground forces, it was decided that the plan for the attack of the oil storage installations should be implemented. In the course of the next four weeks the majority of the leading depots in the distribution system which had not already been overrun by the ground forces were attacked by the air forces, the Tactical Air Forces taking a conspicuous share in these operations.

35. All the synthetic plants except Ruhland were put out of action by the early days of March or reduced to an insignificant output. Ruhland succeeded in remaining in operation longer than the others on account of its remote location but finally succumbed to a special long-range operation by the United States 15th Air Force. By the end of March the whole of the German refining industry, except for some unimportant production by small Geilenberg plants, was out of action and with the return of fair weather the United States 15th Air Force made short work of the remaining producers in Central Europe.

36. From December 1944 onwards few statistics are available of the output of oil products. Production was rapidly diminishing to a trickle. The manufacture of aviation fuel components was almost at a standstill. There was

(13) C.O.S. (45) 78 (O), 25.1.45.

(14) At this time consideration was again given to the desirability of attacking other oil target systems, namely lubricants and tetra-ethyl-lead. A study of this subject is recorded in Appendices 14 and 15

some production of motor gasoline from dispersed distillation units, although this was quite insufficient to alleviate the shortage to any useful extent.

37. A secondary effect of the dislocation of the synthetic plants was the disruption in the output of methanol and nitrogenous products. By September the sharp fall in the production of methanol was limiting the manufacture of propellants, explosives, rubber, plastics and other chemical and synthetic products At the same time the decline in the production of fertilisers was of an extent that made certain an impairment of the 1945 harvest.(15) The loss in the output of these products did not, however, decisively affect military efficiency. Stocks of most munitions were sufficient for those available to consume them up to the final stages of the German collapse.

38. By the beginning of April practically the whole oil industry was immobilised, and during the month most of its constituent units were rapidly overrun by the ground forces. Most of the attacks on oil in the concluding weeks of the offensive were directed against depots in the path of the advance which might have furnished the means for protracting local resistance. Moreover, the whole war production machine was on the point of collapse so that repair and dispersal schemes could no longer be implemented. The task of dislocating the enemy's oil resources had, in fact, been completed.

39. In the last few weeks of the War the main effort of the Strategic Bombers was directed against communications in the small area of Germany which still remained in enemy hands. The fall in industrial output, which had begun in the autumn of 1944, had by now reached a catastrophe. Finally, their territory overrun from East and West, their Armies defeated in the field, their transportation in chaos and their oil output non-existent, the Germans capitulated on the 5th May, 1945.

The Attacks upon Oil Storage.

40. The activities of the *WIFO* Organisation have already been described.(16) Until its operations became hampered by the Allied air offensive this organisation had been efficiently blending, storing and delivering all the liquid fuels for the *Luftwaffe* and a part of the supplies for the Army.

41. The distribution organisation for liquid fuels to the armed forces was based primarily upon ten main depots. These were large, underground installations and their storage tanks and pumping facilities were not only deeply buried and widely dispersed but they were also concealed in many cases by pine trees that had been planted upon them a number of years previously. In the case of the depot at Stassfurt, south of Magdeburg, the storage tanks had been constructed in a salt mine at a depth exceeding 900 feet. In addition to the husbanding of reserves these depots served another important purpose. Receiving bulk shipments of oil products from the producing plants, these installations were responsible for blending these products to the required specifications for all *Luftwaffe* fuels and, to a lesser extent, army gasoline.

42. These ten main depots served some 50 lesser depots in Germany and in addition, in the Western theatre, there were some 150 depots and dumps in France and the Lowlands.

43. Before the Allied landing in Normandy there had been no concerted attempt to bomb fuel storage centres. Between the 6th June, 1944, and the 8th May, 1945, however, the Strategic and Tactical Air Forces, including the United States 15th Air Force from the Mediterranean, carried out 395 attacks on reported fuel depots and dumps in France and Germany. These operations comprised some 25,500 sorties and 54,060 tons of bombs were dropped. These attacks covered two periods, represented by the Battle of France and the Battle of Germany respectively.

44. The strategic reserves of the German army in France, exclusive of tactical reserves and supplies in transit, amounted to some 10,000 tons. These stocks were held in the large surface depots at Argenteuil (and they were later moved into the Metro tunnels of Paris) and the Bordeaux and Lyons areas. Tactical reserves were held in a large number of army dumps and commercial depots in army, corps and divisional areas. The quantities of fuel allocated to

(15) See page 120 of Appendix 12.
(16) See page 2 and Appendix 3.

each military unit were probably not more than one or two consumption units.([17]) The strategic reserves of the *Luftwaffe* in France were in a buried installation at Montbartier, in addition to which supplies were held at two or three field depots or issuing stations.

45. Just before the invasion on the 6th June, 1944, and continuing throughout the exploitation of the beach-heads during most of that month, the Tactical Air Force had as a primary objective, apart from close support operations, the isolation of the area bounded by the Seine and Loire Rivers on the North and South and by a line running South from Paris to the Loire on the East. Although all efforts were directed to this end some supplies continued to reach front-line troops from outside the Seine–Loire triangle. The movement of supplies not only made substantial demands upon road transport but also enhanced the military importance of such supplies of motor fuel as were available.

46. Concurrently with the attacks upon supply lines, the Tactical Air Forces began a programme for the systematic searching out and smashing of the tactical fuel reserves in France. Although this programme did not reach its peak until July, some 29 attacks, using 1,676 tons of bombs, were directed against these targets in June.([18]) In addition the heavy bombers did high-level attacks against the principal strategic depots, those at Gennevilliers, Rouen and Montbartier being heavily damaged. During this period the United States 15th Air Force made three attacks on depots in the Vienna area with inconclusive results.

47. With the isolation of the Normandy battle area practically completed by the end of June the programme to smash all tactical fuel reserves gained momentum in July. At least 50 attacks on such targets were made during the month by the United States 9th Air Force and the 2nd Tactical Air Force. These operations were supplemented by several heavy attacks by the Strategic Air Forces.

48. In the course of August other commitments detracted the Tactical Air Forces from continuing their sorties against oil supplies. However, the Strategic Air Forces stepped into the breach to throw their weight against these targets and with excellent results. On the 4th and 5th R.A.F. Bomber Command attacked depots in the Bordeaux area, and on the 6th the United States 15th Air Force from Italy attacked two depots near Lyons. These attacks destroyed 7,000 tons of the 10,000 tons which comprised the German Army's strategic reserves in France.([19]) The military consequences of this loss were far-reaching. The dislocation of the German forces caused by the attacks upon communications, by the lack of fuel and, above all, by the successful progress of the Allied armies, culminated in the Battle of the Falaise Gap.

49. Another effective attack, which has already been commented upon, was the dislocation of the large *WIFO* storage at Nienburg by the United States 8th Air Force on the 5th August. This was the major supply depot for the *Luftwaffe* in the West.

50. By the end of September practically all oil targets in France had been captured and the following months marked the opening of the Battle of Germany. During the remainder of the year the United States 15th Air Force inflicted slight to moderate damage to a number of active depots on the Danube, the other heavy bomber forces being engaged upon programmes that did not include oil storage targets.

51. On the 8th January, 1945, the United States 8th Air Force successfully attacked the important *WIFO* depot at Derben. This installation, situated on the River Elbe, was a major distributing centre for army fuels to both the Eastern and Western fronts, and damage resulted in some interruption in the supply of fuel to the fighting areas.

52. In the remainder of January and during February heavy attacks were made upon strategic storage installations at Ehmen, Buchen, Dulmen, Neuburg and Ebenhausen. In addition further damage was done to storage at Hamburg, Dresden and Vienna/Lobau.

53. In March both the strategic and tactical bomber forces were able to intensify the weight of attack upon liquid fuel depots. In these sorties practically

([17]) A consumption unit is the amount of gasoline required to drive a vehicle 100 kilometres.

([18]) It subsequently transpired that a large number of these targets did not comprise fuel dumps. They had been described as such by faulty intelligence.

([19]) Interrogation of John.

no major active depot in Western Germany was left unattacked. The depots at Dulmen, Ehmen, Ebenhausen, Buchen and Neuenheerse were rendered incapable of further operation. Heavy attacks on Derben and Hitzacker caused sufficient damage to interrupt rail movements. A large concentration of hits were scored on the 300,000-ton depot at Farge, although the results of this attack were inconclusive.

54. The success of the attack upon Neuenheerse was of particular importance. This was the army depot that was supplying the Ruhr sector of the front. Owing to the damage to rail communications despatches from this installation had been going by road. The destruction of this depot put a stop to the movement of further supplies of liquid fuel to the forces in this area. Two attacks on the depot at Ebrach also affected supplies to the fronts. This army depot was responsible for supplying the centre sector of the Western front and the second attack upon it put it almost totally out of action for a short time.

55. Early in April the United States 9th Air Force completed the destruction of Ebrach, rendering it inactive for good. The United States 9th Air Force and the 1st Tactical Air Force made similar attacks on several other depots causing both damage and fires. The heavy bombers were also adding their weight to the offensive. Their attacks were of increasing effectiveness as any periods of inactivity at the main fuel distribution centres were now of decisive importance in the frenzied distribution of dwindling supplies to German formations. A further sortie by the United States 8th Air Force against Derben sealed in the remaining 1,500 tons of strategic reserves that were left in this depot and it was later captured before the distribution of its contents could be resumed. A similar attack on Hitzacker prolonged the period of inactivity caused by the damage done in March. A further blow struck at Freiham was less decisive as an alternative rail exit to this large underground installation was left undamaged.

56. Probably the most spectacular achievement in April was the final dislocation of Neuburg, the main *WIFO* depot on the Danube. This was the distribution centre for both Army Group G, that was defending the Southern sector of the Western front, and the elements of the *Luftwaffe* covering Southern Germany. Exclusive of Derben, this installation was holding virtually the last remaining strategic reserves of the *Luftwaffe*, amounting in April to some 2,600 tons.([20]) On the 9th April the United States 8th Air Force attacked Neuburg, causing heavy damage to the rail exits. Although it is questionable whether operations were totally stopped, even temporarily, outgoing shipments were at least drastically reduced and thereby acutely embarrassing the supply position of the Army Group G, and especially the 7th Army. On the 18th April the United States 9th Air Force, employing no less than 445 aircraft carrying 806 tons of bombs, caused widespread damage to the depot facilities, stopping operations for at least two or three weeks. It is not known whether further issues of fuel from this installation were later made possible, but it is confirmed by captured documents that a substantial quantity of motor gasoline was scheduled for shipment to Army Group C during the last ten days of April. It is certain that these much needed supplies were never delivered. In the subsequent collapse of Army Group G and the Allied drives to the East, the attacks on fuel supplies stored in the Southern area must certainly be considered a contributing factor.

57. On the 20th April the oil offensive culminated with attacks by R.A.F. Bomber Command on oil stores at Regensburg and, by the United States 9th Air Force, on the *WIFO* depot at Annaburg, and on the storage to the new Kontinentale Oel refinery at Deggendorf on the Danube. With these three missions accomplished the task of the Air Forces against oil was done.

([20]) On the 20th April this depot contained (exclusive of army fuels which are not reported) 5,100 tons of C.3 fighter fuel, 226 tons of J.2 jet fuel and 5,500 tons of aviation fuel blending components. At the end of April the finished aviation fuels in each of the depots in the other *Luftgau* averaged only a few hundred tons, representing not more than one to two weeks' consumption and provided distribution had been possible. It can therefore be fairly stated that before the war ended the Allied bomber attacks had run the *Luftwaffe* fuel supplies dry. The figures are taken from the official documents of Koller. (A.D.I. (K) Report No. 374/1945.)

SECTION XIII.

THE MILITARY EFFECTS OF OIL SHORTAGE.

The lack of natural oil resources resulted in Germany being handicapped in relation to the Allies. In addition, the lack of foreign exchange was a deterrent to the building up of the strategic reserves that would have been advisable to safeguard military supplies for even a war of short duration. Moreover, the relative paucity of mechanical transport in Germany's civilian economy became reflected in the equipping of the German Army which had to rely upon the locomotive and the horse as its principal means of transport. This adverse factor in the national economy was also in part responsible for the small stocks of oil with which Germany began the War.

2. Although in the long run severely handicapped by this basic disadvantage the war machine was so planned that, during the first years of the War, no arm of the services lacked the fuel it needed for the execution of offensive operations. The system adopted by the *Planungsamt*, whereby the fuel needs of the forces were presented by a representative of the *Oberkommando Wehrmacht Wirtschaftsruestungsamt*, ensured that the services had priority. The organisation for distributing the available oil supplies worked well and the possibility of operations being in any way impaired by fuel shortage was not likely, at least up until 1943, to have been contemplated by the army commanders. The opinion of two members of the General Staff(¹) is probably typical of the views held by these military leaders who were concerned with dealing with the military machine as planned and who were not associated with the difficulties of making the best of the available oil supplies :—

> "The German High Command remained unconcerned about oil supplies until the defection of Roumania. Up to that time the various theatres of operations were adequately supplied if only through radical measures of economy."

3. But the broader picture, as seen by Keitel and others who could view the position in a proper perspective, showed that the parsimonious allocations of oil were proving a handicap to efficient running of the whole military machine. Keitel put the matter in black and white in a letter to Speer of the 25th October, 1943, in which he outlined the fuel requirements of the three services for the following year(²) :—

> "During the last two years the supplies of fuel to the Armed Forces has shown a marked increase. This was sufficient to satisfy a limited demand but not enough to permit a full utilisation of available aircraft, motor transport and warships.
>
> Owing to the development of the armament programme the discrepancy between supply and demand is likely to become more acute during 1944.
>
> Enclosed is a survey of fuel demands from the Armed Forces, approved by me . . . .
>
> The especially high increase in the demands of the Commanders of the *Luftwaffe* is based on the increased aircraft programme for 1944.
>
> Even if the armament programme is not fully carried out a certain fuel surplus is desirable in order to be able to build up a modest reserve to allow for a possible breakdown in fuel production . . ."(³)

4. With the exception of the North African campaign, it is questionable whether this inability to utilise fully the available aircraft, motor transport and warships affected to any important degree the operational efficiency of the Armed Forces in at least the first three years of the War. On the other hand, the necessity to observe the strictest economy in the use of liquid fuel imposed

(¹) Westphal and John.

(²) Files of the OKW *Mineraloel Abteilung*, Frankfurt.

(³) This was the first of a series of letters from Keitel to Speer. The situation at this time has been summed up by the Head of the Liquid Fuel Supply Section of the Supreme Command :—

> "In numerous letters after 1943 Keitel pointed out to Speer the paramount difficulties in the supplies of fuel, but he contented himself with this correspondence, which remained ineffective. . . . .
>
> The discrepancy between fuel consumption and actual fuel production for 1944 was really horrifying. A plan drawn up in the beginning of 1944 for the current year on the basis of the aircraft and vehicle programme showed a shortage of 5 million tons for that very year. . . . . only at the end of May 1944, *i.e.*, after the beginning of the Allied air offensive, was Hitler informed by his men of the real fuel supply situation." (Dultz. AO. 232/1/Z.)

FIGURE 6

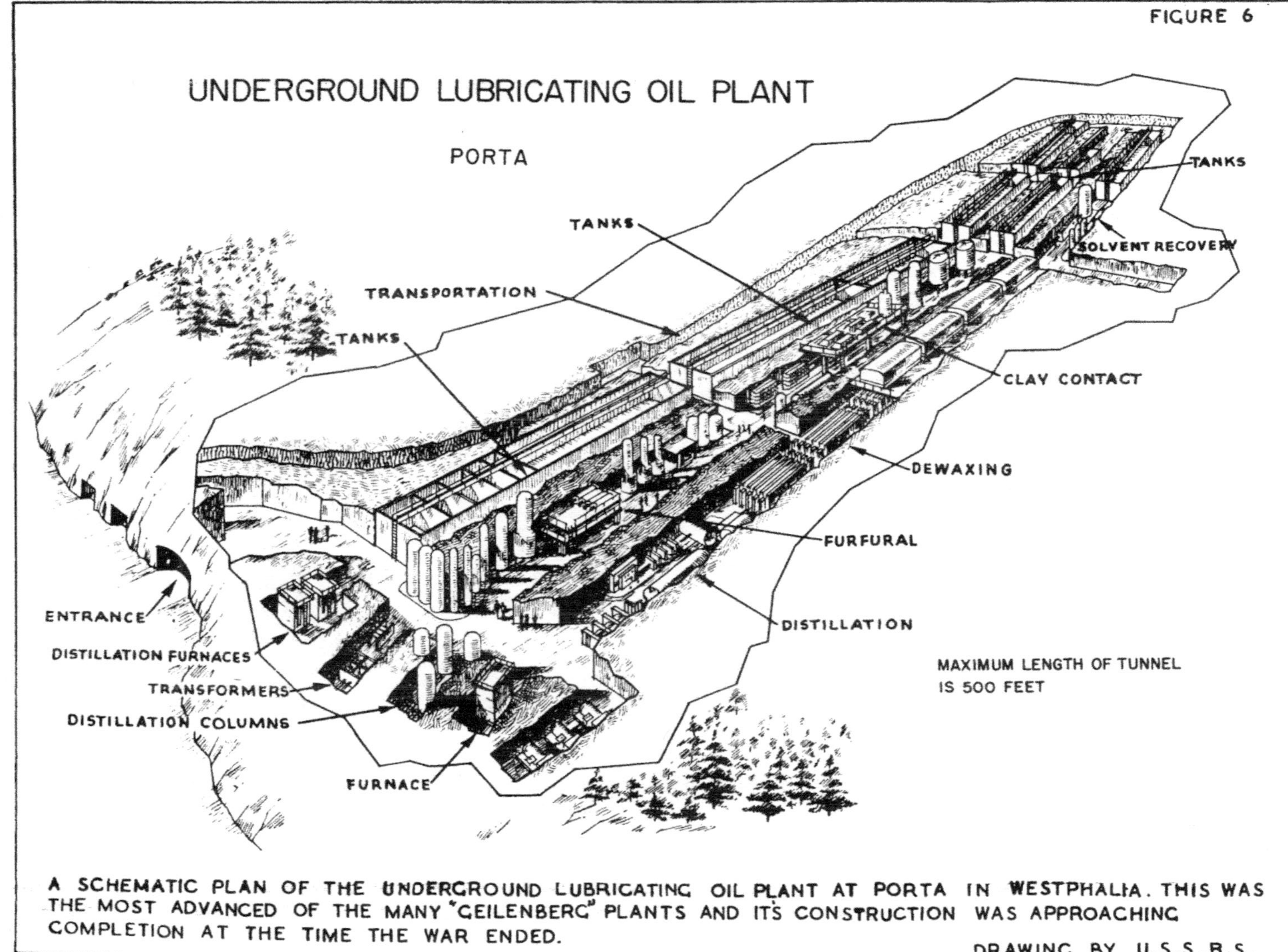

A SCHEMATIC PLAN OF THE UNDERGROUND LUBRICATING OIL PLANT AT PORTA IN WESTPHALIA. THIS WAS THE MOST ADVANCED OF THE MANY "GEILENBERG" PLANTS AND ITS CONSTRUCTION WAS APPROACHING COMPLETION AT THE TIME THE WAR ENDED.

DRAWING BY U.S.S.B.S.

restrictions in training which became increasingly detrimental to military efficiency as time went on. While these economies affected the Navy least of all, a restriction in training in the *Luftwaffe* in 1942 proved a serious matter. It was, however, the Army that bore the brunt of these economy measures.

The Effect of Fuel Shortages on the German Army.

5. The meagre allowances of fuel for the training of drivers resulted in due course in inefficiency at the fighting fronts. There were complaints during the African campaign when the exacting conditions for transport demanded the skilled handling of vehicles. Furthermore, the constant restrictions in the quantities of fuel that could be expended upon manœuvres resulted in reduced effectiveness under battle conditions.(4)

6. The quotas of fuel that were despatched to France from the *Reich* in the months before the invasion were so small that it was impossible for commanders of motorised units to train their men properly in driving. Further, it was impossible for coastal defence divisions to carry out manœuvres which they considered were highly essential.

7. The resistance to the invasion of France in June 1944 caused the largest military demand for liquid fuel since the advance into Russia in the summer of 1941. Von Rundstedt, who was in command of the defences of the West, has confirmed that, at first, he had no anxiety about his gasoline position :—

> " An attempt had been successfully made to build up the largest possible stocks through a policy of radical and strictly enforced economy.
>
> In order to minimise the effect of the anticipated air attacks, fuel dumps were greatly decentralised. It was also foreseen that the attack on railways would cause a diversion to road transport with an attendant increase in gasoline requirements."

8. Von Rundstedt's Chief of Staff has amplified this statement :—

> " Adequate stocks of fuel were built up in anticipation of the invasion. As a result of economy measures and because curtailments could always be expected, requests for fuel (from Germany) were always greater than actual allocations. The strategic reserves for the invasion consisted of 10,000 tons of fuel which was stored in Bordeaux, Lyons and Paris."

9. When Oberst John assumed his duties as quartermaster for fuel supplies for Commander-in-Chief, West, he discovered to his dismay that nothing had been done to change the location of these three strategic dumps, and before action could be taken two were destroyed by air attacks early in August and the third was only saved by removing it into the tunnels of the Paris Metro.

10. The supplies in divisional areas immediately before the invasion were as follows :—

	Sufficient for Kilom.
1. For armoured and armoured infantry divisions ...	500
2. For all other divisions	300–500
3. No fuel stored in Army areas.(5)	

Over two years of fuel stringency had apparently resulted in von Rundstedt feeling satisfied with a strategic reserve of a size that would have been considered disastrously inadequate in terms of the requirements of the invading forces.

11. In the course of July the insufficiency of these supplies became increasingly apparent. The dislocation of communications not only impeded the movement of such supplies of fuel as were available, but the substitution of road transport for rail also added to the demand for fuel. Superimposed upon these difficulties were the effects of Allied fighter bomber attacks on road movements. Units were having to march many miles into battle and were arriving in no fit state for combat.

(4) " The shortage of fuel very badly affected the training of the *Panzer* replacements. These new *S.S. Panzer* Divisions, for instance, had only very little driving experience due to the lack of fuel. The officers and soldiers, therefore, were insufficiently trained in driving technique and in tactics, and thus came to the front in a raw state. Firing with guns and automatic weapons from a moving vehicle must also be taught very thoroughly and we had no opportunity for this. Therefore in actual battle fire was very poor and much ammunition was wasted."

Thomale. (FD. 4641/45.)

(5) U.S.S.B.S. " The Impact of the Allied Air Effort on German Logistics," Chap. VIII.

Difficulties in Italy.

12. Information is lacking upon conditions at this time on the Eastern front but in the case of Italy Field-Marshal Kesselring was faring but little better than von Rundstedt. The position is best described in Kesselring's own words :—(6)

> " In Africa we had a shortage of fuel and it was decisive. But in Italy I made many savings in fuel, even down to the extent that artillery and flak had to be drawn by oxen. I managed to save enough to carry out the necessary movements although the situation was tense.
>
> First of all I ordered a reduction in the number of motor vehicles to the most vital minimum. This was near the end of 1944. Something like that was in effect before but it was not rigid and was according to the situation. I then forced a standstill of all vehicles through non-fighting periods. I introduced the horse-drawn and ox-drawn equipment, as of Cassino time. Heavy trucks were replaced by light ones. In mountains we used cable cars. A special organisation of supply transport handled all the vehicles, using the best drivers and very careful maintenance. It was a strict rule that each vehicle would have to draw a trailer."

13. These regulations were ruthlessly enforced. Unit commanders had to make a daily return of consumption, even under battle conditions, and any irregularites in the submission of this return were punishable by death. The whole conduct of this campaign afforded a remarkable example of the skill with which a defensive battle could be fought with a minimum of motor fuel.(7)

14. The impact of the bomber offensive against the oil plants began to be felt by the armies at the fronts from the beginning of August onwards. The shortage was greatly aggravated by the increased activity in all theatres in June and July. The expenditure of fuel on the four major fronts in July exceeded allocations by 30 per cent. The strain thus imposed became reflected in reduced fighting efficiency.

15. Early in August Army Group North, on the Russian front, reported that their position was becoming jeopardised by the lack of fuel. Four weeks later Army Group South, in the Ukraine, was also in difficulties for the same reason. Their headquarters reported that there was no gasoline that could be spared for moving up reserves nor for changing defensive positions. The capture of Ploesti by the Russians at the end of August put an end to the declining flow of liquid fuel that had been going direct from Roumania to the forces in the South-East and thereafter supplies had to be transported from Hungary and from the depots on the Elbe which were at the same time endeavouring to meet the demands from the Western front.

(6) U.S.S.B.S. Interview No. 61.

(7) The following order, dated 13.6.44, issued by Headquarters, 44th Division, in Italy, illustrates the conservation measures that were taken :—

> " The fuel situation will continue to remain critical as the available fuel is required primarily for the tasks of the Panzer and Panzer Grenadier Divisions. In particular a further scarcity of gasoline must be reckoned with.
>
> In order to preserve valuable vehicles, and those difficult to replace, the strictest measures must be carried out.
>
> In this connection I order :—
>
> 1. All journeys not absolutely necessary must cease. In place of vehicles with a high gasoline consumption, vehicles of economical consumption must be used, *e.g.*, for reconnaissance, primarily *Volkswagen* or motor cycles.
> 2. All drivers must be instructed to switch off their engines immediately when halting, *e.g.*, in traffic blocks. This must be continually supervised by officers.
> 3. All gasoline-driven vehicles except Commanders, despatch riders and signals vehicles, those drawing heavy guns, and ambulances must be withdrawn immediately, loaded with stores which have to be evacuated, and sent back.
> 4. Gasoline vehicles carrying ammunition, fuel and food must whenever possible be replaced by diesel vehicles.
> 5. Gasoline vehicles to be evacuated must be towed by diesel-driven vehicles on level stretches.
> 6. Diesel vehicles must always be held ready for the towing of gasoline vehicles which have exhausted their fuel. This will require foresight in organisation.
> 7. 20 per cent. diesel fuel must be mixed with gasoline.
>
> It is essential in order to avoid the blowing up of vehicles that the discomforts resulting from these measures are put up with. I therefore make it a special duty of all C.Os. to supervise continually the execution of these orders.

16. The same strain was felt in the South-West. A message from Keitel to Kesselring dated the 22nd August stated that the daily average consumption of 800 tons could no longer be met from supplies " in view of the strained fuel production situation and the great demands in the East and the West." Kesselring was instructed to stretch his August allocation to include the first ten days of September, and he was advised that his September quota would be curtailed accordingly.([8])

17. On the Western front the high rate of expenditure of available fuel over the amount allocated had resulted in an almost total depletion of supplies in the combat areas. In addition the situation had been greatly worsened by the destruction of fuel dumps and the interruptions to communications. These factors, compounding one with another, made doubly certain the defeat of the German 7th Army in the Battle of the Falaise Gap.

18. In spite of all efforts to divert as much liquid fuel as possible to the fighting fronts, the position continued to deteriorate from September onwards. It became necessary for the motorised supply columns of all infantry divisions to be converted entirely to horse-drawn transport, resulting in an inevitable reduction in fighting efficiency.([9]) There must have been at this time so many signs of crisis that it is not surprising that by November reports were circulating amongst the German forces of a complete breakdown in oil supplies. To allay any consequent deterioration in morale Speer had to take immediate steps to issue statements emphasising that the Allied air offensive was being frustrated.([10])

19. The brief respite before the Allies launched their offensive in the West on the 16th November enabled some consolidation to be made in supplies and the stocks available to divisions, which included Army stocks, were sufficient for 150 kilometres with the reduced vehicle strength of units at that time.

20. During this month drastic steps were being taken to re-adjust the vehicle establishment of armoured divisions to operate on a minimum of gasoline. On the 7th November Keitel issued a directive instructing that armoured divisions be equipped " in such a way that they can move under their own power, practically without needing any motor fuel in order to avoid their time consuming transporting by railway." This was to be achieved by decreasing the vehicle strength of armoured divisions, by operating about half the remainder on diesel oil or producer gas, and by arranging for diesel vehicles to tow gasoline vehicles when in convoy. Tracked and half-tracked vehicles were, as far as possible, to be towed on trailers drawn by producer-gas units.

The Ardennes Counter-Offensive.([11])

21. A study of the effect of oil shortage upon the outcome of the Ardennes offensive must be made against the background of the circumstances under which this offensive was undertaken. Not only were a relatively small number of below-strength divisions pitted against a much larger number of Allied divisions but the offensive was also handicapped by inadequate preparation and by the lack of confidence of those instructed by Hitler to undertake it. The army commanders rightly regarded the offensive as a military gamble with remote chances of success.

22. By December the supplies of gasoline moving by rail from the central depots in Germany to the armies in the field had become reduced to a fraction of requirements. On the Western front, von Rundstedt had been emulating Kesselring's policy and had been endeavouring, during the lull in the fighting, to set aside some reserves for future operations. The quantities thus accumulated were, however, small and quite inadequate.

23. When the divisional commanders were personally briefed by Hitler for this offensive, Hitler is reported to have stated that he had scraped everything together for this effort and if he did not succeed the War would be lost. However, in spite of the fact that von Rundstedt's quartermaster had laid down that

([8]) U.S.S.B.S. The Impact of the Allied Air Effort or German Logistics, Chap. III.
([9]) Buhle. (U.S.S.B.S. Interrogation, 24.5.45.)
([10]) Hitler Conferences, Vol. 18, 15.11.44.
([11]) The details given of this counter-offensive are largely based on the interrogation of von Rundstedt, von Manteuffel, Blumentritt and Kruse. (C.S.D.I.C. (U.K.), G.R.G.G. 330 (*c*), 1.8.45.)

five fuel consumption units would be necessary, only 1·2 to 1·5 units were available when the attack began. Additional supplies were to be brought up from reserves East of the Rhine. The inadequacy of these supplies is emphasised by the fact that von Manteuffel, commanding the 5th *Panzer* Army, calculated that, on account of the rough terrain and the conditions of snow and ice, a consumption unit was likely to give a range of 50 kilometres instead of the usual 100. Furthermore, von Rundstedt affirms that the attacking forces did not count upon the capture of any Allied fuel dumps, the locations of which were not known. The complications caused by the lack of fuel, both before the battle started and while it was being fought, are illustrated by the report of the officer commanding the *Panzer Lehr* Division.([12])

> "On the 13th December I returned to my Division where I received a few more tanks and some extra fuel. I had been promised enough for a 500 km. advance, but all I got was the normal amount for 200 kms., which in the rough terrain of the Ardennes was scarcely enough for 100 kms. On the 15th–16th December the Division proceeded to a new area. To save petrol the tanks were sent by rail; there were some attacks by two-engined bombers on these trains. Around Christmas Day the Divisional tank repair establishment, which was set up close to the railway station in order to save gasoline, was badly bombed. The long march itself cost about 30 tanks; those which had got bogged down, ran out of fuel or had breakdowns. About this time Model ordered that all stalled vehicles on the line of march should be drained of gasoline which should be used to get tanks up to the front. I refused to do this as I foresaw a situation when I would have insufficient transport to get fuel up to these tanks once they ran dry."

24. Von Rundstedt has also given his description of the difficulties encountered :—

> "I was informed that there was more gasoline available when the original two days' supply had gone. While the bad weather prevented air attack, it also made for bad transport conditions and poorly-trained drivers added to the difficulties. Captured stocks did not materially alleviate the position as not a single big dump was captured. I cannot fix a date when it became clear that gasoline would be insufficient for the attainment of our objective. The stubborn defence of Bastogne and Malmedy was possible because the gasoline shortage prevented German armoured columns from reaching the towns while they were still thinly held."([13])

25. The inadequacy of liquid fuel, combined with conditions of ice, mud and fog, took the punch out of the battle. These difficulties caused a delay in bringing the artillery forward and, when movement was possible, advances had to be made piecemeal. For the same reasons the assault on Bastogne was checked as the divisions from the 6th *S.S. Panzer* Army could not be deployed in time. And when units were finally committed to battle their lack of training placed them at a disadvantage.([14])

26. The distances to be covered added to the worsening of the situation, an aspect that is emphasised by Speer in his account of the battle :—

> "The attack was ordered to begin although the formations had only one or two fuel supply units. The entire supplies of bridge-building equipment still lay in the rear areas, whilst the rest of the supply organisation for the units was insufficient for the distant goal in view. *Generalfeldmarschall* Model and *Oberstgruppenfuehrer* Dietrich called attention to this state of affairs, but the time-table for the attack was persisted in. Without any doubt the lack of supplies was due to the transport difficulties caused by air attack.

([12]) Bayerlein (AMWIS. 303, 23.6.45).

([13]) 12 A.G., M.F.I.U., No. 4, 20.5.45.

([14]) This point was emphasised by Westphal :—

> "The drivers were not well trained and a good many vehicles dropped out. Tank and truck drivers were insufficiently trained because of the drastically enforced fuel economy. The armoured troops remembered bitterly the Senne Lager, where the economy of fuel for training purposes was carried all too far." (12 A.G.I.C., C.I.R., No. 1, 12.6.45.)

In this connection it is recorded in the Speer documents that tanks manufactured at Kassel during the last nine months of the War were tested and run in by means of a producer-gas generator mounted on a trailer and detached when the tank was delivered to the Army. The tanks employed at training schools were likewise run on producer gas.

At the front the effects of the shortage of fuel must have been far-reaching, as the distances which the supplies had to cover were insupportably great. Supplies for Bastogne had to travel a distance of 200 km. by lorry and as neither lorries nor gasoline were available the supplies themselves proved insufficient."(15)

27. Von Rundstedt, in summing up his views upon this offensive, gives four causes for its failure. The first two are the wrong employment of his forces on account of orders from higher authority, and the insufficient number of divisions made available for the task. His third cause is the lack of fuel and the failure to bring up all the fuel available in time. The fourth cause is the inexperience of drivers due to the lack of fuel for training.

28. From January onwards the quantities of gasoline available to the forces in the West became reduced to insignificant proportions. A report by the officer commanding the *Fuehrer* Escort Brigade illustrates the difficulties that were encountered. After stating that the Brigade had only two-tenths of a consumption unit, which was not an exceptional state of affairs but a condition which had persisted for some time, the report went to to say:—

> "Fuel is a means by which operations are conducted. In the present situation, however, fuel dominates the conduct of operations. Tanks with two-tenths of a consumption unit facing the enemy in the front line constitute a risk. The main burden of the fighting is borne by the infantry because, in spite of all concentrations of fuel supplies, our tanks have only been mobile in special circumstances."

29. On the Eastern front the position was no better. Both Speer and Jodl separately confirm that lack of fuel was substantially responsible for the rapid collapse of the defensive front against the Russian break-out from the Baranovo bridgehead in the third week of January. There had been massed at Baranovo approximately 1,200 tanks which were intended to stem the Russian drive into Upper Silesia after the forcing of the Vistula. But the fuel allocation only amounted to one or two consumption units and when the time came these tanks were not capable of proper tactical deployment.(16)

The Offensive in Hungary.

30. Towards the end of January an ambitious counter-offensive in Hungary was planned. The decision to make this offensive was Hitler's and it was made against the opinion of the General Staff that a counter-attack in Silesia was more logical and of greater urgency. Various German leaders were under the impression that the need to safeguard Hungary's oil was a primary reason for the offensive. However, so far as it is possible to arrive at Hitler's motives, it would appear that he was hoping for a decisive military success and that he used the need to retain control of the Hungarian oilfields as an argument to counter that put forward by his advisers.(17)

(15) A.D.I. (K) Report No. 349/1945.

(16) Speer's version of this battle is as follows:—

> "We had utilised here approximately 1,200 tanks for the defence and here it was for the first time that the troops had only one or two fuel consumption units, so that the tanks were practically unable to move when the Russian attack started. . . . . The quick break-through came only because of insufficient supplies of fuel. The quick loss of Upper Silesia, which was practically the last push to break the neck of the entire armament industry, was therefore caused through the attacks on the hydrogenation plants."
>
> (U.S.S.B.S. Interrogation, 19.5.45.)

(17) Various authorities, including Doenitz, Milch and Koller (A.D.I. (K) Report No. 374/1945), have given the need to safeguard Hungarian oil as one of the reasons for this offensive. There is also on record a letter from Speer to Guderian, dated the 15th December, 1944, in which Speer opposes a withdrawal to the *Niebelungen* line and which says in part:—

> "Hungary supplies also (besides bauxite) 40 per cent. of the German mineral oil production and has, in view of the well-known impairment of German synthetic oil production, which is further endangered by coal shortage, decisive importance."
>
> (Hitler Conferences, Vol. 9. FD. 2960/45.)

However, when the time came to launch this offensive, Speer was opposed to it and said so to Hitler:—

> "In my opinion no offensive was necessary at that time, as the fighting line was far enough from these oilfields. After a visit to these oilfields in February 1945 I again confirmed this view to Hitler.
>
> The Hungarian oilfields were, it is true, of decisive importance for our fuel supply as they yielded about four times the quantity of gasoline as the oilfields in Zisterdorf (Austria). Hitler believed a greater military success would be achieved with the offensive in Hungary."
>
> (FD. 4548/45).

The attack began on the 5th March and took the form of a three-pronged thrust from the area North of Lake Balaton, South-eastwards, from the area South of Lake Balaton Eastward and from Yugoslavia Northwards across the River Drava. The attack was strengthened by the crack 6th *S.S. Panzer* Army which had played a prominent part in the Ardennes counter-offensive. The four *Panzer* Divisions that comprised this Army had originally been withdrawn from the West for the protection of the gravely threatened Berlin front.

31. At that time the output of the Lispe oilfields, at the South end of Lake Balaton, was unquestionably important. Moreover, the light crude oil produced was of special value because, by simple distillation, it could yield 30 per cent. of motor gasoline as compared with only 9 per cent. from the German and Austrian crudes. It also yielded a good quality diesel oil which was earmarked for the *Luftwaffe's* requirements of J.2 fuel for jet fighters. The refining of the Lispe crude was being undertaken in whatever refineries in Germany and elsewhere that were capable of operation and, as an additional safeguard, it was being shipped to the various Geilenberg "*Ofen*" that were in process of starting up.

32. In the opinion of Koller, the failure of this offensive was to some extent attributable to the lack of oil. The fate of this part of the front was sealed by the fall of Budapest, which Koller considers was principally due to the impossibility of supplying the garrison by air. A minimum of 200 tons of aviation spirit per day would have been required for flying in adequate supplies, but the *Luftwaffe* could only spare 30 tons a day.([18]) As in the case of the Caucasus an economic need had added to the task of the German armies and had diverted them from following a logically planned strategy aimed solely at the defeat of the opposing forces.

33. Likewise in Italy the supplies being allocated had continuously declined. In the concluding stages of this campaign the immobilisation of mechanical transport contributed substantially to the final collapse.

34. From March onwards the defences of the *Reich* began to crumble with increasing momentum. The dislocation of communications and shortages of supplies were having a paralysing effect. The combined effect of the attacks on rail transport and oil were having decisive consequences. The need for road transport and the fuel for its operation increased in proportion to the extent that the use of the railways became precluded.

35. The encirclement and defeat of the German forces in the Ruhr pocket was facilitated by the destruction by bombing of the army depot at Neuenheerse which deprived the forces in that area of their last remaining stocks of fuel. The few armoured fighting vehicles that remained at this stage became incapable of movement.

36. In April it was apparent that no front line units of the German forces could rely upon the replenishment of such remaining stocks of gasoline that they may have had on hand. The position with Army Group G, facing the American forces on the Southern sector of the Western front, was particularly desperate. The accurate bombing of the remaining oil distributing centres supplying this Army Group denied the supplies that were vitally needed. By this time the disorganisation of the German forces was such that they were incapable of stopping the Allied drive to the East.

([18]) Koller. (A.D.I. (K) Report No. 374/1945.)

SECTION XIV.

THE CRIPPLING EFFECT OF FUEL SHORTAGES ON THE *LUFTWAFFE*.

In reviewing the success of the Allied bombing offensive in depriving the *Luftwaffe* of its striking power it is necessary to survey the events leading up to the start of the attack upon the oil plants in the spring of 1944.(¹)

2. The provision of adequate supplies of fuel had received the close attention of the *Luftwaffe* some years before the War. In 1937 the *Luftwaffe* took over from the *WIFO* a number of *Lufttanklager* (supply depots) that had been specially constructed for the storage of aviation fuel. At that time the greater part of the aviation fuel distribution system had been sited with a view to the possibility of hostilities in the East and South-East rather than in the West. During 1939, however, the prospect of offensive operations against the western Allies led to the planning of a large-scale storage and distribution system in North and North-West Germany which, although primarily designed to meet the immediate needs of the War, was also intended eventually to play an important part in time of peace.(²)

3. In 1937 the German production capacity of aviation fuel was inadequate to provide the anticipated requirements for war and over a period of at least two years substantial purchases were made from importers. The greater part of these purchases were earmarked for what was known as the O.K.W. reserve and by September 1939 this reserve had amounted to 355,000 tons, or equivalent to about three months' consumption under war conditions.

The Depletion of Reserves.

4. During the campaigns in Poland and in the West no restrictions of any kind were imposed on the use of any type of fuel and the demands of all branches of the *Luftwaffe* were fully met until the spring of 1941. At this date stocks were being built up in preparation for the attack on Russia and the first measures to curtail consumption were taken, although these restrictions were designed principally to eliminate avoidable wastage and were not severe.

5. During the first twelve months of the Russian campaign the *Luftwaffe* was able to operate normally and to sustain the maximum effort without any restrictions except possibly in isolated cases where purely local supply difficulties were encountered. However, in the summer of 1942 it became obvious that the unexpected duration of the fighting in Russia, coupled with the great distances which had to be covered, would make stringent economies in oil consumption inevitable in the future. The *Luftwaffe*, with the other services, felt the burden of these economies and as a result the first measures affecting training and operations were taken.

6. As was to happen repeatedly in the future, the flying training branch was the first to feel the pinch and the quota of aircraft fuel allotted to this activity was reduced considerably. Orders were also issued that sorties by bomber and torpedo aircraft in the West should be cut down by allowing the machines to operate only at times when conditions were most favourable. Restrictions were also imposed in the case of transport and communications flights. There was, however, at this time no limitation of operational sorties of any type on the Russian front, of fighter sorties in the West or, in view of the imperative necessity of obtaining meteorological information, of meteorological flights on any front.

7. The supply position began to get increasingly difficult in the late summer of 1942. The growing numerical strength of the *Luftwaffe*, combined with the heavy expenditure of fuel in the operations on the Russian front and in North Africa, was causing a substantial depletion of stocks. A critical position was

(¹) Much of the information on *Luftwaffe* fuel supplies was obtained from the interrogation of Milch, Koller, Galland, Kreipe, Ruhsert and Steinmann. (A.D.I. (K) Report No. 374/1945.)

(²) In its long-term aspects, the scheme envisaged the eventual replacement of land and water transport by main trunk pipe-lines from which subsidiary lines would feed the individual airfields. The subsidiary system, connecting the airfields with existing river port facilities, was to be completed first and the main pipe-lines were to be installed later. The plans also involved the building of considerable additional storage space on the airfields and at the river ports.

Although the details of the scheme are not known it is believed that one pipe-line was to run from Braunschweig to Oldenburg and thence to airfields in that area, and another, starting from the mouth of the Elbe, was to feed the air bases in Schleswig-Holstein.

reached in September, when reserves fell to less than two weeks' requirements. This shortage was not, however, allowed to interfere with operations and the necessary economies were made at the expense of training and non-operational transport and communication flights.

8. During 1942 the allocations to the three services by the *Oberkommando der Wehrmacht* were almost always below the demands submitted.(3) In consequence there was a tendency for consumers to indent for quantities in excess of their real needs in the hope that, as their demand would in any case be cut down, they would thus obtain the quantity required. It was generally the *Luftwaffe* that considered themselves the most hardly treated of the three services in the apportioning of these allocations. More than once, during difficult periods, the production of motor gasoline had to be given precedence and at some expense to aviation spirit. The feeling was that the Navy fared best, but this was probably on account of the fact that the lower-grade fuels were easier to produce, and also the submarine campaign necessarily received a high priority.

9. According to Milch the demands for aviation fuel to the *Zentrale Planung* during the first five years of the War averaged some 250,000 tons per month, rising to a peak of 320,000 tons at one period in 1944. The allocations made by the *Oberkommando der Wehrmacht* were much below these figures. The highest monthly quota ever received, which was allotted in the spring of 1944, amounted to about 198,000 tons, and it was in July of that year that consumption reached the highest peak of the War, being slightly in excess of 200,000 tons.

10. After making allowance for the fact that the *Oberkommando der Luftwaffe* was accustomed to asking for more than it expected to receive, the discrepancy between these figures affords some indication of the extent to which the striking power of the Air Force was limited by the continuous need to conserve fuel.

11. At the beginning of 1943 the position began to be somewhat eased by the increasing output from the synthetic plants, but as operations intensified with improving weather the supply position began to deteriorate in the same manner as at the same time in the previous year. While it did not prove necessary to impose any further operational restrictions there had to be substantial cuts in other directions.

Restrictions upon Training.

12. The first to be affected by these cuts were the flying schools and the savings made in this way were rarely fully restored when the position became easier. The natural result of this policy was that in the end the supply of adequately trained air-crew personnel would have proved insufficient to maintain a normal level of effort if the supplies of aircraft and fuel had remained normal.

13. In June 1943, when Kreipe was put in charge of training, he stipulated before taking over the post that he must be assured of a monthly allocation of 50,000 tons of aviation spirit in order to carry out his duties adequately. He was actually granted an average of 30,000–35,000 tons a month until the autumn and, in fact, he was able to carry out the prescribed programme with this amount. During the late autumn of 1943 and the following winter, the quota was reduced to 20,000–25,000 tons a month, but this had no immediate effect on training owing to the normal reduction of flying hours on account of the weather at this season of the year.

14. In February 1944 the allocation for training was raised to 45,000 tons and from that time until the following July a monthly average of 35,000 tons was maintained. This increased quota was allotted in order to cope with the number of pupils required for the greatly expanded fighter programme and it was made possible by the increased output from the synthetic industry in the spring. Even in June, however, in order to make the best use of the allocation, training had to be concentrated on the production of fighter and ground attack personnel and of about half of the prescribed number of night-fighter crews.

15. At this stage the curtailment in training did not only affect pilots but it also was one reason for abandoning the training of paratroops. This occurred in July when four paratroop-training schools, each turning out 1,250 trained parachutists every three weeks, were disbanded.(4)

(3) In this connection see Keitel's letter to Speer on page 72.
(4) Interrogation of Student. (W.O.I.R., 5.12.45.)

LEUNA UNDER ATTACK.

The *I.G. Farbenindustrie* plant of the *Ammoniakwerke Merseburg G.m.b.H.* under attack by the United States 8th Air Force.

The plant is seen partly blanketed by smoke from bomb bursts and oil fires. In the lower photograph the bursts of A.A. shell fire can be seen.

Leuna was attacked on 42 occasions involving 6,663 sorties carrying a total of 18,316 tons of high-explosive bombs.

16. In August, when the full effects of the damage to the hydrogenation plants began to be felt, the fuel quota available for flying training was reduced once more and in the following month it amounted to only 20,000 tons. From the end of September onwards the supplies underwent a catastrophic decline and by the end of the year they had fallen to negligible quantities, so that training came almost to a standstill.(5)

17. Savings were made in another direction by reducing the quota of fuel to the aircraft industry for bench-testing and flight-testing also had to be cut down. Successive reductions in the allocations for these purposes were made from the summer of **1943** onwards. These reductions proved costly in growing operational inefficiency.(6)

The Establishment of Emergency Reserves.

18. Another cause for limiting consumption at this time was the need to set aside an emergency reserve. The High Command had been seriously alarmed at the critical depletion of stocks that had occurred in the previous summer and there was also the threat of Allied invasion to be faced. Whereas there had always been, at least in theory, an O.K.W. reserve, it was decided that, in addition, there should be set aside the "*Fuehrer's* Reserve," which, nominally at any rate, could only be broached at the express orders of Hitler himself. The designation had a dramatic quality to it that helped to ensure that the stocks set aside in this manner would only be used in exceptional circumstances.

19. There was also a third reserve, known as the *Oberkommando der Luftwaffe* reserve, which had been created by Goering before the War. It was intended to be used for any purpose, operational or non-operational, and it was not of such large dimensions as the other two reserves. It was exhausted by the end of **1942**, but was reconstituted in the autumn of **1944** only to become totally depleted again by the spring of **1945**.

20. In addition, at least several of the *Luftflotte* contrived to build up their own private reserves as a routine precaution and for use in the event of unexpected operations having to be undertaken before the necessary supplies could be obtained through the usual channels.

21. These four systems for making provision against future eventualities had the effect of at least attenuating the length of time in which a stoppage of output could bring about a total stoppage in operations.(7)

22. The reserves set aside in **1943** to meet an Allied invasion consisted of between 6,000 and 9,000 tons in Western Europe and some 12,000 tons in Norway. In the case of France and the Lowlands, the reserves were not larger as reliance was placed on the availability of supplies from the main storage centres in Germany; furthermore, the normal stocks on airfields held by *Luftflotte* 3 were sufficient to enable all the aircraft based at these stations to fly at least a limited number of sorties. In all, it was reckoned that sufficient supplies were available to keep the maximum strength of *Luftflotte* 3 operating for four months and to supply for a period of four weeks any reinforcements that might be sent in. These stocks were maintained at approximately this level until Normandy was invaded some twelve months later.

23. A substantial reserve had been set aside in Norway in anticipation of an Allied landing in that country. Later, when it became evident that danger did not threaten in this quarter, these stocks were partly consumed locally and the remainder was shipped back to Germany although a valuable amount was lost *en route*.

(5) The results of this shrinkage of training facilities on the output of replacements for air-crew personnel can be seen in the extracts from the *Generalquartiermeister's* report on the manning situation which are reproduced in Appendix 16.

(6) According to Milch the curtailment of fuel supplies to the aircraft industry was a serious matter. From 11,000 tons to 12,000 tons a month were needed for testing aircraft engines and for deliveries. Eventually, only one aircraft in five made the proper acceptance flight, the others being flown for twenty minutes and then sent straight to the Front.

(7) This system of carefully managed reserves was sometimes disturbed by the *Fuehrer* himself, who, especially after December 1944, insisted on the execution of certain favourite schemes of his own which made unwelcome inroads on the rapidly dwindling stocks. These schemes, which were usually intended to maintain the morale of the ground forces and as such made little appeal to the *Luftwaffe*, included the maintenance of two bomber *Gruppen* on the Western front solely for the purpose of showing the flag, and courier flights to the isolated garrisons in the Atlantic and Channel ports and on the Greek islands.

24. From the end of 1943 until April 1944 the strenuous efforts that had been made to consolidate the position were achieving results and total stocks, which in December had amounted to 390,000 tons, reached a peak figure at the end of April of 574,000 tons. Thus, when the Allied air offensive against oil began, the stock position was stronger than it had been at any time since the summer of 1940.

The Restriction of Operations.

25. The initial attacks upon the hydrogenation plants mainly responsible for producing aviation fuel caused a heavy slump in output. Whereas the level of production at the beginning of the year had been at the rate of 165,000 tons a month the output in June fell to 52,000 tons. In the following three months the output fell still lower, the figures being 35,000, 16,000 and 7,000 tons. However, when the attacks began, the stocks held by the *Luftwaffe* were sufficient to cover approximately three months of maximum operational effort. As a consequence, the full severity of the crisis was not felt before August

26. In particular, the resistance to the Allied landing in Normandy was not limited to any extent by any lack of fuel. No serious shortages developed in France during the first two or three months. The supply position was also eased by the heavy losses in aircraft and by the fact that in the retreat it proved possible to evacuate a useful proportion of the fuel supplies of forward airfields to bases further in the rear.

27. However, by August, drastic measures had to be taken to conserve available supplies. What were described by the *Oberkommando der Luftwaffe* as "far-reaching limitations upon operations" were imposed. Only fighter operations in air defence were permitted to continue unrestricted. Reconnaissance flights were limited and the support of the Army by bomber and low-flying attack operations was permitted only in "decisive situations." The shortage of gasoline was also one of the causes leading to the disbandment of the bomber units which were broken up at about this time. Shortly after this order was issued night fighter sorties had to be cut down in order to permit of as many daylight operations as possible with the fuel available. Even so, by February 1945 day fighter sorties also had to be restricted to days when it appeared that conditions were most favourable for interceptions.

28. It was in August 1944 that a new aircraft production programme was approved which was to concentrate on fighters and jet aircraft. Goering had at last been convinced that the construction of a big fighter force was essential. Lack of fuel, however, soon made this new production programme meaningless. At a meeting of Armaments Staff of the Ministry of Armaments and War Production on the 7th September it was reported that Hitler had given orders to the combat plane squadrons that they must remain grounded owing to the fuel shortage and that it would therefore be useless to have any more combat planes built. The production of JU.88, JU.188 and JU.388 was consequently crossed off the production programmes from the date of this meeting. The production of jet-propelled "target protecting fighters" was, however, to be continued.(8)

29. As a result of the fuel shortage in the autumn of 1944, the *Luftwaffe* was unable to derive any advantage from the fact that its first-line fighter strength had reached a peak which had never before been attained during the course of the War. In the last quarter of 1944 the production of aircraft of all types had been averaging 3,100 a month. But the decline in fuel production and the diminishing prospect of its satisfactory restoration made necessary a drastic revision in the aircraft production programme. All efforts were directed to increase the output of jet-propelled fighters. The He. 162 was designed, tried out and adopted for large-scale production in the remarkably short time of about four months. Even before the blue prints for it were prepared, the production programme for this fighter was amended on the 23rd September to provide for an output rising to one thousand monthly by the following April. In this same programme it was decided to confine jet-propelled aircraft production to models having one engine only which would therefore consume only half the fuel of twin-jet planes. It would therefore seem that even an adequacy of low-grade jet fuel was not contemplated.(9)

(8) FD. 4955/45.

(9) Buhle and Sauer. (U.S.S.B.S. Interview No. 24.)
The actual output of jet aircraft only reached 130 a month by February. An important proportion of this production had necessarily to be devoted to training.

30. By November the shortage of fuel in all operational areas [illegible] units to be grounded for long intervals. In the West the position was for a time even tighter as stocks were being assembled for operations associated with the Ardennes counter-offensive. But by the end of the month supplies began to get noticeably more abundant. This was not due so much to the reduction in consumption, which had been helped by adverse weather, as to an important increase in production. Geilenberg was momentarily winning in the battle of reconstruction versus destruction. The fact that the strategic bombing forces were unable to maintain the advantageous position gained in September was resulting in production of aviation spirit rising to 18,000 tons in October and to 39,000 tons in November. There was a decline to 24,000 tons in December but these quantities were sufficient to replenish the stocks of operational units.

31. The *Luftwaffe* was thus able to make the sporadic resistance that was maintained in the first months of 1945. These activities included the attack on the Western front airfields on New Year's Day, the intruder operations against Bomber Command airfields in England, and the operations connected with the Remagen bridgehead.

32. No statistics are available of any production after December, by which time the work of the strategic bomber forces in so far as it affected aviation fuel plants, was practically done. By March the remaining stocks of fuel were almost finished although a small additional supply was made possible by the blending of a quantity of iso-octane, amounting to 23,000 tons, with motor gasoline.

33. The last phase in this account of the *Luftwaffe's* fuel supplies was the capture in Southern Germany of the main storage depot at Freiham. It contained a limited quantity of un-blended fuel, but all tanks that had contained fighter fuel were empty.

SECTION XV.

THE OIL SITUATION AS AFFECTING THE GERMAN NAVY AND GERMAN INDUSTRY.

In addition to the restrictions which the Allied oil offensive caused upon the enemy's operations on land and in the air there were also the effects upon the operations at sea and upon the war potential of industry.

The Oil Supplies of the Navy.

2. Until the final stages of the War the German Navy was able to receive its essential fuel requirements but, as in the case of the other services, this was only achieved by ingenuity and by the imposition of economies in other directions.

3. The naval programme resulting from the London Naval Treaty of 1935 formed the basis for the number of large, concealed storage installations that were constructed in subsequent years. These installations were sited in the vicinity of the principal Baltic and North Sea ports and in such a way that their locations would both minimise the risk of losses from air attack and also enable fleet units to draw their requirements from their own bases. The total storage capacity thus provided was rather more than 2 million tons.(1)

4. Precise figures are not available of the oil stocks held at the outbreak of war but the total, including supplies stored in commercial depots, was rather more than 1 million tons. Over 700,000 tons of this quantity was diesel oil and as the estimated wartime consumption of diesel oil was put at only 20,000 tons a month the supplies on hand were therefore sufficient for about three years' requirements and without considering additional supplies from current production. The fuel oil, on the other hand, was equal to only three to four months' requirements. The reason for this unbalanced position is that a large purchase of Mexican diesel oil was concluded in 1939.(2)

5. As soon as war broke out steps were taken to increase supplies of heavy fuel oil. The largest possible allocation was made from German crude oil although this was limited by the need to obtain a large proportion of lubricants from this source. Roumania and Hungary supplied some 20,000 tons a month, deliveries varying from 50,000 tons a month in the summer to 5,000 tons or less in the winter. Purchases from Russia averaged about 8,000 tons a month. Later, an increasing output from Estonia helped to replace the loss of these Russian supplies.

6. Supplies were fully adequate throughout 1940 but towards the end of the following year the position began to give cause for some anxiety. The heavy consumption of diesel oil by the armies in Russia made it necessary for the Navy to hand over a large part of its nest-egg to the Army. This, however, was done without in any way jeopardising the requirements for submarines, which requirements were fully met up to the end of 1944. In the case of bunker fuel consumption was in excess of production and before long supplies became reduced to a hand to mouth basis. There was a falling off in the quantities received from Roumania towards the end of 1941 and by the following spring it was necessary to make increased use of coal tar.

(1) This and much of the following information has been obtained from *Konteradmiral* (Ing.) Adam and *Ministerialrat* Dr. Hans Janszen.

(2) This, and other purchases from Mexico, were made, on behalf of the Navy, by the *Reichstelle fuer Mineraloel*, who, in turn, used a firm of Bremen cotton importers as the purchasers as a blind to prevent detection of the actual recipient of this oil. These transactions with Mexico were done through W. R. Davis, owner of Davis & Co., Inc., of New York. Davis dealt direct with the Mexican Minister of Finance, who was looking for the best market for the oil production of the expropriated British and American properties in Mexico. The German Government disliked dealing with Davis, a foreigner, as an intermediary in these transactions and in 1938 or 1939 the *Reichstelle* sent their assistant director, Dr. Budczies, to Mexico to find out, firstly, whether Davis could be dispensed with and, secondly, whether oil shipments to Germany could be increased. Budczies reported that the services of Davis were necessary, and this resulted in Davis receiving a monopoly agreement for the import of Mexican oil into Germany. Under the terms of this agreement Davis was obliged by the *Reichstelle* to pay a certain percentage of his profits to two German representatives who had formerly been participating in the Mexican-German oil trade. Imports in 1939 are believed to have been about 1,300,000 tons, consisting of crude oil and diesel oil. The former was processed at Davis's refinery, Eurotank, at Hamburg, and the latter went to the Navy. (Interrogation of Hertslet. 11.9.45, AO 230/3.)

Both the high temperature tars from gas works and coke ovens and the tars derived from the low temperature carbonisation of brown coal were brought into use. These tars were mixed with the heavy residual petroleum oils, thus reducing their high viscosity. Although the resultant mixture was contrived to produce a smokeless fuel it was necessary for both storage installations and ships' tanks to be provided with heating devices to render the oil sufficiently fluid for combustion. These pre-heating arrangements were not without their disadvantages. High pre-heating tended towards gas formation and insufficient heating caused the blocking of feed-lines. The steam consumed in raising the temperature of the fuel resulted in some impairment of cruising range, although this was obviated as far as possible by heating the fuel before leaving port. The consumption of these tars eventually reached 70,000 tons a month and with only another 10,000 tons coming from the oil refineries and synthetic plants. Thus it was proved possible to operate the steam-driven units of the Navy very largely by the liquid products from the carbonisation of coal.(3)

7. In the latter part of 1942 there was again a sharp reduction in the export of fuel oil from Roumania and at the same time the Italian Navy was reduced to such straits that the German Navy was forced to divert supplies that they could well have used themselves. While this diversion may not have affected in any way operations in the North Atlantic or Baltic it caused a restriction in the exercising of fleet units and the standard of training became lowered in consequence.

8. The shortage also caused a restriction in the operations of the mercantile marine. Before the end of 1942 approximately 600,000 g.r.t. had been laid up due to lack of bunkers. The pressing need at this time for more tonnage made it necessary for a programme of conversion from oil to steam to be undertaken and this included the costly conversion of a number of partially built diesel vessels to steam propulsion.

9. The shortage of diesel oil was as serious as that of fuel oil. The naval strategic reserve which, in 1941, had been maintained at 200,000 tons, or five months' requirements, had been used up. Additions to the submarine fleet and of small surface craft had increased requirements from 40,000 tons a month in 1941 to an estimated demand of 53,000 tons a month in 1943. The available supplies that could be allocated were not in excess of 40,000 tons a month.(4) The submarine service was given priority and the activities of the surface craft had to be curtailed accordingly.

10. After March 1945, but not before, the disruption of oil supplies finally resulted in a restriction in offensive operations by the U-boats.(5)

11. In the last two years of the War it is probable that the reduced activity of the main fleet units was enforced more by aggressive Allied tactics than by any shortage of oil. The attacks on the synthetic plants actually caused an improvement in the supplies of fuel oil for a short time in the autumn of 1944, and for the reason that it was not possible to process the raw oil feedstocks, these being diverted to Naval consumption, for which they were suitable after blending with other products. However, this phenomenon was of short duration, and for the reason that the attacks on the Ruhr benzol plants yielded an unexpected dividend. The disruption of these plants stopped the supply of the tar oil that was necessary to reduce the viscosity of the heavy residues that were in abundant supply and which were otherwise suitable for Naval purposes.

12. On account of this, and other supply difficulties, Naval operations became directly hampered at the end of 1944 by an insufficiency of fuel to maintain the full-scale activity that was then essential in the Baltic.(6) These operations were necessitated by the retreat of the German Army from Esthonia and the evacuation of the Baltic ports. Not only were supplies insufficient for the naval forces that could have been employed but supply ships were held up that

(3) The Navy attempted to improve their bunker fuel oil supply position by the development of the oil-chalk deposits in Schleswig-Holstein. A loan of RM.8,760,000 was promised to *Deutsche Erdoel A.G.* with a view to receiving in return 35,000 tons of oil a month. The project was approaching completion when it was abruptly stopped by Allied bombing. Details are given in Appendix 17.

(4) Twentieth conference of the *Zentrale Planungsamt*, October 1942.

(5) FD. 4478/45, 1.9.45.

(6) Another serious handicap was the inability of the *Luftwaffe*, on account of the lack of aviation spirit, to provide air support for the fleet in the operations against the Soviet forces.

were needed for the evacuation of troops and equipment. The shortage of fuel oil finally became decisive and Doenitz* has confirmed that it was principally on this account that the Navy was unable to operate properly in the concluding stages of the War.(7)

The Effect of Oil Shortage on Industrial Output.

13. The extent to which the allocation of oil products for the national economy was progressively reduced is shown by the following comparison with the United Kingdom and taking the year 1938 as 100.

	1938.	*1939.*	*1940.*	*1941.*	*1942.*	*1943.*	*1944.*
Germany ...	100	92	49	38	25	20	19
United Kingdom	100	Varying between 70–80.					

(*Note.*—The British figures would have been lower but for the large increase in the consumption of tractor fuel for agriculture.)

14. A much more drastic curtailment was imposed upon the occupied countries. But in the *Reich* the shortage of oil was not sufficient to prevent industrial output from maintaining a high level. Those in the best position to know are quite specific on this point.(8)

15. It is also significant that, in the extensive records of the *Zentrale Planung*, there is no trace of any reference to fuel shortages hampering industrial output, although references to shortages in other spheres are frequent. Where transport was a restrictive factor this was always reported as being due to an insufficiency of vehicles or to their inefficient use.

16. Nevertheless the increasing limitations in the use of liquid fuel driven vehicles, agricultural tractors and river craft added to the difficulties of maintaining output. Moreover the man in the street found life progressively more difficult. The journey to and from work, the delivery of food, the visit of the doctor, and the regularity of the mail services were all rendered more difficult by the lack of transport. When the bombing of the hydrogenation plants cut off the supply of bottled "*Treibgas*," which was the principal fuel of the fleet of vehicles operated by the *Reichspost*, an important proportion of Germany's remaining road transport became paralysed.

17. It can therefore be concluded that the national economy was so attuned to operate on a minimum of oil that the eventual depriving of such quantities as were allocated for non-military use did not directly contribute to Germany's final collapse. Thus the results of the offensive against oil were almost exclusively military in their effects upon German resistance.

---

* F.D. 4478/45.

(7) At the request of the Chief of Naval Staff the Combined Strategic Targets Committee on more than one occasion gave consideration to the possibility of limiting the U-boat offensive by bombing attacks upon the enemy's stocks of diesel oil. As, however, the U-boat offensive was only consuming about 15,000 tons of fuel a month it was correctly judged, in the light of subsequent information, that such attacks could not have been successful in cutting off these supplies.

At the time the War ended the total stock of U-boat quality diesel oil in naval storage only amounted to 9,500 tons. At the same time there were no less than 186,000 tons of heavy unusable residues, resulting from the topping of crude oil for carburants, in Naval storage installations.

(8) The joint opinion of Speer, Kehrl, Schneider and Buetefisch is as follows:—

"There was no reduction in industrial output or in the production of munitions that could be directly attributable to any shortage of oil or of lubricants. The lack of oil did, however, contribute to some extent to transport difficulties which were a retarding factor in maintaining munitions output."

Even after the attack on oil had begun Speer has confirmed that—

"The reduction in the production of motor gasoline and diesel oil for industrial purposes remained within tolerable limits. . . . . The rationing of fuel did not produce serious losses in output in industry. From October 1944 production sank in any case owing to difficulties in rail transport and consequently the demands on motor transport diminished also."

(A.D.I. (K) Report No. 395/1945.)

ANNEX A

The Intelligence Assessment of the German Oil Position

The following notes provide a record of the organisation set up to study the German oil position and of the results achieved.

2. Oil had always been recognised as a weak spot in Germany's war economy, and consequently her means of obtaining oil, in the widest interpretation of the term, had been kept under constant review from 1928 onwards. This study was originally undertaken by the Industrial Intelligence Centre whose work in this connection was later taken over by specialist committees.

The Industrial Intelligence Centre.

3. This was a small staff set up by the Committee of Imperial Defence and, through its Sub-Committee on Industrial Intelligence in Foreign Countries, it was responsible for reporting on the state of industrial and economic preparedness of foreign countries to make war. It received its instructions from the Committee of Imperial Defence but for administrative purposes it was affiliated to the Department of Overseas Trade. In spite of the close secrecy observed by the Germans in the mobilisation of their oil resources for war the Industrial Intelligence Centre was able to report with remarkable accuracy upon the nature and extent of these preparations.

The Hankey Committee.

4. After the Munich crisis and with the imminent prospect of war the study of Germany's oil supplies was intensified. On the 17th October, 1939, there was held the first meeting of a committee (the Committee for the Prevention of Oil from Reaching Germany—otherwise known as the Hankey Committee) under the chairmanship of Lord Hankey, at which he explained that he and Lord Chatfield had been charged by the War Cabinet to keep in touch with the action being taken by various Government departments to prevent oil supplies from reaching Germany.

5. This committee held a number of meetings during 1939, 1940 and 1941 to which representatives of the departments who would have an interest in their deliberations were invited. Among the departments represented were the Foreign Office, Treasury, Service Departments, Mines Department, Department of Overseas Trade, Board of Trade, Export Credits Guarantee Department, Ministry of Shipping, and, at all meetings, representatives of the Ministry of Economic Warfare.

6. The Hankey Committee was responsible for reporting back to the War Cabinet the views and recommendations of all these Departments and especially those of the Ministry of Economic Warfare whose province it was to reduce by any means the quantities of oil at Germany's disposal.(1)

The Lloyd Committee.

7. During the course of its work the Hankey Committee found it necessary to appoint a sub-committee to prepare periodical detailed appreciations of the German oil supply and consumption balance so that the vulnerable points might be more clearly recognised and in order that the strategic implications of German oil problems might be more fully interpreted for the benefit of the Chiefs of Staff and especially for the Chief of Air Staff.

8. A sub-committee was therefore appointed under the chairmanship of Mr. Geoffrey Lloyd shortly after the outbreak of war and it issued its first report on the 13th October, 1939. In all, eight reports were made by this committee, the last of which(2) took the form of the presentation by the Lloyd Committee of a report prepared by a sub-committee under the chairmanship of Sir Harold Hartley.

The Hartley Committee.

9. On the 20th April, 1942, the Prime Minister approved the formation of the Technical Sub-Committee on Axis oil, which was to report, through the Joint

(1) The deliberations and recommendations of the Hankey and Lloyd Committees are contained in the P.O.G. series of War Cabinet documents.

(2) P.O.G. (L) (41) 11. Final.

Intelligence Sub-Committee to the Chiefs of Staff and to the Deputy Chairman of the Defence Committee. This Committee, otherwise known as the Hartley Committee, took over the work of the Hankey and Lloyd Committees.

10. The Hartley Committee, the composition of which is given in Appendix 10, comprised representatives of the intelligence branches of the Services and of other interested Departments, and also two members of the oil industry. In addition the Petroleum Attaché to the American Embassy was invited to be a member and a direct liaison was thus provided with the United States Government in Washington that proved of great value. The Committee, which made a detailed study of both the German and Japanese oil positions, issued reports every six months. In the preparation of these reports much help was given by a sub-committee, under the chairmanship of Lieut.-Colonel S. J. M. Auld, which studied the problems of enemy oil production. This sub-committee co-opted the leading experts in the oil industry, with whose assistance it was possible to assess the output potentials of the enemy's oil plants.

The Enemy Oils and Fuels Committee.

11. Another committee, which was originally sponsored by the Air Ministry for the purpose of target identification and which later became a sub-committee of the Hartley Committee, was the Enemy Oils and Fuels Committee. Its activities were directed by Lieut.-Colonel S. J. M. Auld and its principal task was the study of captured enemy oil products. The analysis of these samples was important both for observing technical improvements achieved by the enemy and also for the intelligence they provided upon the enemy's sources of production.

The Combined Strategic Targets Committee.

12. The last wartime report of the Hartley Committee was submitted in May 1944.(3) Thereafter, in view of the day to day changes in the situation, and the need for immediate appreciation of intelligence for operational purposes, the functions of the Hartley Committee were divided between the Combined Strategic Targets Committee (Oil Committee)(4) and the Foreign Office and Ministry of Economic Warfare.

The First Reports on the Position.

13. Both before and after the Munich crisis the Industrial Intelligence Centre had been keeping the position under continuous review. As subsequent events proved, it correctly appraised the relative weakness of the German oil economy.

14. In a report(5) issued by the Centre in January 1938 the underground tankage that was being secretly constructed for the storage of strategic reserves was listed and described. This information proved of much value in the course of the War. In a further report(6), submitted in June of that year, it was stated that the general level of oil stocks in Germany was too low to justify her risking a war against a strong naval power unless she were confident of forcing an issue within the relatively short time of, say, three months. This conclusion was based on the fact that, although Germany held considerable stocks of certain essential food-stuffs and raw materials, stocks of other necessities, including petroleum, for which no adequate substitutes were available, stood at a very low figure.

15. In a paper produced a year later,(7) bringing the position up to date, it was concluded that stocks of oil products were under three million tons and that these, combined with potential domestic production, might last from four and a half to five months at wartime rates of consumption.

16. Before commenting upon the accuracy of these estimates mention should be made of the difficulties that were encountered by the Industrial Intelligence Centre, and in all subsequent studies, in defining what comprised a "stock," and especially in relation to the oil requirements of a country at war. Not only is oil in storage a fluid commodity in the sense that storage is only part of a pipe-line, but there is also the difficulty in calculating how much oil in the distribution system can actually be made available for a given purpose. The Germans themselves were unable to define to their own satisfaction exactly what usable

(3) A.O. (44) 41.
(4) See Appendix 11.
(5) I.C.F./950 dated 3.1.38.
(6) I.C.F./284 dated 1.6.38.
(7) I.C.F./284 dated 1.6.39.

AVIATION GASOLINE

STOCKS, PRODUCTION, CONSUMPTION AND TONNAGE OF BOMBS DROPPED ON THESE FACILITIES

GREATER GERMANY

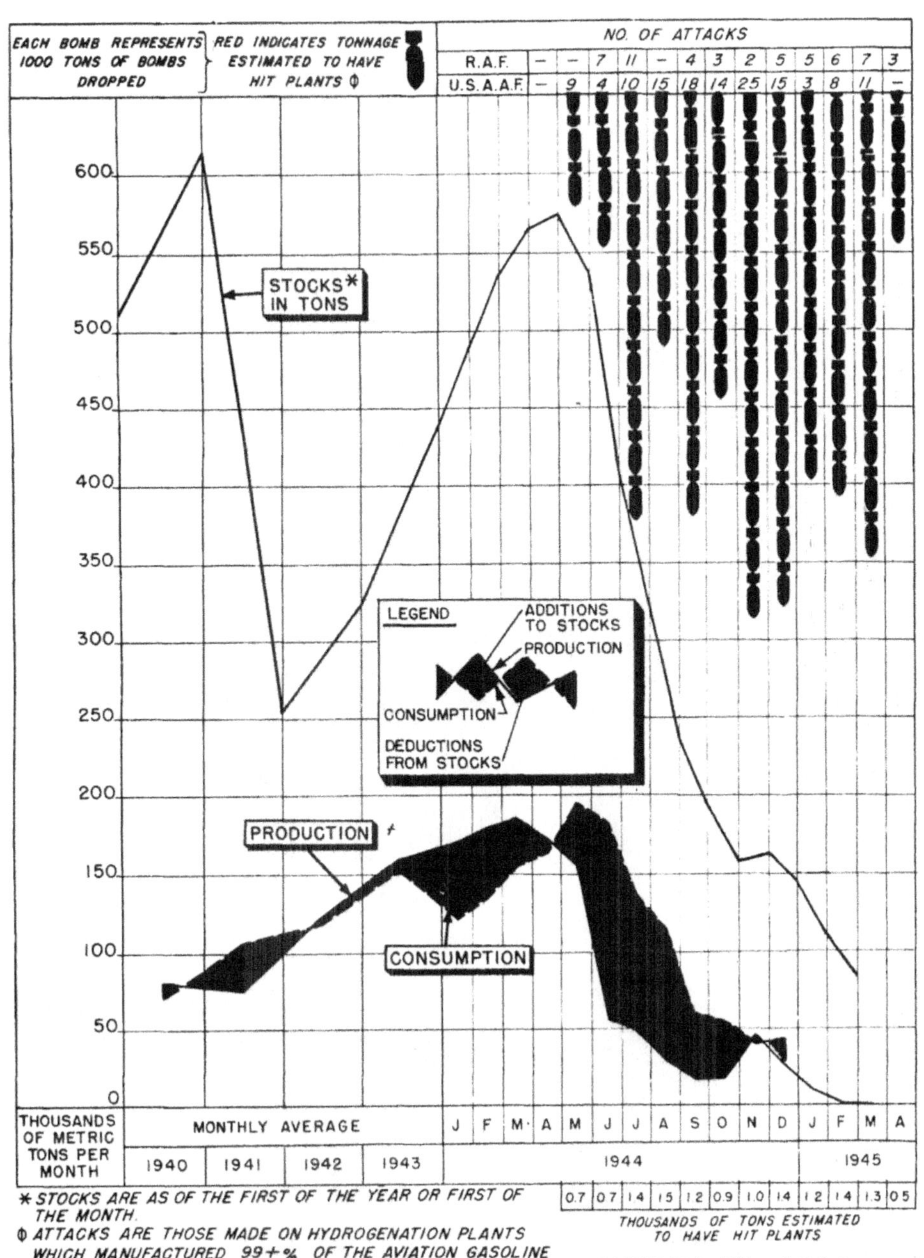

* STOCKS ARE AS OF THE FIRST OF THE YEAR OR FIRST OF THE MONTH.

Φ ATTACKS ARE THOSE MADE ON HYDROGENATION PLANTS WHICH MANUFACTURED 99+% OF THE AVIATION GASOLINE

† INCLUDES IMPORTS AND MISC.

DIAGRAM BY U.S.S.B.S.

Figure 7

stocks they had in hand. It is consequently not possible to compare in sharp focus the figures given in official German documents and those prepared by ourselves.

17. Nevertheless the estimate of the Industrial Intelligence Centre gave an accurate picture both as to the quantity of stocks in reserve and of the extent to which they would suffice under war conditions.

18. In the course of 1940 the statistics were reviewed by the Hankey Committee. By that time sufficient intelligence had been received to confirm that stocks were not large and that oil was giving the Germans some anxiety. At the same time an estimate prepared by the Soviet Government, which was reported to be based on fact, put total stocks at 6 million tons. The indications of shortage combined with this high estimate from Moscow gave emphasis to the possibility that the oil in transit and distribution represented a potential cushion of stocks that should be taken into account as being at least partially available for consumption under stress of war. It was also difficult to believe that Germany could have embarked upon a war with only sufficient fuel to maintain actual operations for a few months.

19. In these circumstances, and on the assumption that there must have been some hidden reserves not indicated by intelligence, the Industrial Intelligence Centre estimate increased from 3 million tons to 5 million tons.

20. However, this revised estimate made provision for 2,800,000 tons to cover the quantities necessary to maintain a distributional minimum; half this quantity was estimated to comprise oil in process (550,000 tons) and in transit (850,000 tons) and the remainder was estimated to be the amount needed to ensure continuous distribution. About one-third of these figures was estimated to comprise fuel oil. A German estimate([8]) of the lowest safety stock margin for aviation and motor gasoline and diesel oil was 800,000 tons, this covering Greater Germany as at June 1941. With the addition of fuel oil and lubricants to the German figure the comparative estimates for "tied" stocks were reasonably close. Nevertheless the upward adjustment of the British estimate for total stocks put the position in a more favourable light than was actually the case.

The Air Ministry View of the Position.

21. It should also be recorded that an independent study of the position was made by the Air Ministry in the summer of 1939. It was concluded that the situation was exceptionally vulnerable. This view was supported by the opinion that, firstly, the annexation of Austria and Czechoslovakia had increased Germany's oil liability and, secondly, that in a major war the oil consumption of a belligerent could be expected to increase considerably over peacetime rates. The Falmouth Report had estimated that in these circumstances there might be an increase in the United Kingdom from 11·5 million tons to 16·75 million tons. It was considered that Germany's needs would rise from 7·5 million tons to 9·5 million tons.

22. It was recognised that the seizure of the oil supplies of other countries would give Germany a potential 5 to 8 million tons a year, but it was considered likely that transportation facilities would limit the amount of oil that could be made use of by Germany. The capacity of the Danube was estimated at from 1 to 1¼ million tons a year only. On the further assumption that Germany and Italy would be allied in war, Italy represented a large added liability.

23. Upon these assumptions it was estimated that stocks would be exhausted after some six months of full-scale effort, and it was concluded that the maintenance of adequate supplies of oil was probably the weakest link in Germany's war potential.

24. Oil was consequently earmarked as a priority bombing target ranking second only to certain military objectives. In the list of targets priority was given firstly to the above-ground stocks of oil at refineries and storage depots, secondly to the hydrogenation plants, thirdly to refineries operating on domestic crude, and fourthly to the Fischer–Tropsch plants. The early attacks were planned on this basis.

([8]) Krauch files.

Subsequent Reports.

25. Except for the divergence in the estimates of the stock position the reports of the Hankey and Lloyd Committees gave a satisfactorily accurate appraisal of the situation. For instance, the report of the 15th March, 1940,([9]) correctly stated that, while there had been a shrinkage in reserves, this was not sufficient to restrict offensive operations and that the position would improve so long as the German armies remained quiescent. The report of the 16th December, 1940,([10]) gave a reasonably accurate study of the trend of production and consumption, and of the influence of Roumania and Italy upon future developments. However, the lack of any means of assessing bomb damage resulted in an over-optimistic evaluation of the results of the R.A.F. attacks at that time upon oil targets and led to the recommendation that these attacks should be continued. These proposals were repeated in the report of the 28th July, 1941,([11]) which foresaw the difficulties that Germany would be in if the Caucasus were not secured and if the Russian campaign became prolonged.

26. By the end of 1941 it had become clear that the war in Russia would be of long duration and a detailed review of the situation resulted in the War Cabinet being informed on the 22nd December that Germany's oil position was now at a crucial stage.([12])

27. During 1942 the study of the potential weaknesses in Germany's oil economy became intensified. The co-ordination of the intelligence gathering agencies of the Services, combined with the co-operation of a Committee set up by the United States Government in Washington, resulted in a steady increase in the amount of information becoming available on the subject. The Washington Committee, known as the Enemy Oil Committee, was formed as a Sub-Committee of the Combined Chiefs of Staff and was made up of representatives of the respective United States Services and of other interested agencies. A close liaison between the Washington and London Committees was maintained by means of an official([13]) of the Ministry of Economic Warfare attached to the British Embassy.

28. In a report submitted in December 1942 it was pointed out that the future trend of German oil supplies was dependent both upon the degree of future military activity and upon the rapidity with which new plants could be brought into operation. It was considered possible that an additional production of 1 million tons of synthetic oil might be obtained in 1943. This possibility was correct to the extent that the revised Four-Year Plan anticipated an increase in output from all sources of 920,000 tons and, as events transpired, the actual output from all sources increased by slightly more than 1 million tons, of which 640,000 tons was synthetic oil.

The Assessment of the Oil Target System.

29. With the approach of the time when a combined strategic bombing force would be ready to make continuous attacks upon Germany the question of target selection became a matter of considerable importance. The primary problem was to define the time period in which a decisive blow, if capable of achievement, upon a source of production would be reflected in a reduction of military striking power. This involved an accurate determination of the quantity of manufactured stocks and of the rate of their consumption. There also had to be taken into account such factors as the susceptibility of the industry to damage and its potential powers of recuperation, the possible results of attacks upon raw materials and ancillary industries, and the ability of the enemy to substitute and improvise.

30. All industrial target systems that offered prospects of limiting military effectiveness within an acceptably short time were analysed in detail. In addition to oil, these systems included ball-bearings, synthetic rubber, chemicals, power plants and weapons manufacture.

31. The onus of this study, as far as oil was concerned, fell upon the Hartley Committee and upon its American counterpart, the Enemy Oil Committee.* On

([9]) P.O.G. (40) 38. ([10]) P.O.G. (L) (40) 18. ([11]) P.O.G. (L) (41) 7.
([12]) W.M. (41) 133rd. Based upon the Eighth Report of the Committee on the Enemy Oil Position (P.O.G. (L) (41) 11). ([13]) Mr. S. Kilbey.

* Before the War an Industrial Section of Air Ministry Intelligence had prepared a priority list of oil production centres as targets. This work was carried on independently of the Hankey and Lloyd Committees and was later merged with the Hartley Committee.

two occasions([14]) representatives of the Hartley Committee had visited Washington for discussions and to resolve the minor differences between the two committees. This liaison was still further cemented by the Anglo-American Oil Conversations of July 1943, for which a delegation of the American Enemy Oil Committee came to London.([15])

32. Liaison with Washington was not confined to personal visits. There was constant interchange of views and of intelligence which went far towards obtaining that uniformity of outlook and recommendations whereby added weight and insistence were given to the value of the enemy's oil economy as a vulnerable and important target system.

33. The assessment of vulnerability depended in a large degree upon the extent of the reserves of oil in stock. No figure that could be relied upon was obtainable through intelligence channels, and the estimate of current stocks had consequently to be deduced by means of a mass of statistical calculations covering production and consumption and upon the basis of a series of six-monthly balance sheets starting with 1939.

34. The calculation of production was less difficult than that of consumption. It was, however, a complex task involving a study not only of the operations of over 150 plants, processing a wide range of raw materials by a variety of methods, but as these plants were capable of producing more or less of any one finished product in accordance with changing demands, there was scope for considerable error in the final computation of the total output of each particular product. There was also the difficulty that some of the plants that had been newly erected had never been seen except by the lens of the aerial camera.

35. The estimation of plant output, which was the responsibility of a committee under the chairmanship of Lieut.-Colonel S. J. M. Auld, was largely the work of a group of experts whose collective knowledge of Germany's petroleum and coal resources, and of the processing equipment employed, resulted in a commendably close appreciation of the output of the majority of the plants.([16]) The centres of aviation fuel production were largely ascertained by the work of the Enemy Oils and Fuels Committee, which, by means of the analysis of a large number of captured samples of gasoline, was able to confirm that the *Luftwaffe* was relying upon certain plants for its supplies.([17]) This information was of great value in the framing and execution of bombing policy.

The Estimates of Consumption.

36. The calculation of the consumption side of the balance sheet involved a study of the changing rates of consumption by the civil economies in all the European Axis countries and also the fluctuations in consumption by the Armed Forces. The statistical work in connection with civil consumption was done by the Ministry of Economic Warfare and, by means of both reliable intelligence and deductions from less reliable intelligence, reasonably accurate figures were obtained. The calculation of Armed Forces consumption was undertaken by the

([14]) Mr. R. H. W. Bruce in 1942 and Mr. O. F. Thompson in 1943.

([15]) A full account of these Conversations is given in J.I.C. (43)340 of 18.8.43. These Conversations also covered the question of Japan's oil position which was studied by the two Committees no less comprehensively. These studies provided the basis for the attacks later made against Japan's oil supplies.

([16]) The capacities of the hydrogenation plants were accurately estimated except for the two new plants at Heydebreck and Auschwitz which were assumed from aerial photographs to be plants of a conventional type, whereas in fact their fuel production capacity was considerably smaller than was estimated. It was largely on this account that the total production from the hydrogenation plants was overestimated by 23 per cent.

The performance of the Fischer Tropsch plants was greatly overestimated, and this was because intelligence did not reveal that, on account of various difficulties, output was much below the designed capacities of the plants. Assessment was also handicapped by the fact that there had been no experience in the operation of these plants in either Great Britain or America.

Crude oil production was also overestimated, the rate of output for the *Altreich* in the spring of 1944 being high by 30 per cent.

Except possibly in the case of Heydebreck and Auschwitz, these errors did not adversely affect the correct allocation of bombing priorities.

([17]) The examination of these samples also resulted in the perception of the advantages being gained by the Germans by the use of aromatic fighter fuels with rich mixture performances, which information was of value in the improvement of Allied fuels.

intelligence departments of the appropriate services and, as the statistics involved were in a number of ways more intricate than the calculation of civilian consumption, some account should be recorded of how these estimates were prepared.

37. Calculations of the consumption by the Army, which were the responsibility of Section M.I. 10 (*c*) of the War Office, were based upon the enemy's order of battle as known to Military Intelligence. The fuel consumption of different classes of divisions was then calculated and a running record maintained of the activities of each division insofar as these were known. The quantities of liquid fuel consumed by each division was then estimated in accordance with the current degree of activity of that division. In the case of a division engaged in active fighting it was assumed, for example, that all the effective vehicles might be averaging a liquid fuel consumption equivalent to 50 miles a day. The estimates were adjusted by a number of factors to cover losses, weather conditions and the nature of the terrain, &c. Account was also taken of the consumption by non-divisional vehicles and by quasi-military transport. Some clue to the general accuracy of these estimates was occasionally afforded by the capture in battle of documents recording the consumption data of individual units. From information obtained since the War ended it is seen that the trends in consumption as estimated by the War Office closely followed the actual fluctuations that occurred. The figures themselves were generally high by a small percentage.

38. The consumption of aviation fuel was also calculated on the basis of the known order of battle combined with a complex appreciation of fluctuations in activity. The co-ordination of this study was undertaken by Section A.I. 3 (*c*) of the Air Ministry Intelligence Branch. Figures were available of the number of operational aircraft by types and these were divided into those units that were operationally engaged and those not operationally engaged. For each type of aircraft an average length of sortie was estimated on the basis of Allied experience and this gave the number of engine hours per operational sortie. Another 25 per cent. was added to cover non-operational engine hours. The average fuel consumption for each type of engine was computed and, after various factors had been taken into account, a total monthly consumption figure was reached. Consumption by ground units and by the Flak organisation was likewise calculated upon the order of battle and by such information as became available upon activity. Some reliable intelligence obtained in **1943** resulted in an upward adjustment in the calculations previously made of consumption by the *Luftwaffe* and the subsequent estimates proved reasonably accurate.

39. In the case of the Navy the estimates were computed by Section 7 of the Naval Intelligence Division. The calculations were based upon the fuel consumption capacity of each vessel and the amount consumed was worked out in accordance with the movements of each vessel as recorded by the Operational Intelligence Centre of the Admiralty. Photographic reconnaissance was of considerable importance in maintaining a record of activity. Consumption in harbour and the requirements of store establishments were worked out on the basis of Allied experiences. The official German Admiralty figures that have since been obtained have shown that the estimates have been correct to within a reasonable margin of error.

40. The possibility that Germany might have overcome the worst of her liquid fuel difficulties by means of substitutes such as producer gas was also studied in detail. The subject was not one that could be reported upon with precision by intelligence sources and estimates of Germany's actual success in this direction were coloured both by the German plans, which were known and seemed capable of achievement, and also by the successes of the neutral countries in their handling of the same problem. In consequence the estimates that were made of the saving in gasoline were invariably too high.([18])

Liaison with Washington.

41. All these calculations of military and civil expenditure of oil were considerably assisted by independent estimates that were made in Washington. These were mostly prepared on a different basis and they consequently provided a useful check on the probable accuracy of the figures as a whole. These very

([18]) In J.I.C. (43) 253 (The Axis Oil Position in May 1943), p. 30, it was reported that the saving in the use of liquid fuel by the use of substitutes might be over 1 million tons a year. The actual figure was probably not more than about half this quantity.

detailed studies of production and consumption resulted in an approximation of the level of stocks. As mentioned earlier in this Report the original estimate of stocks made by the Industrial Intelligence Centre was approximately correct and an arbitrary increase in this figure resulted in the assumed stock level being placed too high. However, the fluctuations in stock levels were estimated with sufficient accuracy to report correctly the changes in the situation. How closely these estimates of the trend in stocks conformed with the true position is shown graphically on page 40. The calculations of the Hartley Committee correctly showed the point at which stocks fell below the " distributional minimum " and the duration of this period until equilibrium had been restored.

42. In June 1943 separate reports on the position were submitted by the Washington Enemy Oil Committee and by the Hartley Committee.([19]) Although these reports contained differences of detail the conclusions reached were almost identical. It was agreed that whereas there had been some recovery in stocks the position was uncomfortably tight. It was foreseen that an expansion in synthetic output would increase supplies and that the defeat of Italy would make more oil available to Germany. In the meantime it was considered that the position would continue to be critical until the autumn of 1943. These two reports were discussed with the American delegation that visited London in July and this resulted in a joint recommendation from the two Committees that, during the next six months, effort should be directed—

(1) to destroying synthetic oil plants and refining capacity, particularly in Ploesti;
(2) to producing the maximum interference with the transportation, distribution and storage facilities on which the Axis oil economy is dependent.
(3) to forcing the maximum consumption and dispersal of available oil supplies through direct or threatened military action.

43. In November 1943 the Hartley Committee submitted a report([20]) that reviewed the position in considerable detail. It was concluded that, although stocks had perceptibly improved, the position as a whole was more vulnerable than at any previous time. The basis for this contention was, firstly, that the Allied advances had brought all the principal sources of oil within range of air attack and, secondly, that the refining capacity of South-Eastern Europe was at that time barely sufficient to meet requirements; refining facilities as a whole appeared particularly sensitive to dislocation and the importance of refineries as targets had become greatly enhanced. It was submitted that the insufficiency of oil stocks was still causing Germany grave anxiety and that any substantial interference with her oil supplies would seriously weaken her ability to continue the War. The report ended with the following deduction :—

> " We should, as far as circumstances will permit, use every endeavour this winter to destroy as much of the enemy's oil resources as possible."

44. The Hartley Committee submitted one further report([21]) before the attack on oil began. This was in May 1944, and it was correctly appreciated that if Roumanian oil became lost to the Germans there would not only be a serious shortage of liquid fuel but their reliance upon the synthetic plants would be even greater than before. It was the opinion of the Committee that if, combined with the loss of Roumania, an Allied offensive was opened on both the Eastern and Western fronts, Germany would then be faced with her consumption requirements being over fifty per cent. in excess of total production. Under these conditions it was concluded that—

> " Within two to three months, depending on the circumstances, military oil supplies would have to be cut to an extent that would cause a most serious contraction in operational mobility."

45. These detailed studies of the German oil industry, and of the potential effects of its dislocation, provided the planning staffs with the material necessary to decide upon the course of future action.

The Co-ordination of Intelligence.

46. It was at this juncture that the need for the immediate appreciation of intelligence for operational purposes made it necessary for the functions of the

([19]) J.I.C. (43) 266 and J.I.C. (43) 253. ([20]) J.I.C. (43) 463. ([21]) A.O. (44) 41.

Hartley Committee to be divided between the newly-appointed Combined Strategic Targets Committee (Oil Committee)([22]) and the Foreign Office and Ministry of Economic Warfare.

47. The Ministry of Economic Warfare, operating in close liaison with the Economic Objectives Unit of the United States Economic Warfare Department, was made the clearing house for all intelligence upon oil and oil targets. This work was assisted by the close co-operation of the Intelligence Departments of the Services, the Ministry of Fuel and Power and last, but not least, the Central Photographic Interpretation Unit (later A.C.I.U.) of the Air Ministry.

48. The Germans maintained elaborate precautions to prevent the leakage of information about their oil position. Plants and depots were closely guarded and all communications on the subject were accorded a high security grading. In spite of these precautions, intelligence reports, amounting on occasions to the number of several hundreds weekly, were regularly received by the Ministry of Economic Warfare. A large proportion of these reports were from the interrogation of prisoners of war and of persons who had been in enemy territory and, while the bulk of such information was not objective, it was possible to draw a reasonably clear picture of the enemy's position.

49. To assist in the sifting and dissemination of this information a weekly meeting was held at the Ministry of Economic Warfare of a small group known as the Enemy Oil Intelligence Committee. The bulletins issued by this committee were given as wide a circulation as their interest warranted and their security grading allowed.

50. In addition to the information obtained from the numerous agencies engaged in the gathering of intelligence there was the invaluable work of the aerial reconnaissance units. The skill and daring of those responsible for photographing enemy activities was ably supported by the personnel interpreting the photographs. The technique that was developed in assessing enemy activity by these means was, especially in the later stages of the War, by far the most valuable single source of intelligence. In the course of time a remarkable degree of accuracy was achieved in judging the state of activity of oil plants, the speed of construction of new plants and the approximate production of oilfields.([23])

Commentary upon Target Selection.

51. A faulty appreciation of the operational capabilities of an oil plant would have resulted in a mis-direction of offensive effort. From the facts available the priorities for attack recommended by the Target Committees were, in general, listed in the correct order. These priorities were based upon the importance of each plant as a producer of gasoline. Although the capacities of the Fischer-Tropsch plants were greatly over-estimated, they were nevertheless listed in their appropriate positions in the target lists, and this was due to the fact that their gasoline output, whether rightly or wrongly estimated, came below that of the hydrogenation plants and above that of the greater number of oil refineries.

52. On account of the many operational factors that decide the selection of a target to be attacked it was not possible for bombing operations to follow precisely the order of target priority recommended. Moreover, the weight of bombs dropped on a given target tended to be in inverse proportion to the distance to be flown to reach the objective. It is consequently difficult to draw comparisons between targets on the basis of bomb density in relation to products output.([24]) Although certain targets were inevitably over-bombed while others were under-bombed, the consequent mis-direction of effort was not on a scale sufficiently great to be the subject of detailed criticism by any of the Allied bombing research teams or by the Germans themselves.

([22]) See Appendix 11.

([23]) A summary of the technique employed is given in Appendix 11.

([24]) In the table given on page 184 the tons of production loss per ton of bombs dropped by different processes is shown as follows:—

Hydrogenation	36
Fischer-Tropsch	10
Oil Refineries	17·7
Miscellaneous	29
Total	26

The unfavourable ratio shown for the Fischer-Tropsch plants is principally due to the fact that most of these plants were in the Ruhr and therefore within short range for heavy night attacks by R.A.F. Bomber Command.

The Results Achieved.

53. Confirmation of the success of the whole intelligence organisation in ensuring that the enemy's oil potential was systematically and accurately attacked is provided by the post-war evidence that there was no significant oil plant that was not located nor attacked in its appropriate order of importance. The Germans themselves, in the course of the oil offensive, were amazed at what they regarded as the uncanny discrimination of the Allied bombers in their selection of targets. Furthermore, the statistical estimates of production and consumption in the concluding stages of the War were sufficiently accurate to ensure that the Chiefs of Staff were correctly advised of the position.

54. The value of the work done by these committees was expressed in a message of appreciation by General C. Spaatz, commanding the United States Strategic Air Forces, of which the following is an extract :—

> "The brilliant work and infinite pains which these organisations have shown in piecing together the multiplicity of intelligence information, have raised the selection of strategic targets to the stature of a science. The consistently sound recommendations which this Committee and its subsidiaries have submitted to the Deputy Chief of Air Staff, R.A.F., have been largely responsible for the decisive execution and successful conclusion of the Strategic Air Offensive in Europe.
>
> The congenial atmosphere and general unanimity of opinion coming from this mixed group of British and American officer and civilian personnel, working together for the defeat of Germany through the determination of sound operational principles and targets, has been an inspiration to all of us and a model for future combined operations."

APPENDIX 1.

Organisation of the German Oil Administration.

Until the general reorganisation of German administration in September 1943, when Speer emerged as the supreme director of war production, the Government machinery for oil administration consisted of the *Reichswirtschaftsministerium* and the *Reichstelle fuer Mineraloel.* This latter organisation, formed from the old *Ueberwachungstelle fuer Mineraloel*, was originally responsible for oil production, consumption, imports, exports and distribution, but its responsibilities had been gradually whittled down, principally by the Office of the Four-Year Plan, which appointed special commissioners for crude production and for synthetic production. On the outbreak of war, in view of the importance of oil, more of the *Reichstelle's* functions were transferred to the *Reichswirtschaftsministerium.*

In September 1943 all questions of oil production, import and export were transferred to the *Rohstoffamt*, a section of the Ministry of Armament and War Production, while the *Planungsamt* of the office of the General Commissioner for the Four-Year Plan became responsible for examining the claims of the armed forces, put forward by *O.K.W. Wirtschafts-Ruestungsamt*, and for general planning.

Under this reorganisation, therefore, oil control was divided into three sectors :

1. *Planungsamt* was responsible for general planning and the control of consumption. Control of civil consumption was vested in the *Reichstelle fuer Mineraloel.*
2. *Rohstoffamt* controlled production and imports.
3. *Reichswirtschaftsministerium* was responsible for market policy, currency questions and price control, principally through the *Reichstelle fuer Mineraloel.*

While the *Rohstoffamt* was in general control of production, technical responsibility was delegated to two Commissioners of the Four-Year Plan. Professor Krauch was *Beauftragter fuer Sonderfragen der Chemischen Erzeugung* and was in technical control of all operations and planning of the synthetic oil industry, and Dr. Bentz was *Beauftragter fuer die Erdoelgewinnung*, with crude oil production as his responsibility.

All firms of the oil industry were compulsory members of the trade federation *Wirtschaftsgruppe Kraftstoffindustrie* and production control was largely effected through the subsidiary sections of the *Wirtschaftsgruppe.* There were four of these main subsidiaries :—

Arbeitsgemeinschaft Erdoelgewinnung und Vorarbeitung.
Arbeitsgemeinschaft Hydrierung, Synthese und Schwelerei.
Arbeitsgemeinschaft Benzolerzeuger.
Arbeitsgemeinschaft Steinkohleteererzeugnisse:

Although not officially part of the governmental machine, these federations were important links in the chain of executive control in that it was only within their competence to convert the general planning directions of the *Planungsamt* and *Rohstoffamt* into a technical production programme.

Oil imports were planned and controlled by the *Rohstoffamt* through the agency of two State importing Companies especially formed for the purpose :—

Rumaenien Mineraloel G.m.b.H. for imports from Hungary and Roumania.
Mineraloeleinfuhrges. m.b.H. for imports from Russia and Galicia.

Questions of import prices and currencies were handled by the *Reichstelle fuer Mineraloel* on behalf of the *Reichswirtschaftsministerium.*

As the execution of the production programme was delegated to the *Arbeitsgemeinschaften*, or trade federations of the various producing industries, so distribution was likewise delegated to trade federations or cartels of oil

A RECONNAISSANCE VIEW OF THE MISBURG REFINERY.

This composite photograph was taken on 30th March, 1945, after the fourteenth and last attack upon the refinery.

distributors. There were eight of these *Arbeitsgemeinschaften*, each varying in importance with the products they were empowered to handle :—

Arbeitsgemeinschaft Mineraloelverteilung, with its executive
Zentralbuero fuer Mineraloel Arbeitsgemeinschaft Schmierstoff-verteilung.
Flugkraftstoffgemeinschaft.
Arbeitsgemeinschaft fuer Petroleumverteilung.
Grossbunkergemeinschaft.
Verteilungstelle fuer Bitumen.
Arbeitsgemeinschaft fuer Test-Benzin.
Arbeitsgemeinschaft White Spirit.

These federations were under the control of the *Reichstelle fuer Mineraloel*, and had a monopoly of the purchase, sale and distribution of their respective products, except in so far as supplies to the Armed Forces were concerned.

The most important of them was the *Arbeitsgemeinschaft Mineraloelverteilung* and its executive *Zentralbuero fuer Mineraloel* which had a monopoly in the distribution of gasoline, diesel oil and liquid gas.

Distribution for service requirements was handled by the service supply corps, assisted, in the case of the Army, by *Zentralbuero* and *WIFO*. Air force fuels were handled almost exclusively by *WIFO* except that motor fuel for ground transport went through *Zentralbuero* channels. The Navy handled its fuel supplies independently and made very little use of non-service organisations.

APPENDIX 2.

Rationing of Liquid Fuels for Civilian Consumption.*

After consideration of the import, production and stock position, the *Reichsministerium fuer Ruestung und Kriegsproduktion*, in consultation with the Supreme Command, handed over monthly a certain quantity of fuel to the *Reichstelle fuer Mineraloel* for distribution for non-military purposes. The *Reichstelle fuer Mineraloel* divided these quantities up between the *Landeswirtschaftsaemter* and special groups of large consumers.

The *Landeswirtschaftsaemter* got their allocations for agriculture and road transport separately and passed them on to their sub-areas, to the *Kreisbauernschaften* and *Wirtschaftsaemter* respectively. These organisations issued coupons to consumers, which coupons were honoured by *Zentralbuero fuer Mineraloel*.

The central administrations of large consumers distributed their global allocation among their provincial administrations, who in turn distributed to their local consumers. Again actual delivery and supply of fuel was made by *Zentralbuero fuer Mineraloel* on the basis of these coupons.

The quantity of fuel distributed each month for non-military purposes depended upon the surplus available after military requirements had been met. When allocations had to be reduced the *Reichstelle fuer Mineraloel* decided, on the basis of statistical reports concerning the general activity of each area, how the consumption cut was to be spread.

The system of allocation was based on the individual necessity of the consumer. On that principle the *Landeswirtschaftsaemter* and *Wirtschaftsaemter* decided on their own responsibility what ration each consumer should get. Only in the case of agriculture and domestic lighting and heating by kerosine was there a system of a basic ration. However, to ensure uniform treatment in different areas the *Reichstelle fuer Mineraloel* issued periodical classifications of consumer groups according to priority and a monthly circular concerning the general policy to be applied in fuel distribution.

* Source: *Zentralbuero fuer Mineraloel*, Hamburg, 23rd August, 1945.

APPENDIX 3.

THE FORMATION AND DEVELOPMENT OF THE WIRTSCHAFTLICHE FORSCHUNGSGESELLSCHAFT.

(Extracted in part from a report by Franz Wehling, General Manager of the WIFO.)

The Company was officially formed on the 24th August, 1934, and was registered in the Commercial Register (Handelsregister) on the 11th September, 1943, at the Amtsgericht, Berlin. The function of the undertaking was the direction and maintenance of a series of works, industrial, commercial and handicraft (Handwerke), and, in particular, the direction and operation of experimental and research plants with the object of research into the aforesaid Economic Branches.* The initial capital was Rm. 20,000. The financing of its undertakings was obtained by means of State finance from the budget of the Ministry of Economics. These funds were administered by the Company on a trustee basis and were subject to audit.

By decision of the Company on the 25th April, 1939, an Advisory Council was formed as from the 1st January, 1939, consisting, in the first instance, of eight members. The members of the Advisory Council were nominated by the Ministry of Economics, which, from its inception, was the Ministry responsible for the operations of the Company, this procedure remaining in force until the end.

The Advisory Council was, in particular, responsible for the entire financial operations of the Company. Already, before their nomination as members of the Advisory Council, the majority of these individuals had been closely associated with the working of the Company in the form of so-called "Technical Commissions" (T.K.). The composition of the Commissions varied as required to deal with matters outstanding so that all concerned could ventilate their views. There was a tendency, during the time in which construction was in progress, for separate Commissions to be set up to deal with mineral oil matters, economic matters, and later, as this matter became of greater importance, the question of installation operations. Even after the formation of the Advisory Council, these Commissions continued in being with, as far as possible, a member of the Advisory Board in the Chair at their meetings.

In general terms, the function of the Commissions remained restricted to that of the section dealing with construction. For the operations of the undertaking, in course of time several clear principles were worked out and gradually brought into effect.

The product remained the property of the principal users.

The principal users themselves closed contracts with the producers in regard to specifications, quantities and prices. Contracts entered into were handed over to the *WIFO* for fulfilment, *i.e.*, for them to obtain delivery of product from works and for storage in the *WIFO* Depots. Where blending had to take place, account was kept of the quantities and values of the component parts, as also of the resulting blends.

WIFO expenses consisted in cost of transport, cost of mixing/blending, depot rent and depot throughout, and were usually calculated on the basis of unit cost per unit weight.

Restriction in Transport responsibilities.

WIFO took delivery of components from producing plants in their own (*WIFO*) rail tank cars and likewise delivered finished products in *WIFO* rail tank cars, to the depots of the users. Users' depots consisted at the beginning of the war of—

(*a*) 50–60 commercial depots for commercial and army use.
(*b*) 10 army depots for commercial and army use.
(*c*) 7 Air Force depots (later a few more) and 30–40 subsidiary Air Force depots for Air Force requirements.

At the beginning of the War considerable use was made of users' depots but later, and with increasing destruction of above-ground commercial depots, the whole throughput tended to be concentrated more and more in the underground depots of the *WIFO*.

* This sentence is an accurate translation of Wehling's description of the purposes of the Company, which were in fact for waging war and for no other purpose than waging war.

In addition to transport in rail tank cars, transport by tank barge on the inland canal system accounted for something like one-quarter to one-fifth of the total tonnage carried. Here the available fleets of the private tank barge owners were used, on average some 35 to 40 tank barges being in use on *WIFO* account.

Restrictions in responsibility for throughput.

Fundamentally, the operations of the *WIFO* at their main storage depots were restricted to receipt from and discharge from storage into vessels and rail tank cars. Only at a later date and at a few places was provisional equipment provided for filling, particularly lubricating oil, into barrels. In the construction projects towards the end of the war, permanent facilities at the main storage depots were planned for barrel filling.

In the case of *WIFO* Depots for the Air Force, equipment was provided for delivery to road tank cars, and in the case of Army Depots, permanent facilities were provided for barrel filling and for the automatic filling of cans and bottles. Of no less importance was the function of these Depots in the blending and ethylising of aviation and army gasolines.

Throughput.

Up to the beginning of the war the capacity of the main bulk storage depots was approximately 820,000 cbm. for all forms of motor and aviation fuels and about 110,000 cbm. for lubricants.

Monthly output potential (turnover) for all depots operating on the basis of a ten-hour day could be put at around 175,000 tons in and out. This capacity was sufficient until the early part of the year 1942, at which time stocks had fallen to around 150,000 tons.

Tied Stocks.

This working stock figure could not be further reduced since it consisted of a large number of individual components of intermediate mixtures and finished products distributed over the whole network of depots. The responsibility of the *WIFO* now became to the largest possible extent purely one of installation operation (throughput). The production of the hydrogenation plants was invariably promptly cleared as large stocks were not permitted at any plants owing to the danger of air bombardment.

Daily deliveries were therefore subject to strict control. The throughput capacity of the *WIFO* had to keep pace with the planned development of production of the hydrogenation plants. Extension of the main storage depots was therefore put in hand with the object of catering for a monthly throughput of about 350,000 tons in and out.

The raising of the throughput potential was to some extent obtained by doubling the railway siding capacities and partly by the building of new storage units with a considerable extension of pumping equipment.

Increase in Storage Capacity.

The complete development programme before the war was established on the basis of a storage capacity of a total approximate 2 million cbm. At the beginning of the year 1941 the building programme was stopped and only those portions of the programme were permitted to continue which were already well under way. Therefore, at the end of the war, extension of the tankage capacity had only been achieved to the extent of around 1½ million cbm.

After, as a result of the quick turn round of products, an extension of the tank capacity was no longer necessary, but this was, nevertheless, partly put into effect as it was still necessary to provide for the possibility of damage through air attacks, and on that account considerable flexibility was necessary.

At the end of May 1944, a permanent and steady decrease in the deliveries followed as a result of the damage to refineries and other production works, which reached its lowest point around October–November 1944. However, even in the following months, little further improvement was obtained.

Capital Reorganisation.

The continuously increasing importance of installation throughput since the beginning of the war compelled a reconstruction in the financing of the company. The relations as of a trustee which existed between the *WIFO* and the German State in regard to all property was brought to an end and the whole value of all property with effect from the 1st April, 1943, was transferred to the *WIFO*. The capital of the company was thus raised to RM. 100 million and the remaining funds, which had been given to the *WIFO*, were converted into a loan of a total amount of RM. 670 million.

NOTE.—Details of other activities of *WIFO* in connection with the construction of chemicals and explosives plants, are not given in this report.

Principal WIFO *Oil Storage Depots.*[1]

	Capacity in Cubic Metres. Fuel.	*Lubricants.*
Hauptlager (Main Underground Depots)		
Stassfurt (Elbe)	220,000	—
Derben (Elbe)	180,000	18,500
Nienburg (Weser)	100,000	10,200
Munich (Krailling-Freiham)	100,000	5,700
Drugehnen (near Koenigsberg)	100,000	6,000
Neuburg (Danube)	100,000	5,700
Hitzacker (Elbe)	100,000	46,800
Farge (Weser)	320,000	—
Niedersachswerfen (Hartz Mts.)	(Not completed)	
Krumnussbaum	(Not completed)	
Umschlarlager (Transit Underground Depots)—		
Vienna	160,000	
Roudnice	100,000	
Heerestanklager (Army Surface Depots)—		
Ruthen	7,200	
Neuenheerse	,,	
Eickeloh	,,	
Zarrentin	,,	
Muenchen Bernsdorf	,,	
Vorderheide	,,	
Ebrach	,,	
Amstetten	,,	
Heiligenstadt	,,	
Maehrisch-Schoenberg	5,000	

Evaluation of WIFO *Properties.*

(From the 1940 *WIFO* Audit Report.)

	Total Value. (In millions of marks.)
Oil Depots—	
(*a*) Main Depots	286·3
(*b*) Drum Depots	·3
(*c*) O.K.H. Depots	67·5
(*d*) Toluol Depots	1·8
(*e*) Transit Depots	44·5
	400·4
Factories—	
(*a*) Toluol Plants	14·5
(*b*) Sulphuric Acid Plants	6·0
(*c*) Sulphur Plants	26·4
(*d*) Nitric Acid Plants	63·5
	110·4
Raw Material Depot Building	2·5
Strontium Mine	·1
Pipelines	29·0
Tankers (water and rail) and drums	163·4
Equipment for alcohol fuel	0·4
Hutments	1·4
Workshops and Equipment	1·4
Other Installations	1·2
Equipment in reserve for building	5·0
Total	715·2

[1] This list does not include Naval oil storage which was not a *WIFO* responsibility.

Lufttanklager.

Surface Depots built by *WIFO* for the *Luftwaffe* and operated by the *Luftwaffe*—

Annaburg	5,400
Bad Berka	6,500
Duelwen	13,000
Ebenhausen	6,500
Ehmen	5,900
Langenselbold	6,200
Weissenhorn	10,400
Niederullersdorf	?
Loewenhagen	?

Main Underground *Lufttanklager* not built by *WIFO*—

Buchen	20,000
Loccum	13,500
Oldendorf	45,000

Turnover Sales.

	Million Marks.
March 1939–40	**247**
1940–41	**593**
1941–42	**693**
1942–43	**1,255**

WIFO *Personnel.*

1937	**355**
1938	**970**
1939	**2,014**
1940	**3,941**
1941	**5,938**
1942	**6,615**
1943	**7,544**

End of March 1943, 1,218 were inducted into the Wehrmacht.

APPENDIX 4.

Kontinentale Oel, A.G.

Kontinentale Oel, A.G. was formed in **1941** for the purposes of exploiting Germany's oil conquests. It was sponsored by both the Government and industry, the paid up capital of RM. 80 million being provided and held by the following shareholders :—

	RM. million.
A number of German oil companies	20
Several of the principal German banks	30
The German Government, through *Borussia Beteiligungs G.m.b.H.*	30
Total	80

The capital provided by the oil companies came from concerns such as *I.G. Farbenindustrie*, *Elwerath*, *Deutsche Erdoel*, *Braunkohle-Benzin*, *Preussag*, *Wintershall*, *Sudetenlaendische Treibstoffwerke* (Bruex), *Benzol-Verband*, &c. The banks were principally represented by the *Deutsche Bank*, *Dresdner Bank* and *Reichskreditgesellschaft*.

Management.

The management of the company comprised the *Aufsichtsrat*, the *Verwaltungsrat* and the management proper. The *Aufsichtsrat*, which comprised twenty-two persons representing the oil industry, the banks and the Government respectively, was presided over by *Reichsminister* W. Funk. The *Verwaltungsrat* was dispensed with early in the war and the direction of the company was left principally in the hands of Karl Blessing and Dr. E. R. Fischer, with Walther Dihlmann as Managing Director. The company had its headquarters in Berlin, but in March **1945** the staff, under Walther Dihlmann, evacuated to Landshut, in Bavaria. The personnel of the company normally numbered about 3,000 of German nationality, although by the end of the war this number had been reduced to about 300. Approximately forty of the staff evacuated to Landshut.

Activities.

The purpose of the company was to operate in all spheres of the oil business which were not already covered by the German oil industry. The company consequently handled virtually all of Germany's oil conquests. The company was, therefore, engaged in production, refining, transporting and distributing. In the execution of these functions, the *Kontinentale Oel* usually acted as a holding company, the activities in the various fields being undertaken by subsidiary or affiliated companies.

At the time the war ended, the company held equipment in Germany, stored in 26 depots, which was worth, according to the estimate of the Managing Director, RM. 12 million.

The activities of the company are briefly covered by the following notes respecting each country in which *Kontinentale Oel* was operating.

Rumania.

Rumania was a primary objective for the company in its work of economic penetration. Before the war German interests held less than 1 per cent. of the total capital of the Rumanian oil companies. In the course of the war *Kontinentale Oel* succeeded in controlling, by direct participation in various companies, no less than 25 per cent. of the production and refining of petroleum in Rumania. The dealings of the *Kontinentale* with Rumania were principally handled through a subsidiary, *Kontinentale Oel G.m.b.H.*, which had a capital of RM. 1 million.

This latter company was formed not only for the purpose of acquiring financial interests in Rumanian companies, but it was also the medium through which German oilfield equipment was supplied to Rumania. This company also entered into a contract with the *Astra Romana* for participating in exploration work.

The following summarises the participation of *Kontinentale Oel* in Rumanian oil companies up to the time of the collapse of that country:—*

(a) *Concordia S.A.R. Bucharest.*

Participation of *Kontinentale Oel*: 54·59 per cent.
Prime value of participation: RM. 23,900,000.
Adm. Délégué: Osterwind.
Activity: Drilling and refining of petroleum, electric power stations, plants for construction of machinery and apparatus, lignite mines, gold mine.

(b) *Colombia S.A.R. Bucharest.*

Participation of *Kontinentale Oel*: 85·53 per cent.
Prime value of participation: RM. 13,100,000.
Adm. délégué: Dr. Gramsch.
Activity: Drilling and refining petroleum.

(c) *Explora S.A.R. Bucharest.*

Participation of *Kontinentale Oel;* 100 per cent.
Prime value of participation: RM. 850,000.

This company did not get beyond the formation stage.

(d) *Sarpetrol S.A.R. Bucharest.*

Participation of *Kontinentale Oel;* 50 per cent.

(e) *I.R.D.P. Bucharest.*

Participation of *Kontinentale Oel* via *Suedostchemie Ges. m.b.H.* (25 per cent. ?) (see below).
Activity: Drilling operations.

(f) *Petrol Block S.A.R. Bucharest.*

Participation of *Kontinentale Oel* via *Suedostchemie Ges. m.b.H.* (see below).
Activity: Refining.

(g) *Transpetrol, Bucharest.*

Participation of *Kontinentale Oel* via *Suedostchemie Ges. m.b.H.* (see below).
Activity: Forwarding.

(h) *Suedostchemie G.m.b.H.*

This was a wholly owned subsidiary of *Kontinentale Oel*, with its headquarters in Berlin, and with a capital of RM. 12,000,000. It held participations in the companies named above.

* In this connection the following extract from a Foreign Office, E.I.D. paper ("German Business Penetration under the New Order," ref. E. 1b, dated 19.6.45) is of interest:—

"The (Rumanian) Government was also opposed to any extension of foreign capital penetration. It brought the Bourse under rigorous control in 1941, appointing commissioners to supervise individual share transactions and to exercise, if thought desirable, State pre-emption rights over any securities on offer. The Germans had consequently to confine their purchases largely to shares already foreign-owned, which meant in effect those held by French and Belgian interests. These interests had already before the war been anxious to sell out but could not do so because of exchange restrictions on the repatriation of capital. The Germans were now able to offer them high prices and payment at home in their own currencies (evidently out of occupation levies). The principal controlling interests so acquired were in *Concordia* and *Colombia*, which had been German-owned before the last war. The *Kontinentale Oel A.G.*, Berlin, bought control of the *Concordia* from the *Compagnie Financière Belge des Pétroles* (*Petrofina*) and other Belgian interests associated with the *Société Générale de Belgique*, and of the *Colombia* from a French group represented principally by the *Omnium Français des Pétroles*, the *Banque Paris et des Pays-Bas* and *Mirabaud et Cie.* In the same year another French group, the *Société Industrielle des Pétroles Roumains*, sold the *Sudostchemie Hand. A.G.* its quarter-interest in the *Industria Romana de Petrol* (*I.R.D.P.*) This undertaking was linked with a refinery enterprise, *Petrol Block*, in which a year earlier the *Erste Brucnner Maschinenfabrik A.G.* (a *Reichswerke Goering* concern) had obtained a dominating interest in return for refinery equipment. Minor German acquisitions included the *Foraky Romaneasca*, which was taken over by the *Concordia* in 1941."

Russia.

The activities of the company in Russia are largely of historical interest, but the company's plans to exploit Russian oil were as follows :—

A wholly owned subsidiary, the *Ost Oel G.m.b.H.*, which had a capital of RM. 4 million, was formed to plan and prepare operations for the exploitation of Russian oil.

Another company, the *Ostland Oel Vertriebs G.m.b.H.*, with a capital of RM. 50,000, was engaged in supplying the *Reichskommissariat Ostland* with mineral oil products, which, in 1943, amounted to about 150,000 tons.

A similar company, the *Ukraine Oel Vertrieb G.m.b.H.*, with the same capital, supplied the *Reichskommissariat Ukraine* with mineral oil products which, in 1943, amounted to about 300,000 tons.

Poland.

Activities in Poland comprised, firstly, a participation with other German companies in production and refining and secondly, in the direct operation of refineries. *Kontinentale Oel* had a 50 per cent. interest in *Karpathen Oel A.G.* which had a capital of RM. 15 million, the remaining capital being held by a number of German oil companies. The company had its headquarters in Cracow, and was engaged in the exploitation of Polish oil. *Kontinentale Oel* was interested only financially in the company and had no direct control in its commercial and technical activities.

Erdoelraffinerie Trzebinia G.m.b.H., which was wholly owned by the *Kontinentale Oel*, and which had a capital of RM. 4 million, was formed to operate the Trzebinia refinery which was the property of *Polski Zwiazkowe Rafinerje Oleyow Skalnych* (*Malopolska*). This refinery was substantially enlarged in the course of the war and it is probable that equipment taken from French refineries was used for this purpose. The company was also constructing a refinery at Galatz, although this plant was captured by the Russians before it was completed.

The Baltic Area.

Baltische Oel G.m.b.H., was formed, with a capital of RM. 20 million, for the purpose of resuscitating the Esthonian shale oil industry. The company set about the reconstruction and enlargement of a number of shale oil plants and was the sole agent of the German Government in the production of Esthonian shale oil.

A company by the name of *Sapropel G.m.b.H.*, with a capital of RM. 20,000, was started in Lettland for the purpose of digging and treating Sapropel (a sort of peat) for the production and distribution of peat coke. The peat coke was produced at Papenburg.

Bulgaria.

Bulgarische Mineraloel A.G. was formed, with a capital of Lewa 3,300,000 and with headquarters in Sofia, for the importation and distribution of oil products in Bulgaria.

Greece.

Kontinentale Oel G.m.b.H. (see Rumania above) had an office in Athens which was engaged in the importation and distribution of mineral oil products in Greece in collaboration with Shell-Hellas, Ltd. In this connection a company entitled *Mineraloel G.m.b.H. Suedost* was formed with a capital of RM. 20,000. Fifty per cent. of this capital was held by the *Bataafsche Petroleum Mij* and the company acted as a clearing cartel for the oil business of *Kontinentale Oel G.m.b.H. Suedost* and *Shell-Hellas*. (It would appear that the B.P.M. holding in this company is without legal justification as the administration of the B.P.M. was not within reach of German control.)

Yugoslavia.

The distribution of oil products in Yugoslavia was handled by *Mineralœl Vertrieb Serbien A.G.*, which had a capital of 1 million dinar. The share holding in this company was as follows:—*

	Per cent.
Kontinentale Oel	30
Jugoslavensko Shell	30
Petrol, Belgrad	20
Refinery, Semendria	20

Albania.

Albanien Oel G.m.b.H. was formed, with a capital of RM. 20,000, for the purpose of refining the crude oil in the Devoli area.

Italy.

Kontinentale Oel took advantage of the collapse of Italy in 1943 to make use of certain refineries in Northern Italy. Cracking capacity was urgently needed for the processing of heavy crude oil residues and the *Kontinentale Oel* made an arrangement with the *Azienda Generale Italiana Petrolii* at Mestre (near Venice) whereby topped crude oil was cracked and the resultant light products were used to alleviate the acute shortage of fuels in Northern Italy. Arrangements were also made with the refineries at Trieste and Fiume for the production of lubricating oils.

France.

The activities of the *Kontinentale Oel* in France comprised the acquisition of equipment from French refineries and an attempt to expedite the production of crude oil in Southern France.

The removal from certain French refineries was based upon an agreement between the German Government and Vichy. The French companies were then forced to conclude an agreement with the *Kontinentale Oel* by which certain equipment was handed over and upon the understanding that it would be replaced after the war.

The participation of the *Kontinentale Oel* in oil exploitation activities in the St. Marcet area in the South of France was in order to expedite the production of crude oil and no doubt also to overcome any possible reluctance on the part of the French to produce a national asset during the period of the occupation. *Kontinentale Oel* concluded an agreement with the French Government in September 1943 under which the former would undertake drilling operations for the *Regie Autonome des Pétroles.* The *Kontinentale Oel* was to provide seven drilling rigs with personnel, and was to receive in return 10 per cent. of the profits on the activities in addition to drilling expenses. The *Kontinentale Oel* did not participate in the oil produced.

Refining in Germany.

The *Sueddeutsche Bau G.m.b.H.* was formed with a capital of RM. 1 million for the purpose of constructing and operating the refinery at Deggendorf on the Danube. This plant had only reached an advanced stage of construction when the war ended. The refinery was to consist of a cracking plant (5,000 barrels of crude per day) and a pipe still (topping 7,500 barrels of crude per day, or vacuum distilling 6,000 barrels of residuum). There were also to be units for the production of white products and for "*Kybol.*" This refinery was being constructed principally of equipment taken from plants in occupied countries.

Forwarding Companies.

Kontinentale Oel Transport A.G., with a capital of RM. 2 million, was formed to undertake the transporting of both crude and finished products. An affiliated company, *Kontinentale Betonschiffbau G.m.b.H.*, with a capital of RM. 20,000, was formed for the purpose of constructing concrete tankers for the German Government.

Source.—Based upon a report prepared by Dihlmann and Donn, Landshut, 6th June, 1945.

* Another source gives participation in this company as follows:—

	Per cent.
Mineraloel Raff. Smederevo A.G.	30
Deutsche Gruppe	30
Kontinentale Oel A.G.	25
Shell	15

APPENDIX 5.

The Levant Plan.

Lord Hankey's Committee directed that, as part of the general policy of acquiring as much Roumanian oil as possible, efforts should be encouraged to maintain the normal markets for that oil in the Mediterranean, even if transactions resulted in a loss. The maintenance of normal markets was difficult owing to the great rise in prices of Roumanian oil. For example, the price of kerosine and gas oil, which were most in demand, had risen about 2½ times in the course of 1939, while prices of supplies from the U.S. had increased by only about 20 per cent. in the case of kerosine and 40 per cent. in the case of gas oil. The Oil Companies therefore could not continue to supply markets which had no clearing or barter arrangements with Roumania, except at a heavy loss and therefore the only means of disposing of Roumanian oil to such markets was to compensate the Companies.

2. In the early months of 1940 a scheme was worked out in consultation with the British purchasing companies engaged in the export of Roumanian oil to non-clearing countries in the Mediterranean. The Companies concerned were the Asiatic Petroleum Co., the Anglo-American Oil Co., Steaua Romana (British), Ltd., and the Anglo-Iranian Oil Co. The Socony-Vacuum Oil Co., of America, also agreed to co-operate in the scheme by shutting out their supplies from the U.S. and purchasing in sterling at Constanza from the Anglo-American Oil Co. Their co-operation was an essential feature of the scheme inasmuch as they contemplated absorbing about half of the total of 750,000 tons which it was hoped would be disposed of under the scheme. The countries concerned were Greece, Egypt, Syria, Palestine, Cyprus and Malta.

3. The basis of compensation was the difference between the actual laid-down cost of Roumanian oil at the port of destination and the hypothetical cost of landing at that port the same quantity of an equivalent grade of U.S. oil, prices including freight, war risks and insurance. The guiding principles laid down by the Treasury were, firstly, that the Companies should not profit more by this scheme than if they had drawn normal supplies from the U.S. and, secondly, compensation should be reduced to the extent that it was possible to increase prices in the consuming countries or secure part of the difference in price from the local Governments concerned. In fact, no such increase in price or relief from local Governments was found possible before the scheme terminated.

4. By the middle of May 1940 conditions underlying the plan had altered fundamentally and it was necessary to reconsider the arrangement. The shipping position in the Mediterranean had changed owing to the closing of that area to British shipping and restrictions on the movement of vessels. Three months' notice of termination of the plan, as provided in the agreement, was therefore given to the Companies.

5. During the period of operation of this scheme from the middle of January to the 8th June, 1940, 165,000 tons of oil were shipped from Roumania to the countries covered by the scheme. The total compensation payable to the Companies, after scrutiny by an independent firm of auditors, amounted to £454,000.

6. Although the amount of oil removed from Roumania under this scheme was only a small proportion of that intended, it may be said that the plan, in its limited scope, secured some definite advantages by maintaining the normal Roumanian markets. It thus encouraged the Roumanians to put up a stiffer resistance and delayed their surrender to German demands. There was also some saving in tanker tonnage and from the use of ships not suitable for other than Mediterranean traffic.

APPENDIX 6.

The Operations of the Goelánd Transport and Trading Company, Ltd., during the First Year of War.

1. The company obtained control, by purchase or charter, over a total of 390 Danube vessels, made up of 51 tugs, 4 motor tank-barges, 54 tank-barges, 270 other barges, 8 elevators and 3 motor boats. This fleet comprised 20,825 horse power, 42,548 tons of tanker tonnage and 271,300 tons of other barges.

2. Apart from vessels which were already enemy-controlled or were owned by the State Navigation companies of Hungary, Yugoslavia and Roumania, this fleet represented a high proportion of the independently-owned shipping available for purchase or charter, amounting to 47 per cent. of the horse power, 74 per cent. of the available tanker tonnage, and 48 per cent. of the total number of vessels.

3. Had these vessels been able either to trade with Germany or to set free others for the same purpose, it is estimated that the quantity of merchandise which they could have carried would have amounted to 203,500 tons of petroleum and 666,000 tons of other goods.

4. By arranging for the transport to Switzerland of petroleum which would normally have been transported by the sea-route, the company was enabled to divert from German service further tank barges capable of carrying an additional 60,000 tons in the period under review.

5. The company was partly instrumental in seducing from their ordinary employment 22 Iron Gates pilots and 80 Lower River pilots and captains in key positions. This scheme caused considerable congestion and delay to German transports.

6. Thirty-five vessels, including 3 tugs, 4 motor tank-barges, 16 tank-barges and 12 other barges, flying the British flag were safely evacuated to Istanbul.

7. On the collapse of France the company was successful in purchasing the 50 vessels of the French fleet already evacuated to Istanbul, and also obtained legal control of the 26 vessels remaining in the Danube. Similar steps were taken to protect the position of Belgian and Dutch vessels so far as was possible.

8. The company was also concerned with the chartering of Greek sea-going tankers which might otherwise have been employed in carrying oil to the enemy across the Black Sea or to Italy, and with arranging for the French tanker *Phenix* to be transferred to the British flag at Istanbul after the collapse of France.

9. Numerous other schemes which gave great promise of causing congestion on the Danube and embarrassment to the enemy were rendered impossible by the miscarrying of Naval plans culminating in the " Giurgiu incident," which gravely compromised the company's position on the river, and for which neither the company nor its employees were in any way responsible.

10. The net cost of these operations to the 11th November, 1940, amounted to some £635,000, of which about £245,000 represents capital expenditure on vessels still owned by the company, and over £40,000 may in certain circumstances be reclaimable from other Departments.

11. By complicated financial transactions at unofficial rates, the company was able to save the Treasury £274,000 on the purchase of lei, and to effect very considerable economies in the proportion of free currency to be provided.

(Signed) M. R. BRIDGEMAN.

December 1940.

Second Report.

The first year of the company's operations was dealt with in an earlier report dated December 1940. In the initial phase, from October 1939 to January 1940, the company was in process of formation. The next few months represented a period of great activity, when British influence in Roumania was still strong enough to enable the company to carry out valuable work in denying to the enemy the use of a considerable proportion of the Lower Danube rivercraft. During this second phase the company, in addition to other activities, succeeded in obtaining control of no less than 328 barges and tank-barges of a total of about 304,000 tons, and 55 tugs and self-propelled barges, aggregating about 21,500 horse power.

Following the Giurgiu incident in April 1940, however, the position became progressively more difficult, and after the collapse of France obstruction was the only practicable weapon, though by then about 75 Allied vessels had been safely evacuated to Istanbul.

The period covered by the second report falls into two distinct parts. During the first, which lasted until February 1941, His Majesty's Minister remained at Bucharest, though in circumstances of increasing difficulty, and the company, while unable to exercise any active control over the vessels which it owned or had chartered, could still, by delaying tactics, carry out the negative function of postponing the date on which they became of use to the enemy.

Since the withdrawal of the British Legation, the efforts of the company's staff have been mainly directed towards conserving such of the assets as had been removed to Istanbul, and ensuring that they were put to the greatest possible use, both from the financial and practical aspect.

While this second report cannot be regarded as of much importance from the standpoint of economic warfare, which constituted the original purpose of the company, it has been thought desirable to bring the record up to date, if only in order to simplify the task of the eventual liquidator.

In conclusion, it may be said that the greatest credit in due to Mr. W. Harris-Burland, the former general manager of the company, for his handling of its affairs. While attached to the British Legation at Belgrade he was captured by the Italians, but he was fortunately released after a short internment and is now serving the Government in another capacity.

The affairs of the company at Istanbul are now being managed by Mr. T. Walton. Owing to the destruction of most of the company's papers before leaving Bucharest, his position has been one of exceptional difficulty, and the Board of Directors is greatly indebted to him for the capable and efficient manner in which he has discharged his duties.

(Signed) M. R. BRIDGEMAN.

May 1942.

APPENDIX 7.

The Attempt to Block the Iron Gates.

Translation of an article in the Hamburger Fremdenblatt *of 9th April*, 1940.

The Outrage on the Danube.

The following particulars have so far transpired regarding the unprecedented attempt by the English at terrorism and sabotage, on a big scale and worked out to the smallest detail, with a view to making the navigation of the Danube impossible, and thereby deprive the countries of south-east Europe of their most important artery.

Already some time ago the accumulation of many English Danube lighters in the port of Giurgiu became suspect to the Roumanian authorities. It was at first thought that the manœuvre was only intended to obstruct shipping movements at Giurgiu, Roumania's most imporant river port. Owing to the watchfulness of a Roumanian navigator the ruthless plan was exposed and by the quick and energetic action of the Roumanian authorities it was baulked. The crew of the English lighters, about a hundred pioneers, special troops and pilots disguised as simple sailors were arrested and the cargo seized. Sixteen military motor lorries were required to remove the explosives.

It is not yet explained how the Englishmen were able to proceed to Giurgiu unhindered after they had declared the cargo in Braila as preserves and spare parts of machinery. From special information, however, it may be taken that it was made possible by the English shipping firm of Watson and Youell, whose head office is in Galatz, and who have their own office in Braila, Sulina and Constanza, with the help of their wide-spread connections and agents. This firm, whose close connection with the political principals in London and the Secret Service has been known for months, appears also to have undertaken the transhipment of the explosives and weapons from the sea-going ship into the Danube lighters and to have declared them as preserves in tins, &c. It is now evident that the stroke was directed against the Iron Gates where the river bed is narrowed between rocks on both sides.

APPENDIX 8.

SUMMARY OF GERMAN OIL AND GAS FIELDS, MAY 1945.

Field.	Position.	Owners.	Operating Company.	Number of Wells.		Depth of Wells in Metres.	Producing Formations.	Producing Method.	Oil Production.	
				Producing.	Drilling.				M 3/Day.	Metric tons per annum.
Bentheim	25 km. E. of Hengels	Deutag Elwerath	Elwerath	3 Gas	3	1,500–1,700	Zechstein	Gas Pressure Flowing	...	...
Berkhopen and Edesse-Olheim	30 km. E. of Hanover	Preussag	Preussag	...	...	...	...	Pumping	10	3,500
Broistedt	16 km. S.W. of Brunswick	Preussag	Preussag	...	...	...	...	...	1–10	2,500
Ehra	...	...	...	...	...	...	...	...	⅓–3	340 in 1944
Emlichheim	Netherlands Frontier	Wintershall	Wintershall	4	1 (4 units on field)	800–850	Cretaceous	Pumping	40	14,600
Etzel	17 km. S.W. of Wilhelmshaven	Preussag	Preussag	...	...	...	...	...	8	3,000 in 1944
Eicklingen	12 km. S.E. of Celle	Wintershall	Wintershall	16	1	552–930	Valendis, Wealden	Pumping	50	18,000
Fuhrberg	15 km. W. of Celle	Wintershall	Wintershall	134	5	150–500	Cornbrash	3 Flowing 131 Pumping	183	61,000
Forst	Rhine Valley	...	...	...	...	...	...	...	...	...
Georgsdorf	Netherlands Frontier	25% each— Preussag Wintershall Elwerath Deutag	Preussag	9	2	800–850	Cretaceous	6 Flowing 3 Pumping	40	14,600
Gifhorn	55 km. of S.E. Celle	Deutsche Erdöl	Deutsche Erdöl	55	3 (2 exploration)	250	Wealden	Pumping	20	6,900
Hademsdorf	35 km. N.N.W. of Celle	Deutsche Vacuum	Deutsche Vacuum	3	3	1,300	Wealden	Pumping	25	10,000
Hambuhren	10 km. W. of Celle	Itag Brigitta	Itag Brigitta	Nil	Nil	400–1,700	...	...	No Production	Developed
Hänigsen	15 km. S. of Celle	Deutsche Vacuum	Deutsche Vacuum	27	4	730–850	Wealden, Valendis	Pumping	120	39,000
Hohenassel	18 km. W. of Hildesheim	⅓ each— Elwerath Preussag Braunschweig G.M.B.	Elwerath	11	2 and 2 outfits moving	850–1,050	Corallian	Pumping	150	48,000
Heide	80 km. N.W. of Hamburg	Deutsche Erdöl	Deutsche Erdöl	40	11	800–1,300	Cretaceous	Pumping	350	120,000
Lingen (Dalum)	N.N.E. of Lingen	50% Elwerath 50% Salzgitter	= Production = Drilling	38	10	1,000	Valendis	Pumping	50–80	24,000
Meckelfeld	15 km. S. of Hamburg	Wintershall	Wintershall	8	Nil	400–1,200	Cretaceous	Pumping	12	4,500
Mölme	45 km. S. of Celle	Elwerath	Elwerath	40	2	...	...	Pumping	35	11,000
Nienhagen	12 km. S. of Celle	Elwerath	Elwerath	110	3	300–1,500	Valendis, Wealden	70 Pumping 40 Bailing	357	114,000
		Wintershall	Wintershall	48	1	625–1,454	Valendis, Wealden	Pumping	75	24,000
		Deutsche Erdöl	Deutsche Erdöl	28	...	600–1,300	...	Pumping	22	7,000
		Deutsche Vacuum	Deutsche Vacuum	8	...	900–1,400	...	Pumping and Bailing	11	3,600
Oberg	7 km. S. of Peine	Deutsche Vacuum	Deutsche Vacuum	80	...	300–600	Wealden and Dogger	20 Bailing 60 Pumping	28	8,700
Reitbrook	15 km. S.E .of Hamburg	Deutsche Vacuum	Erdöl Betrieb Reitbrook A.G.	80	8	480–730	Eocene and Upper Cretaceous L'st	50/60 Flowing 30/20 Pumping	120	41,000
Rodewald-Stiembke	15 km. E. of Nienberg	Brigitta	Brigitta	200	8	250–650	Wealden 5% Serpuline 30% Cornbrash 65%	Pumping	150	48,000
Sottorf	15 km. S.S.E. of Hamburg	Preussag Vacuum	Preussag Vacuum	1–5	...	200–300	Cretaceous	Pumping	Closed down	110 in 1944
Thören	23 km. W. of Celle	50% Deutsche Erdöl 50% Deutsche Vacuum and Brigitta	Deutsche Erdöl	50	5 (2 rigging up)	500–900	Wealden, Dogger	Pumping	84	27,000
Wesendorf	40 km. E. of Celle	50% Deutsche Erdöl 40% Elwerath 10% Preussag	Deutsche Erdöl	10	8 (2 erecting)	870–1,200	Dogger	Flowing	200–210	64,000
Wietze	20 km. W. of Celle	Deutsche Erdöl	Deutsche Erdöl	363	...	180–365	Wealden, Rhaetic	Bailing Pumping	73 by mining 22 from wells	31,000
Wiengarten	Rhine Valley	...	...	...	...	...	...	...	...	...

APPENDIX 8—(*continued*).

Field.	Crude Characteristics.	Refinery.	Gas Production M³/Day.	Gas Treatment.	Gas Products Production.	Power.	Fields Storage M³.	How Crude Removed.	Notes.
Bentheim	No Oil	...	320,000	Nil	Nil	...	None	None	Gas piped to Ruhr; no oil discovered so far.
Berkhopen and Edesse Olheim	... ...	...	Small	...	...	...	...	...	... ...
Broistedt	... ...	...	...	...	...	...	...	...	... ...
Ehra	... ...	...	...	...	...	...	...	...	... ...
Emlichhein	Benzine 5%, kerosene 5%, Diesel 6%	Salzbergen ...	Small	Nil	Nil	Electric ...	Small	Field R.R. to main line	Main Rail Road Tank Cars bottleneck to shipment crude Salsbergen.
Etzel	... ...	...	...	...	...	...	...	...	... ...
Eicklingen	Paraffin 0·885	Misberg, Dolbergen, Peine	5,000/4,000	Combined with Nienhagen Gas (Wintershall)	...	Electric ...	Small. Crude Piped to Nienhagen	Pipeline to Nienhagen then Rail	Eicklingen Field in fairly rapid decline.
Fuhrberg	S.G. 0·915 Paraffin Base	Dolbergen, Peine	Small	Nil	Nil	Electric and Diesel	5,800	Rail from Oldau	Storage full. Railway will be repaired 15.6.45.
Forst	... ...	...	...	...	...	...	...	...	Production negligible. No details available.
Georgsdorf	Benzine 5%, Kerosene 5%, Diesel 6%	Salzbergen ...	Small	Nil	...	Electric ...	Small	Trucked to R.R. Rail from Veldhausen	Main R.R. tank cars bottleneck to shipment crude to Salsbergen.
Gifhorn	S.G. 0·945, Asphalt Base	Hamburg ...	Nil	Nil	Nil	Electric Diesel for Exploration	600	Rail from Wilsche	This crude is a useful source of bitumen.
Hademsdorf	S.G. 0·905, Paraffin Base, Gasoline 15%, Gas Oil 11%	Misberg, Dolbergen	Nil	Nil	Nil	Electric ...	600 on field 500 on R.R.	Rail	Crude can be used as boiler fuel without refining. Bounutz Electric Plant have applied for 30 tons a day soonest.
Hambuhren	None	...	...	...	...	...	...	...	Though Oil shows obtained no commercial (or war) production.
Hänigsen	S.G. 0·892, Paraffin Base	Misberg, Bremen	Small G.O.R. 8·1	Nil	Nil	Diesel Gas Engines and Electric	800	Rail from Hänigsen	Hänigsen Field almost drilled up. Production expected to decline.
Hohenassel	S.G. 0·875, Paraffin Base	Misberg ...	Small	Nil	Nil	Electric ...	500	Rail from Osterlinde	This is a new Field in early stages of devolopment.
Heide	Paraffin 0·867	In field, Hemmingstedt	4,800	Gas pipe to Heide for domestic use. No treatment		Electric ...	Destroyed	To R.R. at Refinery or Pipeline to Kiel Canal	Pipeline to Kiel Canal will mostly solve transport problem.
Lingen (Dalum)	Paraffin	Salzbergen ...	120	Nil	Nil	Electric ...	1,500	Pipeline to R.R.	Pipeline to R.R. now ready. Lack of road transport has restricted production of this Field so far.
Meckelfeld	... ...	Hamburg ...	Nil	Nil	Nil	Electric ...	Small	R.R. to Hamburg	Emulsion troubles require Dismulgan and centrifuge.
Mölme	S.G. 0·866, Paraffin Base	Misberg ...	Small	Nil	Nil	Electric ...	...	Rail from Steinbrück	... ...
Nienhagen	S.G. 0·876, Paraffin Base	Misberg, Peine, Dolbergen	4,000	Charcoal Plant	Gasoline 4–8 tons/day Propane and Butane 2–4 t/d	Electric ...	11,000	Rail from Field	High Gasoline production obtained by recirculating dry gas through crude. The Nienhagen Field as a whole is fully developed, and a decline of the order 3% per month is expected.
	S.G. 0·880, Paraffin Base		3,000/4,000	Charcoal Plant	Gasoline 25 tons/month	Drilling Diesel	4,000	Rail from Field	
			Small	Nil	Propane and Butane 10 t/m	Electric ...	...	Rail from Field	
	S.G. 0·884, Paraffin Base S.G. 0·875, Paraffin Base		Small	Nil	Nil	Electric ...	...	Rail from Field	
Oberg	S.G. 0·855, Paraffin Base	Hamburg ...	2,000/3,000	Charcoal Plant	8 tons Gasoline month	Electric ...	1,100	Rail from Grosse Ilsede	The oil is of good quality and can be used unrefined in tractors.
Reitbrook	S.G. 0·910, Mixed Base S.G. 0·930, Asphalt, Base	Bremen ...	6,000	Charcoal Plant	Gasoline 8 tons Propane and Butane 8 tons/m	Electric, Diesel and Gas Engines	11,000	Barge from Warwisch Rail from Bergedorf	Cretaceous production fully developed. Eocene sand is being drilled in S.E. sector of Field and wells produce about 1 ton per day.
Rodewald-Stiembke ...	S.G. 0·935. Mixed Base	Harburg ...	6,000	Nil	Nil	Electric ...	...	Rail to Hagen	When visited Field had been shut down for 8 weeks owing to lack of transport to empty field storage.
Sottorf	... ...	Hamburg ...	Nil	Nil	Nil	Electric ...	60	Road to Hamburg	Field seems to have closed down during 1945.
Thören	S.G. 0·942 Asphalt Base; S.G. 0·847 Paraffin Base	Hamburg ...	Small	Nil	Nil	Electric ...	500	Rail from Wietze	This Field is close to and operated from Wietze.
Wesendorf	S.G. 0·847 Paraffin Base	Hamburg, Misberg	2,000	Charcoal Plant	Gasoline 600/800 Litres/day	Electric ...	3,500	Rail from Warenholtz	This is a young Field. It is under a water drive, but water has not yet been struck. Wells are flowing through 2 mm. to 4 mm. beans.
Wietze	S.G. 0·942 Asphalt Base S.G. 0·878 Paraffin Base	Hamburg ...	Nil	...	...	Electric ...	16,000	Rail from Wietze	Present production by mining can be maintained for 10 years.
Wiengarten	... ...	...	...	...	...	...	...	...	Production negligible. No details available.

APPENDIX 9.

Development of the Oil Shales of Southern Germany.

The productive oil shales of Southern Germany are confined to the uppermost Lias member (Epsilen Lias) of the Jura Formation. These shales have a specific gravity of about 2·1 and contain from 3·5 to 7 per cent. by weight of a distillable shale oil in the exploitable areas (more than four metres in thickness). The total oil potential is estimated at approximately 40,000,000 m. tons.

In 1937, Lurgi of Frankfurt initiated a limited research programme, and late in 1938 the Portland Cement Co. of Stuttgart did likewise, not only for shale but as a source material for the manufacture of Portland Cement. In April 1942 an auxiliary oil shale distillation unit within the cement plant was completed and put into operation. Daily capacity was 480 tons shale. Daily oil production from this unit varied from 25–30 tons.

The Lurgi Company chose the Frommern area. Construction of a large two-unit plant, designed for a shale capacity of 1,000 tons per unit daily, was started late in 1942, but was only 70 per cent. completed at the time of the Allied troops' occupation in April 1945. The Lurgi plant was designed on the basis of peace-time conditions and encountered difficulties in procuring many critical items incorporated in its design. All buildings and equipment were of a permanent character, and designed for long-term service rather than emergency war production. Accordingly, the Lurgi plant at Frommern did not contribute to the German war effort.

Emergency War Production Programme, 1943–45.

The further possibilities of developing oil production from the low grade oil shales of Württemberg received strong Government support early in 1944, and two large-scale projects in this field were approved and incorporated in the Geilenberg programme. The first of these to enter production was the attempt of the *Kohle Oel Union*.

They chose the area adjacent to Schoerzingen for its underground operations. Because of the urgent necessity of the shale oil itself, and the character of the mining operations not being amenable to laboratory scale practice, very limited preliminary testing work was carried out prior to the actual full-scale working itself.

The process consisted of driving several parallel working tunnels in the shale bed underground about 60 metres apart, then drilling horizontal cone holes connecting the parallel tunnels, loading cone holes with dynamite, and shooting so as to create a predesigned underground "chamber," filled with broken shale, and offering a channel through which hot gases could pass by reason of controlled suction draft, maintained at one end of the chamber. The fire which supplied the circulating hot gases is started at one end of each chamber and allowed to advance toward the other end (60 metres distant) at the rate of approximately two metres daily. Each chamber contains approximately 250 tons of broken shale.

The Deutsche Oelschiefer Forschungs-Ges. was organised late in 1943 with strong Government backing. They had to construct and install 10 oil shale distillation plants in the South Württemberg area, each capable of treating 1,700 tons of shale daily. The motive was the necessity for producing 5,000 barrels daily for the Wehrmacht, without regard for operating efficiency.

Construction began in July 1944. A thousand Estonian workers and shale oil technicians were sent to the site. Ten plants were placed under construction between the 1st July, 1944, and the 1st October, 1944. The first plant started working on the 22nd February, 1945, followed by three other plants in operation by the 18th April, at which time production was suspended by reason of the occupation by the French Army. On the 18th April four plants of 1,700 tons daily capacity each were working. Three were within three weeks of completion and the three remaining would have been ready by the 15th June.

Although *Deutsche Oelschiefer* was given the task of installing ten plants having a total daily output of 17,000 tons of shale, they were also engaged in intensive research work at the request of the Government.

The Portland Cement Co., coincident with their routine oil shale production, were active in research development of methods for treatment of shale in places underground and in special retorts for surface operations. They were recently engaged in the construction of a 240-ton pilot plant using molten aluminium as a direct contact heat source for shale distillation. This was scheduled for trial runs on the 15th May, but all work was suspended on the 18th April.

In general, the German effort at rapid development of the Württemberg shale failed to disclose any new method sufficient to warrant adoption in other localities by the oil shale industry.

Source: U.S. ALSOS Mission EVF/224. 4.7.45.

APPENDIX 10.

Composition of the various Committees.

Lloyd Committee.

Chairman—

Rt. Hon. Geoffrey Lloyd, P.C., M.P., Secretary for Petroleum.

Members—

Air Commodore A. R. Boyle, C.M.G., O.B.E., M.C., Director of Military Intelligence, Air Ministry.

The Hon. M. R. Bridgeman, C.B.E., Petroleum Department.

Sir Leonard Browett, K.C.B., C.B.E., Permanent Secretary, Ministry of Transport.

Lord Cadman, G.C.M.G.

Captain I. M. R. Campbell, R.N., Deputy Director of Naval Intelligence, Admiralty.

Mr. V. F. W. Cavendish-Bentinck, C.M.G., Foreign Office.

Major-General F. H. N. Davidson, C.B., D.S.O., M.C., Director of Military Intelligence, War Office.

Sir Alfred Faulkner, C.B., C.B.E., Permanent Under-Secretary for Petroleum.

Sir Frederick F. Godber, Chairman, Overseas Supply Committee of Trade Control Committee.

Professor N. F. Hall, Ministry of Economic Warfare.

Sir Harold Hartley, K.C.V.O., C.B.E., M.C., F.R.S.

Sir Cecil Kisch, K.C.I.E., C.B., Director-General of Petroleum.

Air Vice-Marshal C. E. H. Medhurst, C.B., O.B.E., Assistant Chief of Air Staff (Intelligence), Air Ministry.

Major Sir Desmond Morton, K.C.B., C.M.G., M.C., Ministry of Economic Warfare.

Major-General F. G. Beaumont Nesbitt, C.V.O., M.C., Director of Military Intelligence, War Office.

Mr. F. C. Starling, C.B.E., Mines Department.

Captain W. D. Stephens, R.N., Director of Naval Intelligence, Admiralty.

Lieutenant-Colonel Sir Geoffrey Vickers, V.C., Ministry of Economic Warfare.

Joint Secretaries—

Mr. A. F. L. Brayne, C.I.E., Petroleum Department.

Commander E. G. A. Clifford, War Cabinet Offices.

Captain A. D. Nicholl, C.B.E., D.S.O., R.N., War Cabinet Offices.

The Technical Sub-Committee on Axis Oil.

Chairman—

Sir Harold Hartley, K.C.V.O., C.B.E., M.C., F.R.S.

Members—

Captain J. P. Charley, R.N., Admiralty.
Engineer Commander G. W. Cannan, R.N., Admiralty.
Major C. L. Edwards, R.M., Admiralty.
Major T. C. Owtram, M.C., War Office.
Wing-Commander D. A. C. Dewdney, Air Ministry.
Mr. R. M. C. Turner, Ministry of Economic Warfare .
Mr. B. M. C. Trench, Ministry of Economic Warfare (successively).
Mr. R. H. W. Bruce, Ministry of Economic Warfare (successively).
Mr. M. Y. Watson, Ministry of Economic Warfare (successively).
Mr: O. F. Thompson, O.B.E., Ministry of Economic Warfare (successively).
Mr. E. A. Berthoud, C.M.G., Ministry of Fuel and Power.
The Hon. M. R. Bridgeman, C.B.E., Ministry of Fuel and Power
Mr. G. D. Macdougall, Prime Minister's Statistical Branch.
Mr. S. P. Vinter, Prime Minister's Statistical Branch (later).
Mr. D. A. Shepard, Petroleum Attaché, American Embassy (successively).
Mr. C. E. Meyer, Petroleum Attaché, American Embassy (successively).
Mr. S. D. Turner, Petroleum Attaché, American Embassy (successively).
Lieut.-Colonel S. J. M. Auld, O.B.E., M.C., Petroleum Board.
Mr. L. A. Astley-Bell, Shell Petroleum Company, Limited.

Joint Secretaries—

Colonel D. Capel-Dunn, O.B.E., War Cabinet Offices.
Mr. A. F. L. Brayne, C.I.E., Ministry of Fuel and Power.
Lieut.-Colonel T. Haddon, War Cabinet Offices (later).
Wing-Commander A. E. Houseman, War Cabinet Offices (later).

APPENDIX 11.

THE WORKING COMMITTEE (OIL) OF THE COMBINED STRATEGIC TARGETS COMMITTEE.

Terms of Reference.

The Working Committee, as originally set up in July 1944, had the task of advising the Joint Oil Targets Committee in fulfilment of its responsibility :—

(i) to keep the Axis oil position under continuous review;
(ii) to assess the effectiveness of attacks;
(iii) to determine priorities.

With the formation of the Combined Strategic Targets Committee in October 1944 these terms of reference were altered slightly in form, but not in substance.

The Working Committee interpreted its task as being concerned first and foremost with the determination of target priorities for current operations in such a way that the aim of reducing the enemy's liquid fuel supplies to nil should be achieved with the maximum speed and the minimum expenditure of effort, and that in so doing the maximum impact should be obtained at all times on the supplies of fuels (and especially gasoline) to the Armed Forces. This involved in practice a close and continuous study of all developments in production and distribution, including particularly the status of individual installations as influenced by bombing.

The Working Committee did not accept responsibility for estimating the trend in the enemy's consumption and stocks, or for assessing the overall effects of the oil offensive on the enemy's strategy and operational capabilities, since this function was already being discharged by the British Joint Intelligence Committee and by the Service Intelligence Directorates individually. A close watch was nevertheless maintained on consumption and stock trends as determined by other agencies in view of the changes in the policy for the attack of production and distribution which such trends might indicate as being desirable.

Constitution.

The constitution of the Working Committee was adapted to the above interpretation of its main responsibility and was representative of the principal British and American agencies competent to advise on enemy oil production and distribution, together with the Air Staffs :—

British Agencies—

Air Ministry : A.I.3.c. (1).
Air Ministry : A.I.3.e.
War Office : M.I.10.c.
Economic Advisory Branch (Foreign Office and Ministry of Economic Warfare).

American Agencies—

United States Strategic Air Force.
Enemy Objectives Unit, Economic Warfare Division; United States Embassy.
Petroleum Attaché, United States Embassy.

Combined Agencies—

S.H.A.E.F. : G2 (O.I.).
Allied Central Interpretation Unit (A.C.I.U.).
Ministry of Home Security; R.E.8.

The Working Committee was empowered to co-opt technical experts as required and made extensive use of this power, both to assist in photographic interpretation and for more general purposes. Frequent use was made of the assistance of individual members of the J.I.C. Technical Sub-Committee on Axis Oil (Hartley Committee) and of the United States Enemy Oil Committee, as well as of representatives of the British Ministry of Fuel and Power and of the principal British and American oil companies.

The Committee normally worked in two divisions since the problems and agencies concerned with oil production differed from those arising in connection with oil storage and distribution.

Working Methods and Sources of Information.

(a) *Oil Production.*

The Working Committee inherited from the Hartley Committee, the United States Enemy Oil Committee and other agencies, full data on the nature, functions and capacities of the enemy's principal oil production plants, evolved as the result of four years of intensive research. This provided a complete and detailed picture of the enemy's oil industry at the outset of the offensive.

In facing its task of advising on target priorities, the Working Committee was influenced by the following considerations:—

(i) In view of the great difference in the size and functions of the individual units in the industry, priorities could not appropriately be allotted on the basis of activity alone.

(ii) In view of the differing size of individual plants, it was necessary to attempt to determine the extent to which each plant was operating at any time, since a large plant in partial operation might be a larger producer, and therefore a more important target, than a small plant in full operation.

(iii) In view of the differing functions of individual plants, the target priority of an individual plant was not necessarily always determined by the size of its total output. Since the primary object of the offensive was from the start to obtain the maximum impact on the output of motor and aviation fuels, the current gasoline yield of each individual plant was made the primary criterion in assessing target priorities. This would differ widely both in accordance with the process used (*i.e.*, hydrogenation, Fischer Tropsch, or crude refining) and, within each process, on the feedstock employed.

(iv) In view of the difference both in size and in function, potential production was taken into account as well as actual production. Thus, in view of the continuing nature of the offensive, the current target value of an idle plant which might be in a position to produce large quantities of gasoline in two weeks' time was often judged to be greater than a small active plant which was producing insignificant quantities of gasoline.

(v) Aerial photography was the principal source of intelligence on the condition of plants. But in view of the nature of the information required it was necessary to develop the technique of assessment of productive capabilities from air cover on much more ambitious lines than had been attempted hitherto. This was the more necessary in view of the probability that, despite the best efforts of the photographic reconnaissance squadrons, the large number of targets to be covered and the handicaps imposed by the weather would on many occasions prevent the use of air cover as an automatic check on activity.

It was decided at the outset that the best method of embodying the above considerations in a form which would provide a convenient basis for the allotment of priorities would be to attempt a running assessment of the output of each individual plant, based on the deductions drawn from air cover and from any intelligence that was available. Although the retrospective results of these assessments enabled estimates to be compiled of the enemy's total output from month to month under stress of the offensive, this was an incidental product and was not the main purpose of the assessments. Their main purpose was to provide a considered forecast of the future productive capabilities of the individual plants against which their current target value could be measured.

In view of the emphasis on the attack of the production of motor and aviation fuels it was also necessary to consider the current capabilities of each plant in this connection. It was appreciated that gasoline could occupy the following percentages of the total finished products output of [illegible] according to the process and feed-stock used :—

Process.	*Feed-stock.*	*Gasoline as percentage of total finished products.*
Bergius	Bituminous coal or H.T. Tar	50–90
,, ...	Low Temperature Tar ...	50
Fischer Tropsch ...	... ...	40–60
Crude Refining ...	Roumanian Crude	30*
,, ,, ...	Hungarian Crude	30*
,, ,, ...	Austrian Crude	10*
,, ,, ...	German Crude	10*

As in the case of the overall assessment, the estimation of individual gasoline outputs produced, as a by-product, the means of estimating the total output of gasoline and the other principal finished products.

The principal burden of estimating the current production of individual plants fell upon the representatives of D. Section (Industrial) of A.C.I.U. The technical considerations involved made this a subject of much greater complexity than conventional damage assessment and required the services of interpreters who had made a specialised study of the processes and equipment used in the industry under attack. Frequent use was also made of the technical advice of oil industry experts.

In this work, synthetic plants proved to be less difficult subjects than refineries and the establishment of activity proved to be less difficult than the measurement of production.

Synthetic plants emit a great deal of steam at nearly all stages of the productive process and steam emission consequently provided a ready means of establishing activity in the various sections of the plant. It was, however, less easy to distinguish activity in connection with testing from activity signifying production and the decision as to the time at which a repaired plant actually resumed production was therefore a very arbitrary matter. This difficulty, however, probably affected the validity of recommendations regarding priorities for re-attack less than it affected the accuracy of the estimates of production. In the later stages of the offensive, however, it was suspected that the enemy may at times have simulated activity in order to provoke unnecessary attacks on plants which he had no intention of bringing back into production.

Since the production of synthetic oil involves the operation of several essential processes in successive stages, the chain of production was known in general to be as strong as its weakest link, and study of the state of the equipment at each essential stage, and of the state of the inter-connections between them, was used as the best available guide to the scale of productive operations. It was, nevertheless, appreciated that any estimate reached on this basis was bound to be of an arbitrary nature, and in an attempt to eliminate systematic errors it was decided in principle not to attempt to judge the operating capabilities of a plant closer than 25 per cent. of its rated capacity. That is to say, plants were usually judged to be 0 per cent., 25 per cent., 50 per cent., 75 per cent., or 100 per cent. operational. This convention was also applied in dealing with refineries.

By reason of the much less complex and more compact nature of the equipment involved, the establishment of activity at refineries gave rise to much greater difficulties than at synthetic plants and on many occasions the interpreters were compelled to put in a non-committal report. Since at most stages of the offensive the enemy was believed to dispose of a supply of crude in excess of available refinery capacity the general rule was observed of regarding any plant which appeared to be operable as probably active. The estimates of throughput which were made under these circumstances were necessarily of an entirely arbitrary nature and the priority recommendations resulting from them were frequently the cause of much anxious debate within the Working Committee and were put forward with much less confidence than the recommendations regarding synthetic plants.

In the case of both the synthetic plants and the refineries, forecasting of future productive potentialities proceeded mainly from consideration of the severity of the damage to key items of equipment. The physical and economic vulnerability of all classes of oil objectives had already been closely studied by the Enemy Objectives Unit of the Economic Warfare Division (United States Embassy) for the purpose of selecting aiming-points for the United States

* Or such higher figures as appropriate in the case of refineries possessing cracking plants.

8th A.F. and United States 15th A.F., and the conclusions necessarily provided the best starting-point for consideration of the effects of the damage actually inflicted. In brief, these conclusions were :—

(i) In the case of synthetic plants, and especially the hydrogenation plants, the continuous nature of the production process necessitated the functioning of a considerable number of separate items of essential equipment (*e.g.*, gas plant, compressors, injectors, stalls, power plant, &c.). Although the gas plant was generally preferred as a primary aiming-point—solely on account of its greater physical vulnerability—it was recognised that serious damage to equipment at any one of these vital stages would involve prolonged stoppage of the plant for repairs. On the other hand, damage to pipework, tanks, &c., could quickly be repaired up to the point where resumption of some production was possible, provided that the essential units, or some part of them, remained intact.

(ii) In the case of refineries, the only effective means of ensuring complete stoppage of production was thought to be the destruction of the distillation units (together with the cracking plants, if any). In view of the nature of the equipment destruction could normally only be expected from a direct hit and, in view of its small size, direct hits would be difficult to achieve. The power houses and boiler plants of refineries were reckoned a poor second as aiming points, since capacity is usually in excess of requirements and, in the event of damage, an emergency supply of power and steam can often be improvised without much difficulty. Similarly, destruction of tanks, pipework, &c., was not expected to do more than cause a temporary stoppage of production, though it might well cause a permanent decline in operating efficiency.

These conclusions, evolved from purely theoretical considerations of the problem, stood the test of experience well and the Working Committee found no necessity to make any radical revision in the principles established by them. Before the offensive was many months old, in fact, confirmation of their validity was obtained by observation of the actual progress of repair and revival of a number of damaged plants. Moreover, by watching through aerial cover the nature and progress of repair measures, it was possible to deduce the main lines of the enemy's repair policy and to establish the time patterns which normally governed the speed of recovery of plants in various circumstances. Analysis of these factors was the particular responsibility of the R.E. 8 representatives on the Working Committee.

The reliance on aerial photography for quick re-assessment of target priorities rendered the Working Committee very dependent on the good offices of the Joint Photographic Reconnaissance Committee, with whom a close liaison was maintained. The Working Committee supplied J.P.R.C. each week with a schedule of priorities for cover, divided into categories in accordance with the urgency of the requirement. During the greater part of the offensive, and notably in its concluding months, these requirements were rapidly and completely met. During the autumn of 1944 (September to December), however, the cover obtained fell substantially short of the minimum necessary for accurate assessment, and the Working Committee's estimates of the activity and production of particular plants during that period often amounted to little more than guesswork. This situation was not due to any slackening in reconnaissance effort by J.P.R.C., who did, in fact, develop unconventional methods, often with success, to obtain cover under the handicap imposed by prolonged bad weather, and the arrears were rapidly made good with the return of better weather. During this period it was found that oblique photographs, or other cover unsuitable for full assessment, were adequate to establish activity and that photographs taken in the course of actual attacks could often be used successfully for preliminary assessment of damage.

The dependence of the conduct of the oil offensive upon regular air reconnaissance and skilled photographic interpretation cannot be over-emphasised. While the Working Committee considers itself to have been very well served in both respects, it was at all times acutely aware how slender was the margin of resources available to ensure the continuity of this work. In particular, the major burden of photographic interpretation of producing plants during the twelve months of the offensive was borne by a single junior officer* of the R.A.F.

* Flight-Lieutenant P. E. Kent.

Although photographic reconnaissance provided the best means of assessing the operating condition of plants for the purpose of regulating target priorities, intelligence from other sources played a valuable part in disclosing the enemy's productive capabilities and intentions. While its receipt was seldom sufficiently rapid to affect target priority decisions it performed important subsidiary functions, especially in ensuring that major production installations were not overlooked, in providing data by which production capacities could be estimated, and generally in providing information after the event by which the accuracy of assessment methods could be checked. The evaluation of such intelligence was mainly the responsibility of the Economic Advisory Branch (Foreign Office and Ministry of Economic Warfare), and this was discharged in part by the distribution of a weekly series of "Notes on Enemy Oil Intelligence."

(b) *Oil Storage Targets.*

At the outset of the offensive information on the oil distribution system was not as complete as the information on production, since much less attention had been paid to this subject. A good deal of research had, nevertheless, already been done by the War Office (M.I.10.c.), the Air Ministry (A.I.3.c.) and S.H.A.E.F.-G2 (O.I.), in conjunction with A.C.I.U., and the intensification of this research rapidly led to the building up of a fairly complete picture.

The interrogation of prisoners and other intelligence sources provided a great deal of information on the oil distribution system, both as regards location and function and as regards current activity. Aerial photography was regularly used to exploit intelligence, both in the confirmation and identification of reported depots and in supplementing evidence from intelligence on activity.

It was appreciated at an early stage of the research into the enemy's oil distribution system that, while the system served incidentally to provide storage space for strategic reserves, especially at its higher echelons, its main function was to act as a channel of distribution for the continuing output of the production plants to the various branches of the Armed Forces. Hence the activity of the system as a whole was likely to depend not so much on the level of strategic reserves as on the level of current output, while the activity of particular installations, especially in the lower echelons, depended mainly upon the operational status of the consumers whom they were destined to serve. Intelligence was mainly of value in identifying the functions of depots, notably the types of fuel handled and the theatres or commands serviced by them. Aerial photography provided an imperfect, but tolerable, guide to current activity. Provided that cover could be obtained with sufficient regularity, a count of tank wagons present was thought to provide a fair guide to activity, unless the depot should be confined to night working. The minimum tolerable frequency of such cover for major depots (*i.e.*, cover at weekly intervals) was seldom achieved, so that no great reliance was placed on the results.

The attack on the major oil storage installations developed at such a late stage in the offensive that no serious attempt was made to assess damage. Shortly before the offensive came to an end, however, the first steps were taken to this end by the despatch of representatives* of the Working Committee to examine the results of attacks on depots which had, by then, been captured. Since the main object of the attack on depots was temporarily to deny their effective use by inflicting damage on loading installations, &c., it was assumed that the effects would, by definition, never be of very long duration. Moreover, in view of the comparatively low priority accorded to the attack of depots, the Working Committee was, in general, content to await evidence from intelligence of renewed activity before recommending re-attack.

Communication of Recommendations.

The conclusions of the Working Committee were embodied in a Weekly Bulletin for submission to the weekly meeting of the C.S.T.C. and for circulation, after approval, to interested Commands. The priority recommendations of the Working Committee, after consideration and, if necessary, amendment at the weekly meeting of the C.S.T.C., were signalled immediately to U.S.S.T.A.F. (for the United States 8th and 15th A.F.) and to R.A.F. Bomber Command, together with a note on any policy changes. The signals were for the information of the Commanders.

Apart from these arrangements for the formal overhaul of priorities at weekly intervals, direct contact was maintained between the Working Committee and the Commands in the United Kingdom for the communication of new information which might necessitate intermediate adjustments of target priorities in the

* Major C. M. Pollock, Captain A. L. Simon (War Office), Captain M. Riveline (S.H.A.E.F.), Captain J. S. Davies (U.S.S.T.A.F.) and Captain L. J. Simon (A.C.I.U.).

Western theatre. Under this arrangement, A.C.I.U. communicated daily the Secretary of the Working Committee the results of any new air cover. The Secretary, after such consultation as was necessary with other members of the Working Committee, communicated any resultant adjustment of priorities to United States 8th A.F. and R.A.F. Bomber Command, repeating this to Air Ministry and U.S.S.T.A.F. The use of a code system enabled these adjustments to be communicated rapidly by telephone in time to be taken into account in the planning of missions for the following day.

As the attack of oil storage targets proved to be of greater operational interest to the Tactical Air Forces than to the heavy bombers, arrangements were made whereby the Tactical Air Forces were able to share the results of the Intelligence work of the Working Committee. This was achieved by the posting to the Air Ministry of a representative of A.2: United States 9th A.F., who attended meetings of the Working Committee as an observer and drew upon their assistance in accordance with the operational requirements of his Headquarters.

Accuracy of Damage Assessments.

The estimated output of oil plants, as calculated from photographs and from intelligence, and the output as reported in the official records, over the period May 1944 to March 1945, compare as follows:—

	Target Committee Estimate.	*Actual Output.*
	Metric Tons.	
Hydrogenation Plants	1,096,000	650,000
Fischer Tropsch Plants ...	475,500	173,000
German Refineries	533,000	529,700
Austrian Refineries	250,000	228,400
Czechoslovakia Refineries ...	208,000	89,000
Polish Refineries	98,000	69,400
France: Merkwiller Refinery ...	18,000	14,700
Roumanian Refineries	687,000	564,100*
	3,365,500	2,318,700

As every plant was a potential target for priority attack there was an understandable tendency to over-estimate, rather than under-estimate, the production possibilities. This factor is probably responsible for a substantial part of the discrepancy in these two totals. The following notes cover the principal discrepancies in the figures.

Hydrogenation Plants.—The output of the three new plants in Upper Silesia was over-estimated by about 150,000 tons, it not being known that two of these plants were principally intended for the production of methanol. The estimated output capacities of the other plants were satisfactorily close to the actual figures.

Assessments of the effects of damage on production were of creditable accuracy as regards post-raid performance except that they usually over-estimated the rapidity with which production was regained. The renewed activity which was interpreted from the photographs in terms of production was, on a number of occasions, the evidence of the plant undergoing tests before resuming production.

German officials have criticised the attacks on the synthetic plants as not being repeated with sufficient frequency to ensure continual immobilisation, there being occasions when plants were able to produce the equivalent of seven to fourteen days' full output before the attack was renewed. This does not seem to have been due to errors in assessment of damage and recovery and especially as the recovery rate usually gave the Germans the credit for a more rapid achievement than was the case. It was due rather to the influence of weather in preventing re-attack and also to the understandable reluctance of the Commands in making new attacks until satisfied that such attacks were justified.

Fischer-Tropsch Plants.—With the Fischer-Tropsch plants there was the same tendency to over-estimate the recovery factor. A more important error was in the calculation of the normal capacity of the plants, which was based on their designed output which totalled 112,000 tons per month. The actual total output capacity, due to production difficulties, only amounted to 57,000 tons per month. Although this resulted in the plants being given a greater target importance than their output warranted, the difference was not sufficiently large in practice to have affected bombing priorities.

Refineries.—The resistance of refineries to bombing was generally under-estimated, although the sum totals of estimated to actual outputs were remarkably close.

* Estimate based on available crude.

APPENDIX 12.

THE REPORTS TO HITLER ON THE EFFECTS OF THE ATTACKS.

In the course of the offensive against the oil plants Speer submitted five reports to Hitler detailing the damage done and the counter-measures being taken Translations of these reports are given below. In some of them a certain amount of unimportant detail has been omitted. Many of the figures given by Speer check with statistics that have been obtained from other sources, and these letters consequently provide an authoritative account of the effects of the attacks. The italics are as given in the original documents.

1. REPORT OF 30TH JUNE, 1944.

State Top Secret. Berlin.

" My *Fuehrer*,

The enemy's attacks on the hydrogenation works and refineries were intensified during June; his current air reconnaissance and espionage enable him to damage severely, mostly soon after they had started up again, those works which had been hit in May.

Although *Herr Generalfeldmarschall* Keitel reports precisely to you on the weight of the attacks and on the results, I feel it is my duty to send you an overall report on the losses of the German fuel production since May this year and of the vital measures resulting from them.

I.—The focal point of the attacks in May and June was the German aviation spirit production.

In these attacks the enemy succeeded on the 22nd June in increasing the effects on aviation spirit by 90 per cent. Only by the most speedy reconstruction of the damaged works—which was well below schedule—can the effects of this catastrophic attack be eliminated.

Nevertheless, aviation spirit production is at the moment utterly insufficient.

In April the Luftwaffe used 156,000 tons and 175,000 tons were produced.

The average daily production for April was 5,850 tons daily.

In May the daily production was as follows :—

May		*Tons daily.*
1st to 11th	... ...	5,845
12th	Attack on Leuna and Bruex	4,821
13th	... ...	4,875
14th	... ...	4,842
15th	... ...	4,775
16th	... ...	4,980
17th	... ...	4,839
18th	... ...	4,920
19th	... ...	5,010
20th	... ...	4,975
21st	... ...	5,025
22nd	... ...	5,075
23rd	... ...	5,051
24th	... ...	5,073
25th	Leuna again in production at 20 per cent. ...	5,487
26th	... ...	5,541
27th	... ...	5,550
28th	Second attack on Leuna, result 100 per cent.	5,526
29th	Attack on Poelitz, result 100 per cent. ...	2,775
30th	... ...	2,743
31st	... ...	2,794

In May altogether 156,000 tons aviation spirit were produced, compared with an essential production of 180,000 tons.

In June the following production is shown :—

June		*Tons daily.*
1st	... ...	2,476
2nd	... ...	2,535
3rd	... ...	2,580
4th	... ...	2,555
5th	... ...	2,511
6th	... ...	2,226*
7th	... ...	1,823
8th	... ...	3,718
9th	... ...	2,756
10th	... ...	2,873
11th	... ...	3,052
12th	... ...	2,120
13th	Gelsenberg drops out 100 per cent. Welheim slight damage	1,078
14th	... ...	1,587
15th	Scholven slight damage	1,527
16th	... ...	1,275
17th	... ...	1,214
18th	Scholven again attacked, slight fall-off in production	1,323
19th	... ...	1,278
20th	Through the attack on Poelitz re-opening of the plant postponed till August	1,392
21st	... ...	1,268
22nd	Scholven fall off in production 20 per cent. Wesseling 40 per cent.	632
23rd	... ...	868
24th	Leuna again in production at 20 per cent. ...	1,268
25th	... ...	1,223
26th	Moosbierbaum production fall-off 100 per cent.	1,204
27th	... ...	1,252
28th	... ...	1,241
30th	... ...	1,218

Total production in June therefore only 53,000 tons aviation spirit as against the requirements of 195,000 tons in May.

After the first attack of the 12th May this year a production figure of 126,000 tons aviation spirit was reported to you in the Obersalzburg for June.

This quantity would certainly have been exceeded, due to the increased speed of reconstruction, but owing to the continuous attacks in June production was well below the estimated figure. Attention is drawn to the fact that production during the second half of June again decreased considerably and only corresponded to a monthly production of 42,000 tons; should the attacks continue an extra ordinary falling off in July can already now be foreseen with certainty.

II.—*Reich* production figures for carburettor fuel are :—

	Tons.
In April	125,000
In May only	93,000
In June estimated at only	70,400

Including imported fuel the June result should be 96,000 tons, with an April consumption of 205,000 tons.

Diesel fuel in Germany was :—

	Tons.
In April	88,900
In May	74,000
In June	66,300

Including imports, resources available in June are 94,000 tons with an April consumption of 194.000 tons.

* Figure not clear in original document owing to overtyping.

Attacks on the hydrogenation plants resulted in the following production and consumption figures for *Treibgas*—the most important substitute for liquid fuel in the country :—

	Tons.
April production and consumption	37,600
In June decrease to	10,400

Calculations based on the results in June show when the small reserves of liquid fuel will be used up.

The losses in aviation fuel are especially serious in this connection since practically the entire aviation fuel in Germany has been stopped, therefore imports can result in but little improvement.

It is to be reckoned with, that the capacity of the hydrogenation works at present under attack—under the most favourable conditions—in accordance with the severity of the damage, can at the earliest only come into commission again in 6–8 weeks. Every week that does not bring additional protection for the hydrogenation plants has dire results, because already the deficit in the June production and the reduced production to be expected in July and August in accordance with the present standard of air attacks, will *doubtless consume the major portion of the reserves of aircraft fuel and other fuels.*

If, therefore, we do not succeed in protecting the hydrogenation plants and refineries by all possible means, the reconstruction of these works, as is shown in June, will not be successful. Inevitably by September this year the supply of amounts necessary to cover the most urgent requirements of the *Wehrmacht* will no longer be assured; *i.e.*, from this moment on an impossible situation will arise which must lead to tragic results.

I have, for my part, issued *inter alia* the following orders :—

(1) The most speedy repairs of damaged plants, utilising labour and materials to the best advantage.

(2) Air raid shelter buildings to ensure the safety of the most important entities of the plants and of the workers who must carry on during attacks. Geilenberg, together with Dorsch, have set up an emergency plan for which 800,000 cbm. of concrete will be rapidly used.

(3) The *Wehrmacht* Motorised Transport Commander has authorised a drive for the construction of generators in *Wehrmacht* vehicles. The requisite generators are forthcoming from industry and the timber supplies are ensured by wholesale felling.

(4) My Planning Office have cut down liquid fuel requirements in the home country, including inland and sea shipping and construction work, to 35 per cent. in June, a further 23 per cent. reduction is planned for July, that is 42 per cent. of the May allocation.

Doubtless the *Wehrmacht* have taken steps rendered necessary by the present situation. I believe it my duty, however, to draw attention to the following facts :—

(1) Flights must be curtailed to the minimum essential; every ton of fuel wasted now may in two months be bitterly regretted, since the increasing fighter programme bears no relation to the decreasing fuel production.

(2) The strictest measures are essential to control the use of carburettor and diesel fuel in the *Wehrmacht* itself. In this connection it must be determined how the war can be continued when only a part of the present fuel supply is available.

(3) Fighter protection of industrial plants must be strengthened, since the *Luftwaffe* must realise that a successful continuation of the attacks will mean that in September only a proportion of their fighter planes will be able, owing to lack of fuel, to fly.

(4) The reconnaissance planes make it easy for the enemy to ascertain when plants have restarted work and hence in a short time they can stop production again. The enemy will not realise that we have restarted the hydrogenation works in this comparatively short period and will, without reconnaissance, leave longer pauses between his attacks and we shall then at least be able to resume partial production.

(5) A considerably increased supply of smoke units, even at the expense of other important items. Consideration should be given to ensuring better camouflage by setting up a dummy plant with the same smoke screen, apart from the white smoke which points to the existence of the actual plant.

(6) In spite of the recent increase in the Flak, it should be strengthened still more, even at the expense of the protection of German towns.

This supplementary protection must be provided for all hydrogenation plants and refineries, even those which are under construction and will be in production in 1–2 months (*e.g.*, Heydebreck and Blechhammer in Upper Silesia).

If we do not succeed in protecting the hydrogenation works and refineries better than formerly, then in September of this year an impossible situation in the fuel supply for the *Wehrmacht* and country will arise. The protection ordered up to now has meant in some easing already in the last attacks, without decisively improving the position. It shows, however, that it must be possible to protect the hydrogenation plants and refineries from attacks far more even now by the concentrated installation of all possibilities.

I had to draw attention, My *Fuehrer*, to the dire developments in the production of fuel.

I ask you to order the additional protection of these works by the sharpest measures.

SPEER."

---

2. Report of 28th July, 1944.

"My *Fuehrer*, *State Top Secret. Berlin.*

The attacks on the synthetic oil plants and refineries in July had the most dire consequences.

It was possible for the enemy, in most cases, to destroy the plants so effectively, shortly after work in them had been resumed, that instead of the expected increase there was a decrease in production, although the reconstruction measures taken lead to the anticipation of a substantial increase.

Development of production was as follows:—

I.—*Aviation spirit*—

	Tons.
In April the air force consumption was	165,000
In April the production was	175,000
Daily average production in April was therefore ...	5,850
Production in May was	156,000
Production in June was	53,000
Production in July was	29,000

After the latest attack on Leuna the production figure can be reckoned at only 15,000 tons aviation spirit, although the guaranteed amounts, owing to reconstruction work had been: for August, 43,000 tons; for September, 69,000 tons again, as prior to the attack on Leuna of the 28th and 29th July, a total of 93,000 tons had been hoped for in September.

	Tons.
For October are anticipated	120,000
For November are anticipated	150–160,000

These figures are anticipated with the promise that no further attacks occur or that they are fully averted, which under present conditions is not to be expected.

That the reconstruction of the synthetic oil plants in July would have been a complete success, without the occurrence of further attacks, is shown by the fact that in spite of numerous smaller attacks in July, production on the 17th July had already reached 2,307 tons, but four days later the reconstructed work was again completely destroyed by an attack, so that on the 21st July the record production depth of 120 tons was reached.

Daily production figures for July were as follows :—

July.		*Tons.*
1st	... ...	1,043
2nd	... ...	1,086
3rd	... ...	954
4th	Attack on Scholven	1,065
5th	Attack on Scholven	1,393
6th	Attack on Scholven	1,645
7th	Attacks on Scholven, Leuna, Luetzkendorf, Boehlen, Heydebreck	916
8th	Attack on Scholven	600
9th	Attack on Scholven	870
10th	Attack on Scholven	961
11th	... ...	751
12th	Resumption at Leuna	1,133
13th	... ...	1,278
14th	... ...	1,271
15th	Increase at Leuna	1,714
16th	Switch-over to special fuel at Leuna ...	1,588
17th	Increase at Leuna	2,307
18th	... ...	1,378
19th	Attacks on Wesseling and Scholven ...	856
20th	Attack on Leuna	970
21st	Attacks on Wehlheim and Bruex	120
22nd	... ...	140
23rd	... ...	140
24th	Resumption at Leuna	600
25th	... ...	417

II.—*Carburettor fuel.*

Production figures are :—

	Tons.
For April	125,000
For May	93,000
For June	76,000
For July	56,000

Anticipated production figures in the *Reich* after reconstruction measures, provided there are no more attacks :—

	Tons.
For August	84,000
For September	101,000

III.—*Diesel fuel.*

Production figures in Germany are :—

	Tons.
For April	88,900
For May	74,000
For June	66,000
For July	62,000

Estimated production figures :—

For August	93,000
For September	110,000
For October	125,000
For November	142,000

IV.—*Bottled Treibgas.*

Production of this, the most important substitute for fuel at home, reckoned in fuel-tonnage, shows the following figures :—

	Tons.
For April	37,600
For June	10,400
For July	5,000

Estimated production owing to Leuna being non-productive :—

	Tons.
For August	3,800
For September	8,200
For October	17,400

With these production figures—mainly in the field of aviation spirit supply—it is impossible to adapt the production to the reserves which have been used up, in September, if the allocation of fuel is to remain the same for August.

If, on the other hand, further attacks are made on the synthetic oil plants, and the enemy succeeds in throttling the aviation spirit production as hitherto, then a planned use of the air force in September or October will be impossible.

The strengthened protection of the synthetic oil plants, through A.A. and artificial fog units, did not prevent the most successful attacks in the last few days.

The fighter protection, which alone is decisive for the protection of the synthetic oil plants, did not increase since the 1st June, but *decreased !*

This is shown by the following contrasts :—

In the Reich.	*1st June, 1944.*	*1st July, 1944.*	*27th July, 1944.*
Fighter (*Jaeger*) total ...	788	388	460
Ready for action	472	242	273
Destroyer (*Zerstoerer*) total	203	156	94
Ready for action	83	64	42

This means, therefore, that in spite of the high production figures of fighters and destroyers in the months of May, June and July the number of planes available, or ready for action, has not risen at home, but *considerably decreased;* the result of this is that the enemy can reach his targets at will and with very small losses only.

The total number of *all* fighters on *all* fronts was :—

On the 1st June, 1944	1,789
On the 27th July, 1944	1,754

so that increased production did not make itself felt appreciably at the front.

I cannot judge if it would not be possible to allocate a larger part of the new fighter production to the home country, despite the emergency position at the fronts and if it would not be of greater advantage to use the fighters at home, as the losses at the front are much greater (destruction on the ground, fighting with enemy fighters, &c.) and therefore it is perhaps better to use the fighters at home.

I can only state that with the continuation of the attacks, judging by our experiences in June and July, the air force can reckon at the very most, in August and September, with *a new production of 10–20,000 tons of aviation spirit.*

Is it not more to the point to protect the synthetic oil plants for the moment so well with fighters that a part production, at least, will be possible, instead of the usual method where one knows with certainty that the air force, at home as well as at the front, will be ineffectual owing to lack of petrol and that there is not the slightest possibility of rebuilding a substantial fuel production in a short time. Irrespective of the far-reaching results which an unprotected home country would have in other spheres of industry and war production (Nitrogen, Synthetic Rubber, Powder, Explosives, Electricity Power Plants, &c.) . . . . So the absolute necessity remains of protecting the German synthetic oil plants, in the next few months.

At the same time it is necessary to reduce the consumption of aviation spirit in August and September still further, with the greatest energy, and indeed only to allocate fuel for the training of fighter pilots and the use in action of fighter and pursuit planes. With the reduced areas now at our disposal, it should not be necessary to maintain a passenger and courier air service, as long-distance telephones and teleprints are available.

On my part the further installation of producer-gas generators is being carried on with the greatest pressure, as well as the rebuilding of the synthetic oil plants and the building of the underground installations carried on with all energy. . . . .

In order to obtain success it is necessary :—

(1) To strengthen the fighter protection at home, in order to increase the losses during the flight to the synthetic oil plants.
(2) To increase the A.A. protection and the smoke units.
(3) To reduce to the minimum, at once, the use of planes.

Hail, my *Fuehrer* !

SPEER."

---

3. Report of 30th August, 1944.

State Top Secret. Berlin

" My *Fuehrer*,

The last air attacks have again hit the most important chemical works heavily. Thereby the three hydrogenation plants, Leuna, Bruex and Poelitz, although only recently in commission again, have been brought to a complete standstill for some weeks.

As the home defence against enemy air attacks promises no appreciably greater results in September as against August, chemical (oil) production in September must now be considerably lowered.

Nevertheless, no effort will be spared to restore the hydrogenation plants so that past production, at least, can be made possible in a short time.

The effect of these new raids on the entire chemical industry are extraordinary as severe shortages will occur not only in liquid fuels but also in various other important fields of chemistry.

(1) *Methanol Production.*

The production of Methanol dropped, as a result of the air-attacks, from an estimated essential production of 34,000 tons in August to 8,750 tons, and will at the outside only attain this figure in September.

The reserves of Methanol will only be 9,000 tons at the end of August therefore heavy inroads will be suffered in the following essential chemical fields in September if this reserve is completely used.

In the powder and explosives sector, the estimated production of precious explosives (Hexogen and Trinitrotoluol) will drop 30 per cent. in spite of the use of emergency measures and what is more, this notwithstanding that Methanol will be reserved for powder and explosives and use of it greatly reduced in other branches.

A particularly severe inroad is to be expected in artificial resins and plastics, the production of which will drop from 4,000 tons a month to 2,700 tons a month due to the shortage of raw materials.

The production of melamine glue will reach about half the requirements, while solid fuel for the *Wehrmacht* (for spirit stoves, &c.) will now be completely counted out in September.

(2) *Buna Production.*

By the failure of hydrogen from Leuna for Schkopau and the air raid damage in Ludwigshafen, Buna production sank from an estimated possible total of 13,000 tons to 5,400 tons in August.

After the new attack on the 24th August on Leuna, this figure cannot be improved in any way in September.

The Buna reserves, which we were able to increase in the monthly production during the last months before the attacks, stood at 9,000 tons on the 1st October so that October is secure as far as Buna is concerned. About one-third of the anticipated production for November will not be sufficient.

(3) *Nitrogen Production.*

Here also the new attack on Leuna has meant a considerable decrease in production, so that against an anticipated output of 85,000 tons for September, the highest possible figure will be 45,000.

This decrease in production will hit agriculture which at the moment holds only about 45 per cent. of its last year's allocation, which means that next year's harvest will suffer unusual losses.

Even worse are the effects in the field of liquid fuel as the hydrogenation plants and oil refineries have again been heavily hit in the last few days in the Protectorate and round Hanover.

I.—*Carburettor Fuels.*

Before the April attacks the carburettor fuel production in the *Reich* stood at 125,000 tons. In August, as a result of air attacks, at the most 60,000 tons were produced in August instead of the 84,000 tons estimated at the end of July.

If similar attacks continue the highest figures to be expected in September and October is 40,000 tons.

II.—*Diesel Fuels.*

Before the April attacks the diesel fuel production figure stood at 88,900 tons.

Due to air attacks only about 65,000 tons were produced in August as against 93,000 tons estimated at the end of July. If similar attacks continue the highest figure to be reckoned with for September and October is 60,000 tons.

III.—*Bottled Treibgas.*

In April the production of bottled gas stood at 37,600 tons (reckoned in terms of gasoline.)

Only about 3,000 tons were produced in August due to air attacks.

If similar attacks continue a figure of 2,500–3,000 tons is the highest to be reckoned with in September and October.

IV.—*Aviation Spirit.*

While 175,000 tons of aviation spirit was produced in April, production fell to 12,000 tons, *i.e.*, to two normal days' production in August due to the destruction of the aircraft fuel installations at Leuna, Poelitz and Bruex which had only recently been restored to working order. For September production, because of the re-building measures, was still estimated at 101,000 tons even on the 15th August.

After the new attacks production will not rise above 10–15,000 tons because of insufficient home defence.

With these results the enemy has hit the chemical industry so heavily that only by abnormal changes in the conditions is there any hope for the retention of the bases for powder and explosives (Methanol), Buna (Methanol) and nitrogen for explosives and agriculture. At the same time the loss in carburettor and diesel fuels is so widespread that even the severest measures will not be able to hinder encroachments on the mobility of the troops at the front.

The possibility of moving troops at the front will therefore be so restricted that planned operations in October will no longer be able to take place. With this fuel situation offensive moves will be impossible.

The flow necessary for the supply of the troops and the home country will therefore be paralyzed in the late autumn of this year, since substitute fuels, such as producer gas, are also inadequate to provide the essential help in all sectors.

There remains only one possibility, and this only with a large amount of luck :—

If the enemy—

(1) As was his former custom, begins his new attacks only when the plants, at present damaged, are again in commission, *i.e.*, in about three weeks, when

(2) The German fighter weapon at home can be so considerably strengthened in this three- to four-week breathing space as to inflict heavier losses on the enemy and to hinder the compact carpet bombardments by splitting up the bomber formations.

(3) In the coming autumn months, operations are restricted through bad weather conditions and both enemy and German air weapons are more restricted in operation.

We shall do the troops a bad service by sending pursuit planes from home to the front and thereby allow the vital materials for the front (powder, explosives and fuel) to be battered.

If it were possible to combat the attacks with some good measure of success in September, then it is feasible that there will be—

Only a 10 per cent. production drop in powder and explosives in October;

A rise in Buna from 5,000 tons in September to 10,000 tons in October;

A rise in nitrogen from 45,000 tons in September to 60,000 tons in October;

A rise in carburettor fuel from 40,000 tons in September to 65,000 tons in October;

A rise in diesel fuel from 60,000 tons in September to 90,000 tons in October;

A rise in aircraft fuel from 10–15,000 tons in September to 75,000 tons in October.

If, however, the homeland is protected only by Flak, then, despite the greatest concentration, no substantial results from defence can be obtained as the attacks on Leuna, Bruex and Poelitz have proved. In this case the production level in October will remain the same as in September but will not exceed it.

The *Luftwaffe* must be ready for this last great stake by the middle of September at the latest. They must include their best strength, their flying instructors and their most effective pursuit planes in this undertaking. The most modern machines must be ready for this attempt at an item figure of not less than 1,200 items.

If this course is taken it will, if successful, mean the beginning of a new air force or it will mean the end of the German air force.

If the attacks on the chemical industry continue in the same strength and with the same precision in September as in August the output of the chemical industry will drop still further and the last stocks will be consumed.

Thereby those materials which are necessary for the continuation of a modern war are lacking in the most important spheres.

Hail, my *Fuehrer*,
Always yours,
SPEER."

---

4. Report of 5th October, 1944.

"My *Fuehrer*, State Top Secret. *Berlin.*

After the last attacks on the hydrogenation plants and refineries repair of those works is still found to be possible in relatively short time as the number of men employed on this work has been increased.

If no new attacks take place we may count in October on the following quantities, which include the fuel gained from the German and Hungarian mineral oil production.

	Tons.
Aviation fuel (September production: 9,400 tons)...	64,400
Carburettor fuel (September: 48,400 tons)	60,600
Diesel fuel (September: 77,300 tons)	100,300

The following quantities could be produced in November and December:—

	November. *Tons.*	*December.* *Tons.*
Aviation petrol	91,900	106,900
J.2 (Fuel for the jet fighters, which is composed of carburettor and diesel fuel)	20,000	24,000
Carburettor fuel	65,000	66,200
Diesel fuel	71,700*	87,100

These production figures include the requirements of industry and agriculture.

An exchange of aviation fuel and of J.2 against carburettor fuel is, of course, possible.

These figures represent the quantities *theoretically* possible after rebuilding and reconstruction, *if no further successful air attacks take place.*

As, owing to the insufficient air defence, further air attacks of equal importance are to be expected, only the following production can be relied on:—

	October. *Tons.*	*November.* *Tons.*	*December.* *Tons.*
Aviation fuel ...	12,000	10,000	9,000
Carburettor fuel ...	40,000	40,000	45,000
Diesel fuel	75,000	80,000	80,000

* The reduced production of Diesel fuel is due to time required for refining of further Mineral Oil Stocks.

As far as the figures for aviation fuel are concerned these might fall rather than rise in November and December, as the continuous new attacks disorganise the system in the plants and thereby make rebuilding considerably more difficult after every attack.

No higher production can be expected in the month following these three months, on the supposition of further air attacks on the hydrogenation plants.

The underground plants for aviation fuel are not yet in operation, while the protected small plants for carburettor and diesel fuel are already producing the following :—

	October.	*November.*	*December.*
	Tons.	*Tons.*	*Tons.*
Carburettor fuel ...	5,000	10,000	10,000
Diesel fuel	20,000	40,000	40,000

These figures are included in the preceding estimates.

That the estimated higher production is possible is shown in the table for the month of September 1944 (see supplement), which gives the estimated *Daily Production* of Aviation Spirit and the actual quantities produced.

The table shows that on the 10th September the forecast figures were reached as several plants were working again, but it shows also that the enemy succeeded in stopping all fuel production *completely* between the 11th and 19th September.

By changing the method of attack, which has so far always been timed shortly after the restarting of the plants, allowing us thus always a few days of production, to a time shortly before the restart of work, the enemy could, without further ado, bring the aviation fuel production *completely* to a stop.

As the following plants will recommence work on the dates given below :—

Poelitz DHD* on 20th September;
Moosbierbaum DHD on 2nd October;
Poelitz on 6th October;
Leuna on 10th October (DHD 2nd October);
Bruex on 1st November;
Blechhammer on 8th October;

it will be necessary to build up in good time before above dates the strengthened fighter protection in such a way that at least 1,000 fighters can ward off successfully the attack which is to be expected shortly.

If this is not carried out the most *we can count on will be the production quantities given for continuation of air attacks.*

Simultaneously with the insufficient production of fuel at the hydrogenation plants, the picture of the production of the chemical industry so essential for powder and explosives, for Buna, &c., has continued to deteriorate correspondingly, so that already difficulties of the severest kind in these branches can be foreseen, if we do not succeed in protecting the chemical works more efficiently.

Admittedly, orders have been given to erect concentrated A.A. protection at some of these plants, which were constructed with particular care (such as Leuna, Poelitz, Bruex, Blechhammer, Ludwigshafen, Oppau). Experience has shown, however, that only the fighters in spite of heavy losses are in a position to inflict equal punishment on the enemy.

The troops will forgo fighter support, which cannot give them essential relief nowadays, if they know that in this way their fuel basis is secured and that munition supplies will not cease owing to lack of powder and explosives.

Front officers in the West, whose supplies of weapons, tanks and munition have improved during the last fortnight, know only one concern and question : Will it be possible to supply the fuel for future operations or will the air attacks of the enemy prevent this?

The employing of all fighter forces at our disposal for the protection of home production has become even more vitally important since the transport situation in the Ruhr regions has deteriorated quite considerably.

Whereas in September 1943 an average of 19,900 waggons of coal could be transported daily in the Ruhr region, this transport fell off during these last days owing to air attacks to 8,700 to 7,700 waggons daily. This means that after 8–12 weeks the stocks within the industry, which amount to four weeks' supply, will be exhausted, so that during this winter an exceptionally serious coal and consequent production crisis will arise, while, on the other hand, the dumps of coal in the Ruhr region mount up continuously. It must be stressed that these

* DHD is a process for making a high octane component for aviation fuel.

figures include the circulation of waggons within the Ruhr region, so that the figures for coal actually sent out of the region must be reduced correspondingly.

There is, therefore, for the next months only one problem: to raise the effective fighting capacity of the German fighter force to such a height as is absolutely possible, to add all available machines to its strength, and then to concentrate this fighter force for the protection of the home armaments and war production.

Hail, my *Fuehrèr*,
SPEER."

(Copies of this letter, and the figures attached, were sent to Goering, Keitel, Doenitz, Galland, Krauch and Geilenberg.)

Aviation Spirit Production, September 1944.

	Production Target.		Actual Production.		
	Daily.	Total.	Daily.	Total.	
Sept. 1 ...	230*	230	...	0	* Poelitz, Ludwigshafen, Huels, Schkopau commenced work.
„ 2 ...	...	460	80	80	
„ 3 ...	...	690	...	80	
„ 4 ...	...	920	...	80	
„ 5 ...	...	1,150	720	800	Ludwigshafen, Oppau, air attack.
„ 6 ...	...	1,380	38	838	
„ 7 ...	...	1,610	529	1,367	
„ 8 ...	...	1,840	375	1,742	
„ 9 ...	...	2,070	427	2,169	
„ 10 ...	...	2,300	303	2,472†	
„ 11 ...	...	2,530	...	2,472	
„ 12 ...	...	2,760	...	2,472	Boehlen, *Gelsenberg*, air attack, Scholven attacked.
„ 13 ...	...	2,990	...	2,472	
„ 14 ...	...	3,220	...	2,472	
„ 15 ...	...	3,450	...	2,472	
„ 16 ...	670	4,120	360	2,832	Planned change over from motor to aviation fuel. *Gelsenberg, Scholven.*
„ 17 ...	...	4,790	...	2,832	
„ 18 ...	...	5,460	...	2,832	
„ 19 ...	...	6,130	...	2,832	
„ 20 ...	...	6,800	176	3,008	
„ 21 ...	...	7,470	186	3,194	
„ 22 ...	...	8,140	385	3,579	
„ 23 ...	...	8,810	216	3,795	
„ 24 ...	...	9,480	...	3,795	
„ 25 ...	...	10,150	260	4,055	
„ 26 ...	1,570	11,720	...	4,055	
„ 27 ...	...	13,290	261	4,316	Anticipated start of *Welheim* and Moosbierbaum. *Welheim* attacked.
„ 28 ...	...	14,860	...	4,316	
„ 29 ...	...	16,430	258	4,574	
„ 30 ...	...	18,000	626	5,200	
	...	18,000	...	5,200	
	...	4,200 (Benzol)	...	4,200 (Benzol)	
	...	22,200	...	9,400	

† All production was stopped on this day.

REPORT OF 19TH JANUARY, 1945.

"My *Fuehrer*, *State Top Secret.* *Berlin.*

The following comparative figures of production results discussed in my report of 5·10·44 and actually shown in the fourth quarter, prove how enduring the effects of the continued air attacks on the hydrogenation plants and the refineries have been.

	Theoretically possible. *Tons.*	*Estimated if inadequate Air Defence.* *Tons.*	*Actual.* *Tons.*
October—			
Aviation spirit ...	60,000	12,000	18,000
Carburettor fuel ...	61,000	40,000	57,000
Diesel fuel (with J.2)	100,000	75,000	66,000
Total	221,000	127,000	141,000
November—			
Aviation spirit ...	92,000	10,000	41,000
Carburettor fuel ...	65,000	40,000	50,000
Diesel fuel (with J.2)	92,000	80,000	73,000
Total	249,000	130,000	164,000
December—			
Aviation spirit ...	107,000	9,000	25,000
Carburettor fuel ...	66,000	45,000	51,000
Diesel fuel (with J.2)	111,000	60,000	75,000
Total	284,000	134,000	151,000

It was possible to improve slightly supplies for the *Wehrmacht* by finally exhausting small reserves. Moreover, Army Group South took direct from Hungarian production :—

	Carburettor fuel. *Tons.*	*Diesel fuel.* *Tons.*
October	8,900	2,300
November	16,300	2,400
December	9,800	3,300

Since 13·1·45 a new series of heavy attacks have been made on the mineral oil industry, which have, up to now, led to the elimination of the large hydrogenation plants of Poelitz, Leuna, Bruex, Blechhammer and Zeitz for a considerable period: this after the last quarter of the previous year when all the plants situated in the West, especially Scholven, Wesseling, Welheim and Gelsenberg fell out completely. After each attack, owing to the need or the destruction of reserves of machinery and apparatus, the repair of the plant becomes more difficult and takes longer. Moreover, it has now been determined that the attacks which take place so often at night now, are considerably more effective than daylight attacks, since heavier bombs are used and an extraordinary accuracy in attaining the target is reported. Consequently, even if during the first quarter of 1945 the repair work and the plants are completely untouched, the theoretical production figures, which seemed possible in the last quarter, will not be reached. The underground plants for the production of aviation spirit and carburettor fuel now under construction will not yet be in production in the near future. The small plants, as substitutes for non-productive refineries were tried out to some extent in December.

For January–April 1945 the following total supplies can, according to the present position, be reckoned with, that is—including the Hungarian production :—

	Theoretically possible. *Tons.*	*Estimated if inadequate Air Defence.* *Tons.*
January—		
Aviation spirit	13,000	12,000
Carburettor fuel	68,000	60,000
Diesel fuel (with J.2)	73,000	65,000
Total ...	154,000	137,000

Of this, from Hungarian production :—

	Theoretically possible.	Estimated if inadequate Air Defence.
	Tons.	Tons.
Carburettor fuel	15,000 = 22·1%	15,000 = 25·0%
Diesel fuel	8,000 = 11·0%	8,000 = 12·3%
February—		
Aviation spirit	22,000	9,000
Carburettor fuel	73,000	53,000
Diesel fuel (with J.2)	93,000	66,000
Total ...	188,000	128,000
Of this, from Hungarian production :—		
Carburettor fuel	18,000 = 24·7%	18,000 = 34·0%
Diesel fuel	10,000 = 10·8%	10,000 = 15·2%
March—		
Aviation spirit	56,000	12,000
Carburettor fuel	85,000	50,000
Diesel fuel (with J.2)	100,000	68,000
Total ...	241,000	130,000

Of this, from Hungarian production :-
(Figures are omitted from text.)

April—		
Aviation spirit	68,000	12,000
Carburettor fuel	85,000	50,000
Diesel fuel (with J.2)	108,000	68,000
Total ...	261,000	130,000
Of this, from Hungarian production :—		
Carburettor fuel	18,000 = 21·2%	18,000 = 36·0%
Diesel fuel	10,000 = 9·3%	10,000 = 15·0%

It must be emphasised that the Hungarian crude oil is of special importance in connection with the supplies of carburettor fuel for the Army, since it contains 30 per cent. of carburettor fuel as compared with only 9 per cent. in German crude oil; from the Hungarian crude oil production of about 60,000 tons, 18,000 tons were obtained as against only 13,000 tons carburettor fuel from the 160,000 tons German crude oil.

The requirements of the Army are considerably higher than the estimated production figures which will decrease somewhat owing to the inadequate air defence :—

Aviation spirit (without J.2) estimated according to present consumption :—

	Tons.
January	45,000
February	40,000
March	45,000
April	45,000

Carburettor fuel estimated according to present consumption :—

	Wehrmacht.	National economy.
	Tons.	Tons.
January	55,000	8,000
February	50,000	7,000
March	55,000	6,000
April	55,000	5,000

Diesel fuel with J.2 (estimated according to present consumption and future consumption of J.2) :—

	Wehrmacht. *Tons.*	*National economy.* *Tons.*
January	60,000	24,000
February	60,000	23,000
March	65,000	25,000
April	70,000	25,000

In the *Wehrmacht* requirement put forward, the following is calculated for J.2 :—

	Tons.
January	20,000
February	20,000
March	25,000
April	30,000

The small reserves of aviation fuel still available will soon be exhausted. There exist no further stocks of carburettor fuel or diesel fuel.

The aviation petrol output is especially exposed and in danger, since production is only possible in four hydrogenation plants, whilst other fuels are on a wider supply basis. It is technically possible to guide production within certain limits, so that the maximum become aviation petrol. The result, of course, is a corresponding limitation of the supply of carburettor fuel and diesel fuel, including J. 2.

The deficit of the Leuna, Poelitz, Blechhammer and Bruex works is outstandingly lasting. In spite of the especially strong *Flak* defence of Leuna, it has been heavily hit several times. Fighter protection has not, at any attack, been sufficient to enable the enemy to be beaten off.

Even now it will still be possible to cover approximately the *Luftwaffe's* requirement from April onwards this year, if the reconstruction and the works of the four plants named can be completed without substantial interruption by further air attacks.

Since production from underground or other dispersed plants cannot be reckoned with during the next months, the fuel supply must depend entirely upon above ground installations. The undisturbed repair and running of the above ground plants is therefore an essential for further supply. The past months have shown that this is impossible under the present conditions of plant production.

Hail, my *Fuehrer*,

SPEER."

APPENDIX 13.

The Results of the Bombing of the Roumanian Oil Refineries.

The following summarises the findings of a bombing research mission which spent three weeks in Roumania in October 1944 :—

1. The bomber offensive from the 5th April to the 19th August of 1944 resulted in a loss in Roumanian refinery throughput totalling 1,139,000 tons of crude for the period in question and 640,000 tons for the period of reconstruction, thereafter—a total of 1,779,000 tons. Of this total, a small percentage was able to be exported for refining elsewhere, leaving about 1,500,000 tons of crude shut back in the fields. This could have represented a loss of about 1,000,000 tons of products to the Axis oil economy and about 400,000 tons to the Allies consequently. Actually exports of products to Germany for the period were found to have been only 800,000 tons less than normal,* unusually large withdrawals having been made during this time from the Roumanian economy.

2. The previous assessments of overall loss sustained by the Axis through this bomber offensive have been found approximately correct. These previous assessments, based on air cover and intelligence information, were in error as regards a few specific instances, and refinery throughput estimates were from 2 per cent. to 35 per cent. too high for the eight main refineries (90 per cent. of the total capacity) in the period May to August 1944.

3. The effects on refinery operation per ton of bombs falling in the refinery varied from negligible losses to losses of over 4,000 tons of crude throughput. The average for the period was 2,000 tons of crude denied to the Roumanian refineries per ton of bombs falling within refinery limits. This corresponds to 126 tons of crude per ton of bombs dropped.

4. The loss in production was caused, in only a minor percentage of the cases of damage, by direct hits or near misses on vital units (crude stills, cracking coils, or boiler houses); also in a minor percentage of the cases, by fires affecting vital units; in large percentage of the cases, by hits, fires, or near misses on units of secondary importance (lines, tanks, treaters, pumphouses).

5. The bombs used were 500 lb. GP except in a few cases where 500 lb. RDX, 1,000 lb., 2,000 lb. or 250 lb. GP bombs were used. It was found that per unit of weight, the 500 lb. bombs were the most effective. It was also found that explosions occurring at or above ground surface were more effective than those below ground.

* Normal exports for the period would have been one million tons.

APPENDIX 14.

ALTERNATIVE TARGET SYSTEMS : LUBRICANTS AND ETHYL FLUID PRODUCTION.

Throughout the offensive the attack of oil production targets was directed primarily against the major producers of gasoline in order to secure the maximum impact on the supply of aviation and motor fuels. In the later phases the attack was extended to producers of benzol, in view of its extensive use in motor fuels, and to the distribution system handling supplies of motor and aviation fuels to the armed forces. Although the primary aim was never changed, the possibility of crippling enemy resistance by the denial of tetra-ethyl-lead fluid or lubricants was considered at various times.

Tetra-ethyl-lead fluid is an indispensable constituent of high octane aviation and motor fuels. While the elimination of the limited number of sources from which this product was obtained would have ultimately caused a serious restriction in gasoline supplies of adequate quality, decisive attacks upon these plants would not have been possible before the end of 1943. By that time the precautions taken by the Germans against such attacks would have probably ensured an adequate continuity of supplies. These precautions included the maintenance of a reserve stock equal to six months' consumption requirements.* A study of the vulnerability of tetra-ethyl-lead fluid supplies is given in Appendix 15.

The question of giving priority to the attack of lubricating oil production was also considered from time to time. In particular the comparative merits were discussed of attacks on aviation lubricants as opposed to aviation spirit as a means of embarrassing the *Luftwaffe*. It was also realised that if a critical reduction could be secured in the supply of other lubricants this would, sooner or later, affect not only all forms of transport, but also the greater part of Germany's industrial production.

However, the decision was reached that primary importance should not be attached to lubricating oil plants. This decision was largely based on the probable time factor which would influence the incidence of the effects of a successful attack on lubricants production. It seemed probable that some time would elapse before the elimination of the principal lubricating oil plants could affect fighting efficiency, and especially as the *Wehrmacht* was likely to have strategic reserves and these could have been conserved by restrictions on non-military consumption. Furthermore, it was to be reckoned that a continuing supply would be partly assured by the reconditioning of recoverable lubricants. Finally, it was evident that successful attack of the synthetic oil plants and crude refineries, besides immediately affecting the production of fuels, would also in due course affect production of lubricants by depriving the lubricating oil plants of their feedstocks.

It is now seen that this decision was the right one. From the beginning of the war, and thanks largely to the useful lubricating oil yield of the German and Austrian crude oils, there was generally an adequacy of essential lubricants. The armed forces suffered no basic shortages up to the early summer of 1944. Civilian supplies were also well maintained throughout this period, although the impairment of qualities of certain grades at times caused difficulties.† Supplies were, however, insufficient to meet the needs of the occupied countries and the lack of lubricants in these areas contributed to economic decay and the consequent liabilities that these conditions imposed on Germany.

In due course the lubricating oil producers were, in fact, comprehensively attacked although principally for the reason that the plants were potentially capable of producing motor fuels and lubricants. The production of lubricants, which in 1943 had been averaging 65,000 tons a month, fell in the summer of 1944 to less than 40,000 tons a month. After a slight recovery in the autumn, output thereafter steadily dwindled to the point at which the figures were too unimportant for inclusion in the statistical records.

* Ahrens (A.D.I. (K) Report No. 399/1945).

† Annual deliveries of lubricants in Greater Germany to civilian users, exclusive of the Reichspost, were as follows:—

	Tons.
1938	519,600
1941	351,912
1942	324,507
1943	324,228
1944	310,685

Records of the *Arbeitsgemeinschaft Schmierstoff-Verteilung G.m.b.H.*

APPENDIX 15.

VULNERABILITY OF ETHYL FLUID PRODUCTION AS A TARGET SYSTEM.

Consideration was more than once given to the desirability of attacking the production sources of ethyl fluid, which is the indispensable constituent of high octane aviation and motor fuels. This objective had attractive possibilities as the successful denial of all supplies would have had drastic effects on the production of suitable gasoline for both the *Luftwaffe* and the army.

Ethyl fluid comprises tetra-ethyl-lead blended with a corrective agent in the form of ethylene dibromide. The plant for manufacturing both these products is vulnerable and both the raw materials and the finished products are highly inflammable. In the case of tetra-ethyl-lead, the plant comprises pressure vessels together with an elaborate system of pressure piping and, in view of the health hazards involved in the process, repairs would take longer than would otherwise be the case in view of the high standard of workmanship required. In the case of ethylene dibromide, the plant consists largely of earthenware equipment, which is fragile, made to order and would take time to replace if no spares were available.

Reasons for Not Attacking.

In spite of these vulnerable characteristics, the target system offered by ethyl fluid manufacture at no time gave practical prospects of achieving decisive results. It was known that, in addition to a pre-War tetra-ethyl-lead plant at Gapel-Doberitz near Berlin, a second plant had been erected and it was not until the later stages of the oil offensive that its location was positively identified by photographic reconnaissance at Froese near Magdeburg. There was also a French plant at Paimboeuf which, according to intelligence reports which appeared to be confirmed by photographic reconnaissance, did not seem to be in operation. In addition there were two small plants in Italy which sufficed for Italian requirements and there were no reports of the Germans needing to make use of them. In the circumstances it was considered probable that substantial stocks of ethyl fluid had been built up and that adequate safeguards had been taken against the possibility of air attacks. Ethylene dibromide production was also not considered likely to offer a decisive target as the number of plants in operation were not known.

The Froese plant near Magdeburg was photographed in February and March 1944, but it was not until December of that year that a portion of this large *I.G. Farben* plant was positively identified as a unit for the manufacture of tetra-ethyl-lead. When, at this time, consideration was again given to the possibility of attacking ethyl fluid production, the reduction in gasoline output had proceeded so far that, whatever the position may have been in the past, excess productive capacity and considerable reserves were considered to be certain.

A further photograph in February 1945 indicated little or no activity and it was assumed that output had been restricted on account of the small production of gasoline.

The Actual Position.

Only two tetra-ethyl-lead plants were in operation in Germany during the war, namely, Gapel-Doberitz and Froese, with a monthly capacity of 100 tons and 300 tons respectively.

In addition the following plants were in operation :—

	Tons per month.
Paimboeuf (France)	120
Bussi (Italy)	50
Trent (Italy)	15

This total output would suffice for the production of approximately 225,000 tons of petrol per month. Taking into account that the German planned production of aviation fuel was to be 200,000 tons per month, and eventually rising to 300,000 tons, and that, in addition, a substantial tonnage of motor gasoline also required tetra-ethyl-lead, this production was barely adequate to meet requirements. This is borne out by the fact that the following plans had been made, but were not consummated, at the time the war ended :—

(*a*) As early as 1942 the construction was projected of a 400 ton per month unit at Heydebreck, in Upper Silesia.
(*b*) It was proposed to enlarge Gapel-Doberitz from 100 to 200 tons per month by moving in equipment from either Bussi or Paimboeuf.
(*c*) In September 1944 it was decided, in view of the intense air attacks, to erect an underground plant in Brixlegg, Austria, with a capacity of 200 tons per month. Although the completion date of this plant was planned for March 1945, erection had hardly started at the time the war ended.

Tetra-Ethyl-Lead Production.—Gapel-Doberitz operated normally up to May 1944. From then on production decreased, due both to lack of coal, on account of the disruption of transportation, and to lack of ethyl chloride, on account of the destruction of plant facilities, mostly in Ludwigshafen.

Froese operated normally up to the end of July 1944, after which time production decreased sharply, due to the lack of ethyl chloride.

Contrary to intelligence reports to the effect that the French plant was not being operated, it was started up in 1942 and produced almost continuously, the output being taken by the German army.

The production of the three major plants was as follows :—

Tetra-Ethyl-Lead Production.

(In Metric Tons.)

	Gapel.	Froese.	Paimboeuf.	Total.	Equivalent petrol in metric tons.
					Per year.
1940	1,200	3,600	603	5,403	2,060,000
1941	1,200	3,600	134(*a*)	4,934	1,880,000
1942	1,200	3,600	315	5,115	1,950,000
1943	1,200	3,600	847	5,647	2,150,000
1944—					*Per month.*
January	100	300	64	464	177,000
February	100	300	64	464	177,000
March	100	300	64	464	177,000
April	100	300	...	400	152,000
May	50(*b*)	300	...	350	134,000
June	50(*b*)	300	...	350	134,000
July	50(*b*)	300	...	350	134,000
August	50(*b*)	240	...	290	111,000
September ...	50(*b*)	119	...	169	65,000
October	50(*b*)	139	...	189	72,000
November ...	50(*b*)	91	...	141	54,000
December ...	50(*b*)	30	...	80	30,000
Total for 1944				3,711	1,437,000

(*a*) Shipped from stock, not production. (*b*) Estimated.

Ethylene Dibromide Production.—Two ethylene dibromide plants were available, namely, the French plant on the Mediterranean coast, which extracts bromine from sea water, and the German plant of the *Brennerei und Chemische Werke Tornesch G.m.b.H.* in Holstein.

The French sea-water plant was never operated and this was probably because the output of the German plant was adequate. Bromine was available to this latter plant in practically unlimited quantity from the salt mines, although the output capacity of ethylene dibromide was dependent on the ethylene-producing facilities and the bromination facilities in Tornesch. Production was adjusted to correspond to that of tetra-ethyl-lead, there being no other use for ethylene dibromide.

Production was as follows :—

Ethylene Dibromide Production—Tornesch Plant.

	Metric Tons.	
1940	2,767	
1941	3,240	
1942	4,685	(744 tons shipped to Japan)
1943	2,574	(500 tons shipped to Japan) (plant shut down 2 months by fire)
1944—		
First half ...	2,430	
Second half ...	1,846	Production decreased on account of curtailed T.E.L. production.

If the tetra-ethyl-lead plants had been put out of action, the effective bombing of the Tornesch and Marseilles plants would have made the restitution of ethyl fluid production doubly difficult.

Stocks.—At the outbreak of war stocks were between 800 and 860 tons, of which about three-quarters were set aside as the national reserve. This quantity was considered sufficient to cover mobilisation needs, together with current production, for only two to three months. No record is available of the subsequent fluctuations in ethyl fluid stocks. It is apparent, however, that after an anxious period during the early stages of the war, a comfortable stock was gradually built up. By October 1943 stocks had amounted to no less than 2,500 tons, the majority of which was dispersed in the underground depots of the *WIFO* organisation.

Vulnerability.

The Germans anticipated the possibility of attacks upon ethyl fluid production and planned accordingly. The stock on hand in October 1943 would have been capable of treating 950,000 tons of aviation fuel, which would have sufficed for at least six months' operations. However, in the plan prepared in the autumn of that year to increase aviation gasoline production, it was estimated that in 1944 ethyl fluid output would fall short of requirements by from 1,800 to 2,000 tons. It was proposed to keep Gapel, Froese and Paimbœuf working to capacity and to have the new plant at Heydebreck in full operation by April 1945. It was consequently estimated that, until the beginning of 1945, production plus withdrawals from stocks would meet the increased fuel requirements. It was also expected that the jet-plane production programme would result in a decline in the demand for high-octane gasoline, which, in turn, would make the new Heydebreck plant redundant but available in the event of the loss of the other plants.

In 1943 these targets would have been difficult to attack on account of the long range, their small size, and the probability of heavy losses of Allied aircraft. If, however, these plants had been put out of action in that year the loss of ethyl fluid supplies would not have been reflected in reduced front-line activity in less than about nine months. It is probable that during this time alternative production facilities could have been erected.

Successful attacks in the Spring of 1944 might have caused an operational shortage of aviation gasoline before the end of 1944, but only if consumption had been maintained at the same rate as during the early months of the year. Owing to the attacks on the synthetic plants consumption was drastically curtailed in the second half of the year and in these circumstances the stocks of ethyl fluid on hand would probably have been sufficient to have met requirements to the end of the war.

If the location and the capacity of the Froese plant had been known at the time the oil offensive began, the Strategic Targets Committee would have possibly included the three tetra-ethyl-lead plants in the list of targets for attack. It is, however, unlikely that they would have been accorded a higher priority than the hydrogenation plants, and especially as the Committee was of the opinion that aviation fuel stocks in the Spring of 1944 were unlikely to exceed two or three months' consumption. As the offensive progressed it became increasingly apparent from intelligence that fuel supplies were dwindling, and this knowledge added to the determination of the Committee to ensure the destruction of those synthetic plants that were still in operation. It is now evident that this was the correct policy and that a diversion to another target system in the form of tetra-ethyl-lead plants would not have had such decisive results.

Conclusions.

1. In the latter stages of the war the production of ethyl fluid was dependent on two tetra-ethyl-lead plants and one ethylene dibromide plant.

2. If these plants had been attacked as soon as such attacks could have been successfully executed, Germany's powers of resistance would not have been impaired, as the reserve stocks on hand would probably have sufficed until production had been restored.

3. As long as high octane gasoline as at present constituted is the principal fuel for aircraft and ground transport it will need to contain ethyl fluid as an essential component. The denial to Germany of production facilities for tetra-ethyl-lead and for ethylene dibromide, and the prohibition of abnormal stocks of these products, would seriously handicap Germany's ability to wage war.

APPENDIX 16.

Effects of Fuel Shortage on Luftwaffe Aircrew Replacements, September 1943–November 1944.

(Extracted from " Verlust-Verbrauchs-und Bestandszahlen der Wehrmacht, einschl. Waffen-S.S.")

20*th September*, 1943.

The aircrew position can be considered as satisfactory in relation to the number of aircraft available. In all types of unit the crew strength was higher than the strength in aircraft without in general reaching the establishment figures, however. Aircrew replacements from the A.O.C. Training were in accordance with the quota of fuel allotted for this purpose.

31*st May*, 1944.

The proportion of aircrews to aircraft is sufficient at the moment. Owing to rationed aircraft fuel supplies, it must be expected that aircrew replacements cannot (in future) be supplied for all types of units to the full extent required by the armaments programme. In training, therefore, the emphasis must be laid on the production of fighter pilots for the defence of the Reich.

A decline in the strength of operational personnel must consequently inevitably be accepted in other types of unit, owing to the insufficiency of material for training (above all shortage of fuel and aircraft).

30*th June*, 1944.

At the moment, the number of crews is satisfactorily in excess of the number of aircraft. Owing to the aviation fuel shortage which has arisen, the bias of training has been altered so as to produce the full number of fighter and ground attack crews and, provisionally, 50 per cent. of the night-fighter crews. All other types of unit must subsist for the time being on their present strength or on the remaining output of final training schools. The resumption of training for these units is dependent on an increase in the fuel allocation. The special recruiting drive for volunteer pilots for defence of the Reich is having good success. There is a particular shortage of personnel with powers of leadership for the fighter units. This will be eased as much as possible by transferring officers from bomber units to fighters.

31*st August*, 1944.

The proportion of aircrews to aircraft available is generally satisfactory, except in the case of ground attack and reconnaissance crews where the position is tense on account of the reduced output of pilots, which is caused by the fuel position. If the allocation of fuel to the A.O.C. Training is maintained at its present low level it will only be possible to concentrate on the training of fighter pilots. As regards bombers, if maximum effort is sustained the aircrew position could be maintained only for four months by utilising personnel in the reserve training units and those who will be released by the proposed disbanding (of certain units). In all other types of unit one must be prepared to accept a reduction of strength corresponding to the operational losses.

30*th September*, 1944.

The demand for pilots arising from (increased) aircraft production was still covered by the transfer of bomber and transport pilots. As a result of ground-attack training having been given lower priority than that of fighter and tactical reconnaissance personnel, by the 1st November the aircrew strength of the ground-attack units will sink to 50 per cent. of establishment. On account of the fuel position this decline cannot be avoided. The shortage of reliable group (*Verband*) and unit commanders for fighters has not yet been entirely remedied in spite of the transfer of experienced officers, particularly from the disbanded bomber units, on account of the extra training required, although the position has improved. . . . .

The reduction of bomber output, due to the fuel position and the need to increase fighter output, has gone so far that in October only Me. 262 and Ar. 234 have been turned out.

The backlog in the ferrying of new aircraft has again increased, mainly owing to weather conditions, though also due to reduced acceptance and ferrying capacity of the operational units (especially in the East), as a result of fuel shortage.

31*st October*, 1944.

During the month of October pilot training declined by 30 per cent. owing to the reduction of gasoline supplies. The planned strengthening of the Reich defence was, however, made possible by the transfer of complete units (*Verbaende*) to the fighter arm. The observers released by the conversion of bomber units to jet-propelled types were transferred to the paratroops, the air gunners were transferred to the railway defence of Luftflotte Reich, while the radio operators transferred to radio control duties with the front-line units of Lw. Kdo. West. . . . .

The decline in the aircraft strength of the ground-attack units is not due to inadequate aircraft deliveries but to the inability of units to accept machines owing to the shortage of crews and of gasoline (C.3) in the operational areas.

30*th November*, 1944.

The output of aircrew personnel has fallen further owing to the small quantities of aircraft fuel available for training. Only through selecting advanced pupils and taking instructors released by the reduction of training facilities, as well as the transfer of bomber crews to fighters, has it been possible to prevent the number of available aircraft from exceeding that of crews. When these reserves of flying personnel have been used up it will be some time before pupils who are now in the preliminary stages of training can become available.

In night-fighters the critical point has already been reached, *i.e.*, the number of aircraft, including Luftflotte reserves, but excluding pools (*Leitstellen*), exceeds by 294 the number of crews available.

APPENDIX 17.

Contracts between Deutsche Erdoel A.G. and the German Navy for Heide-Meldorf Oil.

In the area of the Heide oil field, in Holstein, oil has been obtained from two sources, from oil fields by means of wells and from a mine. In the case of the mine, the oil is contained in a strongly impregnated chalk of Cretaceous (Senonian) age, called for convenience "oil-chalk." From the point of view of the *Deutsche Erdoel A.G.*, who held the concession covering the Heide-Meldorf structure, the drilling and the mining for oil were seen differently. The drilling and production of oil by wells was an ordinary commercial undertaking. Mining of the oil-chalk was not a commercial proposition and it would not have been undertaken but for the interest of the German Navy.

Drilling in the Heide Field.

Before **1914** drilling in Holstein had already been started in the neighbourhood of seepages but without success and nothing more was done until 1930.

Geological surveys then concluded that the presence of oil-chalk, both in outcrop and in wells, might indicate a secondary migration of oil from deeper layers, and that, therefore, commercial prospects were greater than developments to that date had showed. The first wells were drilled near the oil-chalk outcrop and, though they were successful in finding oil, the results were not very promising. Commercial production was, however, obtained in **1935** from the Hoelle and Friedrichswerk fields which continued into **1940** and **1938** respectively.

Drilling was later undertaken on the flanks of the structure and successful production obtained in the Lieth Field in March 1938. The well Holstein **36** secured an initial production of 205 tons a day from the Permian (Zechstein). Continued drilling exploration led to the discovery in March **1939** of the Rickelsdorf Field showing that the Lieth Field was not an isolated occurrence and that further oil pools might be expected in the Heide region. It was the strike of Holstein 38 in the Zechstein with an initial production of 58 tons, that led to the decision to place a refinery at Hemmingstedt.

The production of crude oil in Holstein naturally brought up the question of local markets. The largest consumer in the region was the German Navy, interested, not in crude oil, but in fuel oil. It was to supply this need that the refinery at Hemmingstedt was built. In May **1940** a long-term agreement with the Navy was made for the supply of this oil. A little later, owing to the great need for lubricants, orders were given that topped crude might not be used as a stock for fuel oil. The Navy was therefore supplied with diesel oil instead, and the topped residue railed to other refineries for making lubricating oil. Only since the summer of **1944** was the German Navy again supplied with fuel oil from Heide-Holstein.

It is necessary to stress that these contracts with the Navy were such as might have been made with any commercial firm. The Navy was not financially interested neither in the exploitation of the fields nor in the Hemmingstedt refinery. Neither did the *Deutsche Erdoel* ask for any financial guarantee from the Government.

In one sense only did the Navy contribute to the development of the area. During **1939** and **1941** it built a 19-kilometre pipeline from the Hemmingstedt Refinery to Gruenenthal on the Kiel Canal. Both the line and the pumping station were paid for by the Navy and were thus their property. In 1940 and **1941** it was the Navy that insisted on the field being produced to maximum capacity.

Mining in the Heide Field.

The predecessor of the *Deutsche Erdoel-Aktiengesellschaft*, the *Deutsche Petroleum Aktiengesellschaft* (*D.P.A.G.*) which became a part of the Group of the first in **1926**, and which was absorbed in **1940**, tried in 1920 to exploit the Heide oil-chalk by mining on a commercial scale, and planned to set up a distillation plant for the production of crude oil. Owing to various technical difficulties, these trials were not carried out. The work was stopped in **1926** after the *Deutsche Erdoel* had secured some influence in *D.P.A.G.*

To be assured of a source of fuel oil the Navy had secured from the *Reich* a reserve on the exploitation of the Heide oil-chalk deposits. Discussions to get the work going began in 1939. The start with development was postponed owing to the long period it was seen would be required for construction, the war being always counted a short one. Discussions were continued in 1940. Before work was started on a large scale the project was studied from every point of view. Previous attempts that had been made to exploit the oil-chalk showed clearly that, thanks to considerable technical difficulties, the project was not a commercial one. It could be justified only as a national project supported by national means.

The discussions in 1940 led, in May 1941, to a ten-year agreement between *D.E.A.* and the Navy. It was agreed that all the products obtained from the oil-chalk should be for the exclusive use of the Navy. Under this agreement the *Deutsche Erdoel* undertook to sink the mine and to erect the distillation plant. Of the estimated cost of RM. 6,000,000 the Navy (*Reich*) granted a loan of RM. 4,000,000. Further, the *Reich* guaranteed to refund the investments in full should mining prove impossible on account of technical difficulties, or prove too expensive. The total of RM. 4,000,000 which the *Deutsche Erdoel* or its successors had already spent on attempts in previous years was, however, to remain a charge on the company.

The construction work, which continued in the next three years, proved to take longer and to be far more expensive than had been anticipated owing to war difficulties and rising costs of materials. Consequently, the Navy agreed to increase their loan by four-fifths, namely, from RM. 4,800,000 to RM. 8,760,000 an amount, in point of fact, that has not been paid.

But the contract was never executed. One distillation unit was completed, but the continuous operation of the plant required the operation of two units for overhaul and clearing out and consequently full working experience was not obtained. On the completion of the second distillation unit the plant was damaged by air attacks. The Hemmingstedt refinery area was bombed in all on thirteen separate occasions.

The technical difficulties which accompanied the underground mining delayed the work as much as bombing delayed surface operations. The Heide oil-chalk is a highly plastic formation. It can be squeezed in the hand; a pile of it will settle down under it own weight. Under the pressure of an over-burden in a mine it will "flow" and so will cave in a mine shaft. Gas also brought difficulties. The cost of sinking a shaft or of running a gallery grew out of all proportion to the returns obtained. One metre of gallery at Wietze cost RM. 80 to run; at Heide a metre cost RM. 1,000.

All these difficulties together made fulfilment of the contract impossible. It was agreed between the Navy and the *Deutsche Erdoel* that the company should bring the plant to the productive stage as a large-scale experiment, entirely for the Navy's account. And that this trial should run to the end of March 1945. The Navy insisted on the work continuing. In the end events hindered these arrangements. The heavy bombing attack in April put an end to all aspirations. It would be difficult to conceive an attack more thoroughly executed than the destruction of the Hemmingstedt refinery and storage plant on that occasion. The distillation of the oil-chalk had hardly started, and the company are still owed a large sum of money by the *Reich* for their attempted contribution to the national cause.

APPENDIX 18.

Some German Views upon Allied Strategic Bombing Policy.

The following summarises some of the comments by various German leaders upon Allied bombing policy. As a number of these statements are extracted from their context they should be noted in relation to their application to the oil target system only and not as a commentary upon all target systems as a whole. The fact that these critics can have had no understanding of the operational considerations that largely dictated bombing policy needs also to be borne in mind. Furthermore, some of the witnesses in other interrogations expressed themselves just as strongly about the importance of other target systems.

Reichsmarschall Goering.

Goering claimed that the Germans realised at an early stage of the Allied attacks that the Allied Air Force intended to bomb by systematic selection of related targets. Immediately after the first attack on the oil industry, they were sure that the synthetic oil plants would then become a first priority target. Generally speaking, attacks selected the right targets and did not overlook any installations the bombing of which would have ended the war sooner. . . . .

The attacks on synthetic oil plants were the most effective of all strategic bombing and the most decisive in German's defeat. " Without fuel, nobody can conduct a war."

(A.M.W.I.S. No. 317. P. 11.)

Reichsminister Speer.

" The final decision (Germany's collapse) was due in great part to the elimination of synthetic oil production. Oil and high-grade steel were the focal points but steel was at no time decisively affected by bombing."

(F.D. 2960/45. A.O. 4/Z.)

" In 1942 we undertook a study of the possibilities of economic warfare through aerial attack. Unfortunately the results of these studies were never put into action. While making these studies we discovered to our terror that it would be possible in the case of Germany to paralyse our industry to such an extent in comparatively short time that a continuation of the war would become impossible for us. . . . . The happenings of 12th May had been a nightmare to us for more than two years."

Speer, in his detailed criticism of Allied bombing policy, has emphasised that the attacks upon transportation would not have prevented the maintenance of a constant flow of fuel to the Front provided that the supplies were available. Moreover, " destruction of the chemical (synthetic oil) industry and the refineries would have resulted in a speedier collapse than that caused by a breakdown of transport, which requires more attacks." . . . .

" With only one target (system) you could have brought about the collapse of Germany within eight weeks so that further resistance would have been impossible either in the east or in the west. . . . . For instance, in the attacks on the chemical plants (synthetic oil) only a part of the bomber formation attacked the chemical plants. As far as I know only 20 to 30 per cent. of the total amount of bombers. Had you concentrated 100 per cent. instead of 25 per cent. reconstruction would have been impossible. No labour would have entered the plants because of the continued bomb danger. In that case the war would have practically ended within six to eight weeks. . . .

" With each attack the plants were more and more shaken up so that the time period in which reconstruction was possible was increased more and more, because the connections were then damaged, torn apart or not tight any more. This became more and more apparent even when there were not hits in the immediate vicinity. The situation became more and more critical. Then a point is reached when reconstruction becomes extremely difficult." . . . .

(U.S.S.B.S. Interrogation 19/20.5.45.)

Generalfeldmarschall Milch.

" The attacks on the airframe industry undoubtedly restricted activity but did not destroy production. The attacks on engine factories were more serious, but still more decisive were the attacks on the synthetic oil plants. The *Luftwaffe* had reserves to last about two and a half months with very gradual use, after which production would have to be resumed or all would have been lost. We were not successful in resuming production. The English intelligence system was marvellous. After a plant had been running again for one or two days it was again bombed. . . . . If the synthetic oil plants had been attacked six months earlier, Germany would have been defeated about six months sooner. At that time it would have been impossible to have adapted production within a period of two and a half months to enable the synthetic oil plants to be protected and at the same time to have had fuel for operational purposes.

The importance of oil targets in relation to ball bearing factories was in the ratio of about ten to one. When other targets were being bombed there was always the greatest apprehension lest attacks would be made upon oil targets."

(C.S.D.I.C. S.R.G.G. 1313(C) 4.5.45.)

Dr. E. R. Fischer.

If the air attacks had been concentrated on industry, particularly oil, chemicals, power and transportation, the War would have been over one year sooner. However, the air attacks being dispersed widely over German industry enabled them to make repairs and bring industrial installations back into production.

(U.S.S.B.S. Interview No. 67.)

Dr. Karl Hettlage.

"The War would have ended much sooner if precision bombing attacks had begun earlier than they did—certainly if the German transport system and oil production had been attacked."

(U.S.S.B.S. Interview No. 12A.)

Generalleutnant Galland.

In Galland's opinion it was the Allied bombing of the oil industries which had the greatest effect on Germany's war potential. . . . . Galland wondered why we waited so long to attack German oil production.

Galland considered that the bombing of communication facilities such as railroads, canals, bridges, &c., was the most important single factor in the defeat of Germany. In fact, he termed this bombing "decisive". . . . . He rated bombing targets in the following scale of effectiveness :—

(1) Transport facilities, because of their direct importance to military operations and war production.
(2) Oil targets, because of their relation to the function of air forces, armoured forces and military transport and industry in general.
(3) General industry, including aircraft production.
(4) Attacks on cities to cripple man-power.

(A.D.I. (K). Report No. 311/1945.)

General-Major von Massow.

The attack on oil production was the largest factor of all in reducing Germany's war potential.

(A.D.I. (K). Report No. 311/1945.)

General Ingenieur Spies.

The oil industry was always a weak point in German economy and targets like Leuna easily damaged. Once the bombing had begun it was extremely effective and Allied intelligence was such that they knew exactly when a target was ready to recommence refining and immediately open further attacks. . . . . In his opinion the only mistake we made was that oil was not made the primary target from the beginning.

(A.D.I. (K). Report No. 311/1945.)

Generalleutnant Veith.

The sequence of Allied attacks was logical and effective. Firstly in the attacking of the aircraft industry which had not yet recovered sufficiently by the time the invasion arrived and, secondly, the destruction of the oil industry and the simultaneous dislocation of the communications system were decisive.

(A.D.I. (K). Report No. 311/1945.)

General Thomas.

"The lack of liquid fuel made itself felt in the War against Russia and had a particularly bad effect on reinforcements for the African operations. If the Allies had begun their planned attacks on the oil sources and the synthetic fuel industry at an earlier date, the War would have been considerably shortened."

(F.D. 4503/45).

APPENDIX 19.

Note on Oil Plant Defence Measures.

The Germans had been in constant fear of the bombing of their oil plants and had been anticipating the attacks for some time. In 1942 they realised that serious damage to the hydrogenation plants would have disastrous consequences and in 1943 it was decided, at a meeting of the *Reichsanstalt Luftwaffe*, that submarine plants and liquid-fuel plants were likely to be the primary targets for attacks.

This decision did not apparently cause any major changes in the disposition of the *Flak* defences, but measures for the protection of the plants were speeded up. In December a conference was held at Leuna, under the chairmanship of Krauch, which was attended by the representatives of most of the important plants. Views were exchanged on the means of protecting the plants and the conference terminated with an order from Krauch that extensive protective measures were to be started at once. All available man-power and material was to be co-ordinated in carrying out the task and the work was to be completed at the expense, if necessary, of stopping work temporarily on new projects.

Protection of Equipment.

For various reasons this work did not go forward as planned, and there was considerable discordance between the *Zentrale Planung* and the various departments responsible for ensuring the availability of materials and for getting the work done. On the 21st April, 1944, Goering demanded that Krauch should investigate whether the most vulnerable parts of the buna and hydrogenation plants could be protected by two metres of concrete. On the 5th May Krauch reported that the programme for the elementary protection of the plants against splinter and fire damage was at that time only one-third completed. The cost of this minimum protection for fourteen oil and two buna plants was estimated at RM. 120 million. The labour involved was calculated at 24,000 men for one year and the materials required were as follows :—

Structural steel	22,000	tons.
Cement	220,000	,,
Sand	1,400,000	,,
Wood	20,000	cu. metres.

The expense of providing complete protection whereby essential equipment would be protected by concrete would have amounted to an estimated cost of RM. 1,000 million. An even more ambitious scheme involving the covering of all buildings (but not overhead pipe-lines or railways), was calculated to require 200,000 tons of steel, 1,200,000 workmen for one year and an expenditure of RM. 6,000 million.

Smoke Screens.

Difficulties were also encountered in arranging for protection by smoke screens. Although it was agreed at the *Fuehrer's* Headquarters in October 1942 that forty plants, of which ten were oil plants, should have smoke protection by the following April nothing adequate was accomplished within that time. In May, at a meeting of the *Zentrale Planung*, Milch announced that strict orders had been given that the hydrogenation plants were to have the first priority in smoke screen protection. There was, however, a shortage of chlorsulphonic acid used in the smoke generators, and, as it was decided that the Ploesti refineries must be given preference, the remaining supplies were insufficient to meet the needs of the principal plants in Germany. In spite of the attempts to increase the output of this acid, there was never enough to meet the demand.

Decoys.

A number of decoy plants were erected and especially in the Leuna and Leipzig areas. These were duly observed as decoys by the aerial reconnaissance units. While the illumination of these decoys at night, together with simulations of bomb explosions and of target indicators, succeeded in attracting a number of bombs there is no record of their causing the diversion of any substantial part of an attacking force.

Flak.

Although oil plants had always been regarded as of high importance when deciding the disposition of *Flak* defences it was not until the beginning of 1944 that the defences were more definitely concentrated upon the protection of oil. On the 31st January, 1944, an order was sent from the Chief of the *Oberkommando der Wehrmacht* to the Chief of the General Staff of the *Luftwaffe* :—

> "The *Zentrale Planung* is greatly concerned with the protection of the fuel plants, since there has recently been a shortage of fuel for the *Luftwaffe*. I demand that air defence be strengthened at the most important fuel plants."

This evoked a reply on the 12th February in which it was stated that, although there had been a slight weakening in the *Flak* defences at a few plants, this was being adjusted and at the same time *Flak* protection was being considerably strengthened at fourteen of the hydrogenation plants.

Plants Vulnerable at the Time the Attacks Started.

Although increasing attention was later given to the protection of the plants, the fact remains that these measures were wholly insufficient at the time when the attacks started. If the main items of equipment had been provided with concrete shelters by May 1944, the damage done in the initial attacks would have been greatly lessened. Although, in the words of Speer, "the happenings of the 12th May had been a nightmare to us for more than two years," these vital plants were, nevertheless, allowed to remain in a vulnerable condition up to the time the attacks started.

APPENDIX 20.

Note upon Aiming Error and Weapon Effectiveness in Relation to Oil Plants.

The complex question of weapon effectiveness is the subject of a separate study. However, some reference should be made to the difficulties involved in securing decisive results upon oil plants with the weapons that were available during the course of the offensive.

Without the assistance of precision ground control ("OBOE"), which was only available for certain targets in Western Germany during the concluding phases of the offensive, the oil plants could only be successfully attacked under visual conditions. The importance of visual aiming conditions is emphasised by a study made of three plants, namely, Leuna, Ludwigshafen-Oppau and Troglitz-Zeitz.*

These three plants covered a total area of three-and-a-half square miles, and they represented 22·4 per cent. of Germany's synthetic oil production capacity. The study revealed the accuracy achieved by 146,000 high-explosive bombs. As this number comprised 38·3 per cent. of the total number directed against all the hydrogenation plants, the figures may be taken as representative of these targets as a whole. Of the bombs dropped on these three targets only 1 in 29 hit structures essential to production. In round numbers, out of every 100 dropped :—

87 missed the target, 3 hitting decoy plants.
8 landed in open spaces inside the target area, causing little or no damage.
2 landing inside the fence lines failed to explode.
1 hit pipe-lines or other utilities, doing reparable damage, and only
2 hit buildings and important equipment.

As these statistics exclude the bomb tonnage that failed to reach the target area (aircraft shot down, abortive missions, &c.) the proportion of effectively-directed bombs actually transported was even smaller than these figures indicate.

The effect of weather on bombing accuracy is shown by the following tabulation of the results, under different conditions, against Leuna, Ludwigshafen-Oppau and Zeitz.

	Total Tons of Bombs Dropped.	*Tons of Bombs Hitting Plant.*	*Per Cent. of Hits.*
8th Air Force visual ...	3,993	1,069	26·8
8th Air Force part visual, part instrument ...	4,553	566	12·4
8th Air Force full instrument	11,870	641	5·4
R.A.F. night technique ...	9,540	1,505	15·8
Total or Over-all ...	29,956	3,781	12·6

In any analysis of these statistics various considerations have to be taken into account. There is, for instance, the fact that aiming technique improved greatly during the course of the offensive. At the same time, the physical protection of the plant equipment was materially strengthened. In regard to bombs, the types used were dependent upon the supplies available, which comprised a large proportion of bombs which were not the most suitable for oil-plant attacks.†

---

* This study was made by the United States Strategic Bombing Survey.

† The average weight of bombs dropped on oil plants by R.A.F. Bomber Command was 660 lb. and the average for the United States Strategic Air Forces was 388 lb.

APPENDIX 21.

The Geilenberg Plan for Plant Dispersal.

Although the Geilenberg plan for the dispersal of the oil industry was a failure a summary of what the plan comprised and what it achieved is of interest for two reasons. Firstly, there is the cost of the plan to Germany in terms of the abortive use of materials and man-power and the factor of this cost in hastening Germany's collapse. Secondly, there is the question of whether, if the project had been launched sooner, or the War had ended later, the plan could have successfully frustrated the aims of the bombing offensive.

The difficulties of dispersing one of Germany's major heavy industries were so great that those responsible for taking this step hesitated some three months before deciding that the task should be attempted. But even if there had not been this understandable delay and if the work of dispersing and burying had begun in June 1944, it is certain that the amount of oil that could have been produced from these concealed plants would have only been sufficient to have provided a small fraction of essential requirements.

The plan broadly comprised two major projects. To provide motor gasoline and diesel oil for the Army and the Navy, the largest amount of crude oil was to be distilled in a substantial number of small refining units. For the production of aviation fuel both hydrogenation and catalytic cracking plants were to be erected.*

Distillation.

The distillation of the crude oil was the simpler of the two projects. It was proposed that forty small distillation units, known as "*Ofen*," would be built in pairs at twenty dispersed locations. They were to be sited in mines, caves or concealed in wooded country. Each unit was to have a monthly throughput of 3,000 tons yielding 350 tons of gasoline, 1,000 tons of diesel oil, and 1,500 tons of residues. Initial production from the first plants completed was to begin in the middle of September 1944. In addition, five distillation plants, each with a throughput capacity of 12,000 tons a month, were to be erected in five suitable bomb-damaged factories which could provide the necessary boiler plants. These units were known as "*Roeste*." Initial production from the first plant was to begin in October 1944.

These two types of plant were intended to ensure a minimum production of 14,000 tons of gasoline and 40,000 tons of diesel oil. It was hoped that the normal refineries, by the repair of bomb damage, would produce about 6,000 tons a month of gasoline and 16,000 tons of diesel oil. This programme would thus provide about 20,000 tons of motor gasoline. The inadequacy of the quantities thus to be produced is shown by the fact that in December 1944 the consumption of motor gasoline was 90,000 tons, of which the *Wehrmacht* consumed 79,000 tons.

Cracking.

Another type of plant was proposed for processing crude oil residues. This was a normal cracking plant, known as "*Taube*," which, with a throughput of 25,000 tons a month, was expected to produce 8,000 tons of gasoline with 15,000 tons of fuel oil. Production was not expected to start before April 1945. In addition, eleven small primitive cracking plants, designated "*Jakob*," were planned.

Shale Oil.

The Wuerttemberg shale deposits were also to play their part, and ten primitive carbonisation plants, known as "*Wuesten*," were to produce a total of 20,000 tons of light diesel oil monthly. Production was planned to begin in the latter part of October 1944.

Aviation Fuel.

The erection of aviation fuel plants, involving high pressures and complicated equipment, was a more difficult undertaking. The original plan contemplated the construction of five underground hydrogenation plants known

* The major part of the information on this subject has been obtained from the Krauch files (20k).

as "*Schwalben*." Two were to be in the Elbe Valley, between Dresden and Aussig, a third in Thuringia and a fourth near Buckeburg and Minden. The fifth, which was to include an alkylation unit, was to be placed in the I. G. Farben gypsum mines at Niedersachswerfen, in the Harz mountains. The principal feedstocks of these plants was to be brown coal tar. They were primarily intended for the production of aviation fuel. To speed their erection the underground sites chosen for at least two of them had already been in preparation for other industries.† But, in spite of the priority that was accorded to these "*Schwalben*," production was not expected to begin before August 1945 at the earliest.

Additional aviation fuel, using diesel oil feedstock, was to be forthcoming from catalytic cracking plants known as "*Meisen*" and which were planned to yield 6,000 tons a month of aviation spirit and 8,000 tons of J.2 jet fuel. Output was not expected to begin before March 1945.

One further type of aviation spirit plant was contemplated. This was a reforming plant for converting 25,000 tons a month of motor gasoline into 20,000 tons of aviation fuel. It was known as "*Kuckuck*," and it was intended to start operating in April 1945.

Thus, if the plans had been consummated, an underground production of about 40,000 tons a month of aviation spirit might have been possible. This compares with the consumption of the *Luftwaffe* in December 1944 of 50,000 tons.

Lubricants

To safeguard lubricating oil production it was intended to construct a small number of underground plants known as "*Dachs*."‡ They were designed to produce 38,000 tons of lubricants a month, in addition to some fuel oil and other products. Although first production was expected to start in December 1944, none of these plants was ready to operate at the time the War ended. It was hoped that the repair of existing plants would provide another 16,000 tons a month of lubricating oils.

Fischer–Tropsch.

In the case of the Fischer–Tropsch process a small unit, assembled from the Lutzkendorf plant, was erected in the Leipzig gas works. Output capacity was intended to be 500 tons of liquid products a month but the plant was not ready for operation at the time of the collapse. Although the erection of ten small Fischer–Tropsch plants was contemplated, to be known as "*Karpfen*," none of these projects was completed.§

Ancillary Products.

For the manufacture of ancillary products the construction of the following plants was also planned. None of them came into operation.

Designation.	*No. of Plants.*	*Purpose.*
"*Molch*"	6	Phenol extraction.
"*Steinbock*"	2	Catalyst manufacture and maintenance. Paraflow production.
"*Iltis*"	3	Super-fractionation.
"*Kranich*"	1	Catalytic cracking.
"*Rabe*"	1	Tetra-ethyl-lead.
"*Kybol*"	2	Diethyl benzene.
"*Krebs*"	2	Fischer-Tropsch.
"*Fasan*"	2	Benzene alkylation.

This, in broad outline, was the scheme to save the situation. Equipment for these projects was to be largely provided by bomb-damaged plants and it was also hoped that the looting of further equipment from the disused refineries in occupied countries would help to make up any deficiencies.

Failure of the Plan.

The execution of the project was handicapped from the start. Difficulties over equipment, labour and transport multiplied themselves. These difficulties increased with the dislocation of the transportation system. The ruthless priority

† Saur. (U.S.S.B.S. Interview No. 48).
‡ See illustration facing page 72.
§ C.I.O.S. Report, Item 30, File XXVII-68.

powers granted to Geilenberg defeated their own ends. The materials earmarked for new plants were confiscated to make repairs to old ones. The indiscriminate seizing of equipment added to a state of disorder that eventually made the systematic assembling of these complicated plants an impossibility.

No statistics are available to show what this immense plan cost in terms of effort. Over and above the many thousands of workers employed there was the burden on the transport system of the great tonnage of steel and materials that had to be moved considerable distances. In many cases these movements to erection sites had been completed but the sites themselves were not ready for erection to begin.

The first "*Rost*" plant, in Hanover, came into partial production in September. In October two "*Rost*" plants and eight pairs of "*Ofen*" plants started operating and in November twelve more pairs of "*Ofen*" were in use. Although the general location of some of these plants had been reported by intelligence, aerial reconnaissance was unable to confirm that any of them were actually in operation and none was attacked. Their total production up to January 1945 was some 62,000 tons of finished products of which 19,000 tons was motor gasoline and 43,000 tons was diesel oil. None of the other types of plant was ever completed except certain of the "*Wueste*" shale plants, production from which was negligible.

Even if the Germans had begun their whole programme of dispersal and concealment a year sooner it is unlikely that the resultant production of liquid fuels by the beginning of 1945 could have amounted to more than a small proportion of requirements. There might, however, have been a single exception in the case of the J.2 light diesel oil used by jet fighter aircraft. An adequate supply of this low-grade fuel could probably have been forthcoming from numerous distillation plants and their small size, together with the fact that a considerable number of them could have been erected, would have made them difficult targets for air attacks.||

On the other hand high octane aviation spirit could have only been produced in adequate quantities from large plants of a size well within the scope of high level precision bombing. If these plants had been too deeply buried to sustain damage from bombing then the experience of the attacks upon the underground oil storage installations has been sufficient to prove that, given air supremacy, the normal rail outlets to such plants would be vulnerable to dislocation even if the installation itself could not have been damaged.

|| A corollary is that if military aircraft are propelled in the future by low grade oil fuel, the production of such fuel, whether from crude oil or coal, could be achieved by a multiplicity of small subterranean plants that could be rendered reasonably safe from high explosive bombs. A diverse pipeline system with concealed terminals could safeguard the intake of raw materials and the despatch of the finished product.

Index to References.

Abbreviations.

A.C.I.U.	Allied Central Interpretation Unit (at R.A.F. Station, Medmenham).
A.D.I. (K).	Assistant Directorate of Intelligence, Air Ministry, Section "K."
A.M.W.I.S.	Air Ministry Weekly Intelligence Summary.
A.W.I.R.	Admiralty Weekly Intelligence Review.
CG/U.S.S.T.A.F.	Commanding General United States Strategic and Tactical Air Forces in Europe.
C.I.O.S.	Combined Intelligence Objectives Sub-Committee.
C.S.D.I.C.	Combined Services Detailed Interrogation Centre.
C.S.T.C.	Combined Strategic Targets Committee.
D.C.A.S.	Deputy Chief of Air Staff.
E.O.C.	United States Enemy Oil Committee.
F.D.	Foreign Documents Unit of the German Economic Department, of the Foreign Office.
F.I.A.T.	Field Information Agency, Technical, of the Control Commission for Germany.
I.I.C.	Industrial Intelligence Committee.
J.I.C.	Joint Intelligence Committee.
J.P.R.C.	Joint Photographic Reconnaissance Committee.
J.S.M.	Joint Staff Mission, Washington.
P.O.G.	Committee for the Prevention of Oil going to Germany.
R.E. 8.	A section of the Research and Experiments Department of the Ministry of Home Security.
U.S.S.B.S.	United States Strategic Bombing Survey.
W.O.W.I.R.	War Office Weekly Intelligence Review.

Personalities.

Adam, Konteradmiral (Ing.) u. Dipl. Ing.—In command of the Fuel and Transport Group of the Supreme Command, Navy.

Ahrens, Oberstabsingenieur.—Was employed in the Fuel Supply Section of the Department for Fuel Supplies to the Air Force. When this department was reorganised in 1944, oil matters were handed over to an office under the Chief of Armaments, and with this reorganisation he became Head of the Supply Section.

Bayerlein, Generalleutnant.—Commanded the *Panzer Lehr* Division.

Bentz, Professor Dr. A.—General Commissioner for Crude Oil Production. Was Supervisor of all geological and geophysical work both in Germany and in satellite countries.

Blumentritt, General.—Chief of Staff to von Rundstedt when commander of the Forces in Northern France at the time of the Allied Invasion in June 1944.

Brauchitsch, Generalfeldmarschall Walther von.—Commander-in-Chief, Army, 1938 to December 1941.

Brauchitsch, Oberst Bernd von.—Adjutant to Goering as Commander-in-Chief of the Air Force. Held post until end of war.

Brochhaus, Dr. Hans.—Chairman of the Working Committee for Crude Oil Production and Refining. This Committee came under the Department for the Oil Industry in the Raw Materials Division of the Ministry for Armaments and War Production. Formerly General Director of *Elwerath* and *Deurag/Neurag.*

Buetefisch, Dr.-Ing. Heinrich.—A director of the *I.G. Farbenindustrie* in charge of the Oil Division. Also on the Boards of *Braunkohle-Benzin A.G.*, *Ammoniak Merseburg G.m.b.H.*, *Norddeutsche Hydrierwerke A.G.*, *Stickstoffwerke Ostmark A.G.*, *Donau Chemie A.G.* and *Mineraloelbaugesellschaft A.G.* Head of the Economic Group, Fuel Industry in the Raw Materials Office of Ministry for Armaments and War Production.

Buhle, General. In 194[illegible] was Chief of the Army Staff (a Supply appointment) in the Supreme Command of the Armed Forces. In 1945 he was successively Head of the Office for Armaments for the Army and Head of Armed Forces Armaments.

Cramer, General-Major Johann.—Commanded the *Afrika Korps* under Rommel.

Drescher, Major.—Representative of Griebel on Keitel's staff.

Dietrich, S.S. Oberstgruppenfuehrer Sepp.—Commanded the 6th *S.S. Panzer* Army in the concluding stages of the War.

Dihlmann, Walther.—A director of *Kontinentale Oel G.m.b.H.*

Doenitz, Generaladmiral Karl.—Supreme Commander of the Navy from February 1943 until May 1945, when he succeeded Hitler as Chancellor of the Reich.

Dultz, Dr. Wilhelm.—Deputy to Griebel.

Fischer, Dr. phil. Ernst Rudolf.—A director of the *I.G. Farbenindustrie, Deutsche Gasolin A.G., Ammoniakwerk Merseburg G.m.b.H., Oberschlesische Hydrierwerke A.G., Donau-Chemie A.G., Hydrierwerke Poelitz A.G., Apollo-Mineraloel-Raffinerie A.G., Kontinentale Oel A.G.,* and *Dynamit-Nobel A.G.* Head of the Oil Department of the Raw Materials Division (Kehrl) of the Ministry of Economics, which later became the Ministry for Armaments and War Production.

Friedensburg, Ferdinand.—A publicist who has specialised on the subject of oil. His books and articles have been regarded as authoritative.

Galland, General.—Air Officer commanding Fighters, until January 1945.

Geilenberg, Edmund.—Until May 1944, was Head of the Main Committee for Munitions. He was afterwards appointed Special Commissioner for the Repair and Dispersal of Oil Plants.

Goering, Reichsmarschall Hermann.—*Reichsminister* for Air, Commander-in-Chief of the Air Force, Successor Designate to Hitler, Chairman of the War Cabinet. Member of the Secret Cabinet Council, Prime Minister of Prussia, President of the Reichstag, Trustee of the Four-Year Plan, President of the Prussian State Council, Game Warden of the Reich, Chief Forester of the Reich, Head of the Hermann Goering Works.

Griebel, Kapitaen zur See.—The directly responsible member of the O.K.W. (Supreme Command of the Armed Forces) for the supply of liquid fuels and lubricants.

Guderian, General-Oberst.—Chief of Staff of the Army from July 1944 to March 1945. Inspector-General of Armoured Troops from February 1943 to March 1945.

Halder, General-Oberst.—Chief of the Army General Staff from 1938, until September 1942.

Hertslet, G.A.—German Economic Representative in Mexico.

Hettlage, Prof. Dr. jur. habil. Karl M.—In Central Office of Speer Ministry. Was responsible for the Ministry's financial administration, and the financial control of *Reich* industrial enterprises.

Jodl, General-Oberst Alfred.—Head of the Operational Planning Staff of the Supreme Command of the Armed Forces, from 1939 to 1945.

John, Oberst.—Quartermaster-General to von Rundstedt (Commander-in-Chief in the West).

Kaufmann, Karl.—*Reichskommissar* for Shipping.

Keitel, Generalfeldmarschall Wilhelm.—Head of the Supreme Command of the Armed Forces under Hitler from 1938 to 1945.

Kehrl, Praesident Hans.—Director of the Raw Materials and Planning Departments of the Ministry for Armaments and War Production (Speer Ministry). In addition, a director of eighteen companies associated with steel, oil and textiles.

Kolb, General-Major.—Was in charge of Technical Training at the Air Ministry. Head of the Chemical Division of the Raw Materials Office in Speer Ministry.

Koller, Generalleutnant.—Head of the Operations Staff of the Supreme Command of the Air Force until November 1944, and thereafter Chief of the General Staff of the Air Force.

Krauch, Professor Dr. phil. Carl.—Formerly Chairman of the Board of Management of the *I.G. Farbenindustrie.* Appointed in 1936 to a two-fold position in the Government as General Commissioner for Problems of the Chemical Industry and also as Head of the Office for Economic Organisation. These two administrations were concerned, under the organisation of the Four-Year Plan, with the development of oil, nitrogen, rubber and chemicals. The administration eventually became within the orbit of the Speer Ministry.

Kreipe, General der Flieger.—Air Officer Commanding flying training during the twelve months ending July 1944, thereafter Chief of Staff of the Air Force until November 1944, when he became Head of the *Luftkriegsakademie.*

Lammers, Dr. Fritz.—*Reichsminister* and Chief of the *Reich* Chancellery.

Manteuffel, General Hasso von.—Promoted Major-General in 1943 when he took over command of 7th *Panzer* Division formerly commanded by Rommel. In March 1944 was promoted Lieut.-General, commanding *Panzer* Division *Grossdeutschland.* In the autumn of 1944 commanded the 5th *Panzer* Army and later the 3rd *Panzer* Army.

Martin, Professor Dr.-Ing. Friedrich.—Managing Director of Ruhrchemie A.G. and a director of four other companies. Head of the Propellents and Explosives Department in former Armaments Delivery Office (abolished in 1944).

Massow, General-Major von.—Air Officer Commanding flying training. Succeeded Kreipe.

Milch, Generalfeldmarschall Erhard.—Secretary of State for Air 1933, and Inspector-General of the Air Force from 1939 to June 1944. *Generalluftzeugmeister* from December 1941 to June 1944. Deputy to Speer as Minister of Armaments and War Production and Plenipotentiary for Armament in the Four-Years' Plan from July 1944.

Model, Generalfeldmarschall.—Commanded Army Group " B " in the West from September 1944 to 1945.

Nagel, Dipl.-Ing. Jakob.—Head of the Transport Section of the Ministry for Armaments and War Production.

Ritter, Dr. Albert.—Dr. Krauch's executive. On the Boards of some seventeen oil and chemical companies.

Robertelli, Admiral.—Chief of the Italian General Staff.

Rosenkrantz, Major (Retired).—Deputy Chief of Supplies and their Allocation, Distribution and Rationing in the Ministry for Armaments and War Production.

Rossi, General.—Deputy Chief of the Italian General Staff.

Ruhsert, Oberst.—Head of the Allocations Department, Air Ministry, until end of War.

Rundstedt, Generalfeldmarschall Gerd von.—Commander-in-Chief, West, 1942 to July 1944, and from September 1944 to March 1945.

Sauer, Dipl.-Ing. Karl Otto.—Chief of the Armaments Staff of the Ministry of Armaments and War Production.

Schell, Generalleutnant von [illegible] was Chief of Staff to the Inspector of Armoured Troops. In 1940 was made Under-Secretary of State to the Transport Ministry. Later was Commander of the 25th *Panzer* Division in Norway.

Schneider, Dr. Paul.—Technical adviser to Rosenkrantz.

Speer, Albert.—*Reichsminister* for Armaments and War Production from 1942. Joint Chairman (with Milch) of the Central Planning Office.

Spies, General-Ingenieur.—Chief Engineer of *Luftflotte* 2 until autumn 1943. After a period of ill-health was appointed, in August 1944, Chief Engineer of *Luftflotte* 10.

Steinmann, Professor.—Professor of Science and Electrical Energy at the Berlin University. Was Consulting Engineer to the Air Force and, from July 1944 onwards, Head of the Air Force Technical Administration.

Student, General-Oberst.—Commander-in-Chief Paratroops from 1938 and throughout the War.

Thomale, Generalleutnant Wolfgang.—Chief of Staff of the Inspector-General of Tank Troops.

Thomas, General.—Head of *Wirtschafts-Ruestungsamt* in the Supreme Command of the Armed Forces. Later Director of the Department for Questions of War Economy and Armaments in the Speer Ministry.

Udet, Generalluftzeugmeister Ernst.—Head of the Office for Technical Matters in the Air Ministry from 1936 until his death in November 1941.

Veith, Generalleutnant.—Air Officer Commanding *Flak* Training.

Warlimont, General.—Deputy Chief of the Operational Planning Staff of the Supreme Command of the Armed Forces and Head of its Operational Departments from 1939 to July 1944.

Wehling, Franz.—Managing Director of *WIFO*. Also a director of *Kontinentale Oel A.G., Mineraloel Einfuhrgesellschaft m.b.H., Rohstoffhandelsgesellschaft m.b.H., Rumaenien Mineraloel Einfuhrges. m.b.H., Suedostchemie-Handelsgesellschaft m.b.H.*

Westphal, General.—Chief of Staff to von Rundstedt (Commander-in-Chief in the West).

Zeitzler, General-Oberst Kurt.—Chief of German General Staff, replacing General Halder in 1942, until the 21st July, 1944.

Table 1.

Production in Greater Germany.

(In thousands of metric tons.)

—	Aviation Gasoline.	Motor Gasoline.	Diesel Oil.	Fuel Oil.	Lubricating Oils.	Kerosene.	Liquefied Gases.	Miscellaneous Production.†	Total.
				1940.					
Crude refining*	18	135	202	207	451	124	4	313	1,454
Hydrogenation	612	299	365	51	4	...	169	4	1,504
Fischer-Tropsch	...	223	131	...	6	...	52	37	449
Coal tar distillation ...	...	24	83	470	1	...	...	34	612
Alcohol	...	80	...	...	...	...	...	...	80
Benzol	14	364	...	...	...	...	...	175	553
Total	644	1,125	781	728	462	124	225	563	4,652
				1941.					
Crude refining*	11	157	287	192	545	108	4	288	1,612
Hydrogenation	847	319	620	78	6	...	226	11	2,107
Fischer-Tropsch	...	228	119	...	11	...	68	48	474
Coal tar distillation ...	...	26	88	541	2	...	...	35	692
Alcohol	...	60	...	...	...	...	...	...	60
Benzol	31	360	...	...	...	...	...	206	597
Total	889	1,150	1,114	811	584	108	298	588	5,542
				1942.					
Crude refining*	7	174	377	56	657	152	5	301	1,729
Hydrogenation	1,340	292	722	122	17	...	262	17	2,772
Fischer-Tropsch	...	228	97	...	8	...	52	61	446
Coal tar distillation ...	...	35	89	669	1	...	...	36	830
Alcohol	...	6	...	...	...	...	...	...	6
Benzol	40	302	...	...	...	...	...	243	585
Total	1,387	1,037	1,285	847	683	152	319	658	6,368
				1943.					
Crude refining*	4	150	429	53	767	182	5	343	1,933
Hydrogenation	1,745	386	787	135	35	...	323	20	3,431
Fischer-Tropsch	...	254	99	...	15	...	60	56	484
Coal tar distillation ...	...	34	94	820	...	...	...	37	985
Alcohol	...	18	...	...	...	...	...	...	18
Benzol	35	320	...	...	...	...	...	302	657
Total	1,784	1,162	1,409	1,008	817	182	388	758	7,508
				1944.					
Crude refining*	3	145	466	66	614	124	235	‡	1,653
Hydrogenation	996	293	318	68	24	...	176	‡	1,875
Fischer-Tropsch synthesis	...	160	62	...	10	...	74	‡	306
Coal tar distillation ...	...								
Alcohol	...	} 363	65	753	...	...	352	‡	1,578
Benzol	45								
Total	1,044	961	911	887	648	124	837	‡	5,412

* Includes products from imported crude and unfinished oils.
† Includes solvent naphthas, asphalt and paraffin.
‡ Included with liquefied gases.

TABLE 2.

German Crude Oil Production.

(In Metric Tons per Year.)

	Old Germany.	*Austria.*	*Total.*
1940	1,052,000	413,000	1,465,000
1941	927,000*	635,000	1,562,000
1942	817,000*	869,000	1,686,000
1943	776,000*	1,107,000	1,883,000
1944, 1st Quarter (at annual rate) ...	768,000*	1,195,000	1,963,000

* Includes Pechelbronn from 1st July, 1941, at the rate of about 60,000 metric tons per year.

TABLE 3.

Synthetic Oil Plants' Capacities.

(In terms of Finished Liquid Products.)

Hydrogenation.

Plant.	*Capacity.* Tons/Year.*
Leuna	600,000
Boehlen	300,000
Magdeburg	240,000
Zeitz	300,000
Scholven	240,000
Gelsenberg	420,000
Welheim	150,000
Poelitz	600,000
Lutzkendorf	48,000
Wesseling	204,000
Ludwigshafen-Oppau	48,000
Moosbierbaum	90,000
Bruex	600,000
Blechhammer†	165,000
Heydebreck†	40,000
Auschwitz†	None
Total	4,045,000

† The ultimate designed capacity was Blechhammer 396,000, Heydebreck 90,000 and Auschwitz 30,000 tons per year.

Fischer Tropsch.

Plant.	*Capacity.* Tons/Year.*
Sterkrade	60,000
Castrop	40,000
Homberg	66,000
Wanne Eickel	66,000
Kamen	85,000
Dortmund	48,000
Ruhland	170,000
Deschowitz	40,000
Lutzkendorf	12,000
Total	587,000

* Capacities are based on approximate installed capacity as at the end of the war.

Table 4.

Capacities of Refineries.

(Metric Tons per Year throughput—Capacity before June 1944.)

Name.	Location.	Topping.	Cracking.	Vacuum Distillation.	Dewaxing.	Selective Refining.	Propane-Deasphalting.	Acid Refining.
Germany—								
†Rhenania (Shell)	Harburg (Hamburg)	300,000	18,000 (B)	530,000	150,000 (3)	120,000 (13)	...	90,000
†Rhenania (Shell)	Grasbrook (Hamburg)	...	...	...	...	120,000 (12)	...	80,000
Rhenania (Shell)	Monheim	(70,000)	...	120,000	...	6,000 (12)	...	65,000
†Nerag–Deurag	Misburg	500,000	150,000 (A)	370,000*	150,000 (3)	180,000 (11)	...*	50,000*
†Wintershall	Luetzkendorf	120,000	...	...*	50,000 (4)	40,000 (14)	...*	...*
Wintershall	Salzbergen	60,000	60,000	...*	20,000 (5)	...	...	...*
Deutsche Gasolin	Emmerich	70,000	...	70,000	...	...	...	...*
Deutsche Gasolin	Dollbergen	50,000	...	50,000*	20,000 (7)	...	...	...*
Deutsche Erdoel A.G.	Wilhelmsburg	120,000	...	...*	50,000 (9)	3,000 (14)	...	...*
Deutsche Erdoel A.G.	Heide	200,000	...	...	...	...	...	...
Phoenix–Sengewald	Hannover	20,000	...	...*	...	...	...	...
Mineraloel-und Asphalt A.G.	Ostermoor	120,000	...	...*	...	...	...	...
Oelwerke J. Schindler	Hamburg	...	...	30,000	...	...	...	...*
Oelwerke J. Schindler	Peine	16,800	...	...	...	...	...	...?
Ebano	Harburg (Hamburg)	540,000	...	...*	...	...	...	...
Eurotank	Hamburg	180,000	120,000	...	...	...	...	...
I.G.	Oppau	15,000	...	...*	4,000 (6)	...	...	...*
†Vacuum Oel A.G.	Oslebshausen	120,000	...	...*	30,000 (3)	60,000 (10)	...	...
Vacuum Oel A.G.	Wedel (Hamburg)	...	...	30,000	15,000 (7)	...	...	...*
Pechelbronn	Pechelbronn	120,000	...	...?	...?	...?	...	...?
Ernst Schliemann's Oelwerke	Hamburg	...	...	...	...	...	...	...*
Mineraloelwerke Albrecht	Hamburg	...	...	...	...	...	...	...*
2. Austria—								
Shell-Floridsdorfer	Wien-Floridsdorf	100,000	...	30,000	5,000 (7)	...	...	...*
Vacuum Oel A.G.	Kagran	60,000	...	...*	...	...	...	...*
Dea-Nova	Schwechat	180,000	100,000	40,000	...	...	...	...*
Benzol-Verband	Voesendorf	100,000	...	30,000	...	...	...	...*
Deutsche Gasolin	Korneuburg	80,000	...	30,000	...	...	...	...*
I.G.	Moosbierbaum	150,000	...	...*	...* (6)	...?	...?	...*
Ostmaerkische Mineraloelwerke	Lobau	300,000	...	...	...	...	...	...

TABLE 4 (*continued*).

Name.	Location.	Topping.	Cracking.	Vacuum Distillation.	Dewaxing.	Selective Refining.	Propane-Deasphalting.	Acid Refining.
3. Protectorate and Slovakia—								
I.G.	Presburg	200,000	100,000 (A)	...*	25,000 (6)	...	...	...*
Oderfurter Mineraloelwerke	Oderfurt	70,000	...	...*	20,000 (1)	20,000 (15)	...	...*
†Vacuum Oel A.G.	Kolin	120,000	...	120,000*	30,000 (8)	40,000 (10)	...	...*
Fanto	Pardubitz	150,000	...	90,000	40,000 (1)	...*	...*	...*
"Oderberger"	Oderberg	...*	...	...?	...?	...	...	...?
Galicia and General Government—								
Kontinentale Oel (?)	Trzebinia	270,000	...	165,000	{ 20,000 Pressen 40,000(2) in Bau	} 40,000 (10)	...	32,000
Vacuum Oel A.G.	Dzieditz	100,000	...	...*	Press-Dewaxing	...	...	...*
5 Refineries in	{ Drohobycz	340,000	...?	...*	" "	...	...	...*
	Jaszlo	70,000	...?	...*	" "	...	...	...*

* Denotes capacity not exactly known.

† Denotes modern refineries that could be expected to run to capacity at all times.

(A) Dubbs.

(B) Paraffin cracking for lube oil synthesis.

(1) Barisol.
(2) Edeleanu—
 Dichlorethane.
 Methylenchloride.
(3) B.A.T.
(4) Propane.

(5) Schmitz-Kell.
(6) I.S. (Ester).
(7) Pressen E.P.
(8) Bi-Zentrifug.
(9) Dichlorethane.

(10) Duosol.
(11) Furfurol.
(12) SO_2—Edeleanu.
(13) B-SO_2 in Bau.
(14) Phenol.
(15) S.N.P.

TABLE 5.

Statistics of the Roumanian Oil Industry.

(In Thousands of Metric Tons.)

Year.	Crude Oil Output.	Drilling (Thousand metres).	Refinery Runs.	Domestic Consumption.	Exports.
1938	6,610	288	6,228	1,674	4,495
1939	6,240	256	5,837	1,785	4,178
1940	5,810	235	5,472	1,862	3,493
1941	5,577	253	5,255	1,811	4,072
1942	5,665	339	5,237	2,098	3,374
1943	5,266	344	4,903	2,007	3,150

TABLE 6.

Roumanian Oil Exports By Products.

(Metric Tons.)

Year.	Crude Oil.	Gasoline.	Kerosine and Wh. Spirit.	Gas Oil.	Fuel Oil.	Lub. Oil.	Other Products.
1938	335,437	1,586,357	827,207	754,918	917,267	40,580	32,996
1939	329,168	1,595,341	793,523	626,041	770,202	41,247	22,059
1940	300,414	1,333,500	623,719	664,699	521,098	34,097	15,410
1941	389,278	1,808,965	520,603	806,791	410,666	60,789	75,214
1942	426,208	1,474,036	322,005	587,163	344,747	119,198	100,187
1943	361,552	1,486,639	348,230	421,142	356,662	86,054	98,889

TABLE 7.

Roumanian Oil Exports by Countries.

(Metric Tons.)

—	1938.	1939.	1940.	1941.	1942.	1943.
Germany*	999,240	1,285,153	1,429,807	2,885,229	1,822,207	1,795,555
German Army	...	...	...	34,351	369,452	715,749
Italy†	560,475	629,350	342,943	761,667	862,179	391,354
Bulgaria	79,768	93,744	95,151	53,057	43,394	21,559
Greece	200,215	75,293	187,304	10,161	30,622	25,967
Switzerland...	88,873	113,801	92,481	107,268	87,910	57,605
France	289,338	238,062	87,144	19,412	82,433	14,664
Turkey	53,616	30,424	148,267	57,939	1,380	12,794
Hungary	198,076	165,016	34,643	277	...	...
Other Countries	1,643,604	1,243,136	910,535	125,361	57,693	59,845
Bunker Sales	381,557	303,592	164,662	17,584	16,272	64,076
	4,494,762	4,177,571	3,492,937	4,072,306	3,373,542	3,159,168

* Including Czechoslovakia.
† Including Albania.

TABLE 8.

Hungarian Crude Oil Production.

(In Metric Tons.)

1936	450	1941	421,700
1937	2,200	1942	665,200
1938	42,800	1943	837,710
1939	143,200	1944	809,970
1940	251,400		

TABLE 9.

Actual Output Achieved by the Geilenberg Dispersal Plants.

(Finished Products in Metric Tons.)

	1944. Sept.	Oct.	Nov.	Dec.	*1945.* Jan.	Feb.	Mar.
Rost 1	300	2,000	...	...	60	No figures available.	
„ 2	...	60	200	90	...		
„ 3	...	...	...	400	900		
Ofen 1/2	...	800	2,200	2,700	2,300		
„ 3/4	...	700	1,400	1,700	2,600		
„ 5/6	...	...	400	800	900		
„ 7/8	...	...	400	700	700		
„ 9/10	...	...	800	1,400	1,100	Total planned production for February and March was 11,000 tons and 17,000 tons respectively.	
„ 11/12	...	200	200	800	1,100		
„ 13/14	...	100	200	100	...		
„ 15/16	...	70	30	...	300		
„ 17/18	...	200	400	50	700		
„ 19/22	...	900	1,500	1,700	2,300		
„ 33/34	...	400	800	1,700	...		
„ 37/38	...	...	200	600	1,600		
Total	300	5,430	8,530	12,740	14,460		

TABLE 10.

Approximate Allocation of German Civil Consumption in 1938.

	Motor Gasoline. %	Diesel Oil. %	Fuel Oil. %
Road Transport	88·0	29·0	...
Railways	...	3·5	...
Shipping	0·3	18·8	60·0
Agriculture	1·2	12·3	...
Civil Aviation	0·8	...	...
Industry	9·7	33·0	36·0
Domestic	...	3·4	4·0
	100·0	100·0	100·0

TABLE 11.

Consumption of Alcohol as Motor Fuel.

(In Metric Tons.)

Year.	Ethyl Alcohol.	Methyl and Other Alcohols.	Total.
1935	165,900	21,600	187,500
1936	144,000	51,600	195,600
1937	125,900	63,000	188,900
1938	199,300	12,900	211,200

TABLE 12.

Consumption of Principal Oil Products in Germany.
(Based upon Deliveries ex Stock.)

(In Thousands of Metric Tons.)

Date.	Aviation Spirit.		Motor Spirit.		Diesel Oil.*		Fuel Oil.	
	Military.	Civil Economy.	Military.	Civil Economy.	Military.	Civil Economy.	Military.	Civil Economy.
1936	9†	3†	7†	171†	5†	91†	21†	36†
1937	13†	5†	9†	176†	8†	103†	17†	40†
1938	18†	7†	15†	213†	10†	136†	44†	33†
1939	27†	10†	31†	192†	17†	105†	44†	52†
1940—								
January	27	...	56	72	17	59		
February	36	...	42	71	19	59		
March	41	...	38	76	14	78		
April	78	...	40	84	16	97		
May...	100	...	97	65	34	91		
June	90	...	118	57	37	70		
July...	78	...	76	64	22	90		
August	99	...	74	67	21	98	No data available.	
September	91	...	93	68	24	104		
October	86	...	104	81	29	108		
November	74	...	110	81	29	102		
December	63	...	78	65	25	72		
	863	...	926	851	287	1,028		
1941—								
January	47	...	90	65	26	64	53	14
February	51	...	66	62	20	61	53	14
March	95	...	88	63	23	83	53	14
April	91	...	86	59	35	86	53	14
May...	126	...	91	58	30	89	53	14
June	132	...	268	50	80	67	53	14
July...	137	...	153	55	58	88	53	14
August	129	...	194	55	89	90	53	14
September	122	...	174	45	61	96	53	14
October	115	...	158	44	34	93	53	14
November	88	...	109	43	57	77	53	14
December	73	...	133	41	70	52	53	14
	1,206	...	1,610	640	583	946	636	168
1942—								
January	76	...	90	35	61	42	62	10
February	74	...	84	33	57	37	62	10
March	98	...	86	35	31	57	62	10
April	106	...	118	32	26	62	62	10
May	126	...	117	30	34	60	62	10
June	128	...	102	31	26	46	62	10
July...	110	14	173	30	43	53	62	10
August	118	14	152	25	64	64	62	10
September	103	12	128	22	65	74	62	10
October	94	14	140	27	57	63	62	10
November	98	14	124	22	27	54	62	10
December	91	12	133	23	68	37	62	10
	1,217	80	1,477	345	559	649	744	120

* Excludes Marine Diesel. † Monthly Average.

(In Thousands of Metric Tons.)

Date.	Aviation Spirit		Motor Spirit.		Diesel Oil*		Fuel Oil.	
	Military.	Civil Economy.	Military.	Civil Economy.	Military.	Civil Economy.	Military.	Civil Economy.
1943—								
January	95	14	103	21	68	31	79	6
February	116	15	131	20	69	32	79	6
March	149	18	130	24	78	49	79	6
April	129	16	101	24	66	49	79	6
May...	130	18	114	26	69	49	79	6
June	130	17	108	25	74	38	79	6
July...	176	18	153	26	79	50	79	6
August	149	19	153	28	81	60	79	6
September	157	...	165	27	79	60	79	6
October	155	...	169†	28†	77†	63†	79	6
November	120	...	151†	24†	73†	55†	79	6
December	112	...	145†	25†	83†	34†	79	6
	1,618	135	1,623	298	896	570	948	72
1944—								
January	122	...	147	26	92	23	80	5
February	135	...	139	27	95	42	84	5
March	156	...	140	31	101	33	89	6
April	164	...	130	30	100	58	83	5
May...	195	...	112	31	85	25	93	5
June	182	...	138	44	71	48	49	4
July...	136	...	150	26	98	23	80	4
August	115	...	180	23	84	44	42	4
September	60	...	95	14	47	25	86	4
October	53	...	79	12	41	33	49	4
November	41	...	65	10	47	25	74	4
December	44	...	79	11	56	26	45	4
	1,403	...	1,454	285	917	405	854	54

* Excludes Marine Diesel. † Estimated.

Table 13.

German Civilian Consumption of Automotive Fuels.

(In Thousands of Metric Tons per Year.)

——	Pre-war.	1940.	1941.	1942.	1943.	1944.
Motor gasoline	2,400	851	640	345	289	285
Diesel oil	1,500	1,028	946	649	570	405
Total liquid fuel	3,900	1,879	1,586	994	868	690
Bottled gas*	108	225	298	319	388	210
Generators fueled with wood, anthracite, etc.*	...	...	20	125	245	370
Methane and gas*	...	1	1	2	12	42
Total non-liquid fuels*	108	226	319	446	645	622
Grand Total	4,008	2,105	1,905	1,440	1,513	1,312
Percentage of pre-war consumption ...	100	53	48	36	38	33
Additional wood-generator fuel used by army*	...	...	...	†	75	130

* Non-liquid fuel tonnages given in terms of equivalent gasoline.
† Small amount included in the 125 civilian consumption above.

TABLE 14.

Motor Gasoline Allocations to the Civil Economy.

(In Metric Tons.)

Date.	Regional Economic Offices (*Landeswirtschaftämter*)	Agriculture.	Postal Service and Railways.	War Plants and Construction.	Reserve.	Miscellaneous Consumption.	Total Allocations.	Actual Consumption.
1941—								
January ...	46,400	...	3,700	10,900	2,200	4,200	67,400	65,000
February	41,800	...	3,700	10,100	2,000	1,500	59,100	62,000
March ...	36,600	...	3,300	8,900	1,500	1,200	51,500	63,000
April ...	36,600	...	3,000	8,900	1,700	1,300	51,500	59,000
May ...	36,000	...	3,100	8,900	2,100	1,300	51,400	58,000
June ...	35,200	...	3,200	9,200	2,200	1,600	51,400	50,000
July ...	34,600	...	3,000	7,300	2,900	1,300	49,100	55,000
August ...	34,500	...	3,000	8,000	2,300	1,300	49,100	55,000
September	33,200	...	2,800	7,800	1,800	1,400	47,000	45,000
October ...	30,300	...	2,400	7,600	1,900	1,300	43,500	44,000
November	30,400	...	2,400	7,600	2,300	1,300	44,000	43,000
December	26,200	...	2,300	3,800	2,000	900	35,200	41,000
Total ...	421,800	...	35,900	99,000	24,900	18,600	600,200	640,000
1942—								
January ...	26,300	...	2,200	4,000	1,900	1,100	35,500	35,000
February	25,100	...	1,800	3,700	3,600	900	35,100	33,000
March ...	24,100	...	1,800	3,500	3,400	900	31,700	35,000
April ...	24,100	...	1,800	3,500	1,400	1,000	31,800	32,000
May ...	21,300	...	1,600	3,100	1,300	1,200	28,500	30,000
June ...	20,700	...	1,600	3,200	1,400	1,000	27,900	31,000
July ...	21,400	...	1,600	3,200	1,400	900	28,500	30,000
August ...	18,200	...	1,300	2,600	1,600	700	24,400	25,000
September	17,000	...	1,200	2,500	2,300	700	24,500	22,000
October ...	17,200	...	1,200	2,900	1,900	900	24,100	27,000
November	15,300	2,100	1,200	2,900	1,300	700	23,500	22,000
December	14,300	1,300	1,100	2,500	1,700	600	21,500	23,000
Total ...	245,000	3,400	18,400	37,600	21,200	10,600	337,000	345,000
1943—								
January ...	14,300	1,300	1,000	2,000	1,200	1,000	20,800	21,000
February	13,300	1,300	900	1,700	900	1,000	19,100	20,000
March ...	13,900	1,500	1,000	1,800	1,400	1,000	20,600	24,000
April ...	13,200	1,500	1,000	1,800	1,600	1,000	20,100	24,000
May ...	13,000	1,400	1,000	1,800	1,800	1,000	20,000	26,000
June ...	12,500	1,800	1,000	1,900	1,700	1,000	19,900	25,000
July ...	10,900	1,800	800	1,800	1,500	1,000	17,800	26,000
August ...	10,600	2,100	700	1,700	1,700	1,000	17,800	28,000
September	10,100	2,000	700	1,600	3,100	1,000	18,500	27,000
October ...	9,600	1,500	700	1,600	4,000	1,000	18,400	28,000
November	9,600	1,300	600	2,100	3,800	1,000	18,400	24,000
December	9,100	1,000	600	2,000	3,600	1,000	17,300	25,000
Total ...	140,100	18,500	10,000	21,800	26,300	12,000	228,700	298,000
1944—								
January ...	9,100	1,000	600	2,400	3,400	1,300	17,800	26,000
February	8,600	1,000	600	2,400	3,000	1,100	16,700	27,000
March ...	9,200	1,400	600	2,500	3,100	1,200	18,000	31,000
April ...	9,100	1,600	600	2,600	2,900	1,100	17,900	30,000
May ...	9,100	1,600	700	2,400	2,900	1,400	18,100	31,000
June ...	8,600	1,700	600	2,200	1,100	1,200	15,400	44,000
July ...	7,700	1,600	400	1,900	1,900	800	14,300	26,000
August ...	7,300	1,700	400	1,800	2,300	800	14,300	23,000
September	3,500	900	300	1,330	1,100	1,500	8,600	14,000
October ...	2,800	1,100	300	1,300	1,500	1,800	8,800	12,000
November	2,500	1,100	300	1,200	1,600	1,500	8,200	10,000
11 Months' Total ...	77,500	14,700	5,400	22,000	24,800	13,700	158,100	274,000

TABLE 15.

Diesel Oil Allocations to the Civil Economy.

(In Metric Tons.)

—	Regional Economic Offices (*Landeswirtschaftämter*).	Agriculture.	River and Ocean Shipping.	Postal Service and Railroads.	War Plants and Construction.	Reserve.	Miscellaneous Consumption.	Total Allowance.	Actual Consumption.
1941—									
January ...	29,900	9,800	6,800	4,800	15,900	3,000	1,100	71,300	64,000
February ...	27,800	10,000	4,100	4,800	15,400	4,000	900	67,000	61,000
March ..	26,100	25,900	7,000	4,600	15,300	2,500	900	82,300	83,000
April ...	25,600	26,600	7,500	4,600	15,400	5,800	1,100	86,600	86,000
May ...	25,400	22,400	8,400	4,600	15,400	2,000	1,100	79,300	89,000
June ...	24,000	13,800	8,000	4,500	16,300	2,300	1,400	70,300	67,000
July ...	22,400	23,000	7,200	4,100	15,000	3,700	1,200	76,600	88,000
August ...	22,300	29,400	7,700	4,100	15,400	1,800	1,200	81,900	90,000
September...	21,700	39,000	8,900	4,000	14,700	2,200	1,100	91,600	96,000
October ...	21,000	38,300	8,700	3,700	14,500	2,300	1,000	89,500	93,000
November ...	18,900	30,000	7,800	3,300	13,000	1,600	1,000	75,600	77,000
December ...	17,100	10,000	6,500	3,200	7,500	4,000	900	49,200	52,000
Total ...	282,200	273,200	88,600	50,300	173,800	35,200	12,900	921,200	946,000
1942—									
January ...	17,300	7,400	5,700	3,200	7,800	3,400	900	43,700	42,000
February ...	16,600	7,000	4,700	3,200	7,200	6,300	700	43,700	37,000
March ...	17,000	22,500	6,000	3,200	7,900	4,700	800	62,100	57,000
April ...	15,200	26,000	5,400	2,900	5,400	1,800	800	57,500	62,000
May ...	14,600	24,000	5,300	2,800	5,400	1,700	900	54,700	60,000
June ...	14,100	11,500	6,400	2,600	5,100	1,600	900	42,200	46,000
July ...	13,900	24,000	8,100	2,500	4,900	1,600	1,100	56,100	53,000
August ...	13,200	33,000	7,900	2,400	4,600	1,900	1,100	64,100	64,000
September...	13,200	40,400	6,600	2,400	4,500	3,300	1,100	71,500	74,000
October ...	12,400	36,900	6,000	2,300	4,800	3,000	900	66,300	63,000
November ...	12,400	27,000	6,000	2,300	4,600	2,800	800	53,900	54,000
December ...	12,100	7,500	5,800	2,200	4,300	2,200	700	34,800	37,000
Total ...	172,000	267,200	73,900	32,000	66,500	34,300	10,700	656,600	649,000
1943—									
January ...	12,200	7,000	4,600	2,200	3,400	1,600	1,600	32,600	31,000
February ...	11,500	8,000	4,600	2,100	3,300	2,300	1,800	33,600	32,000
March ...	11,900	18,000	5,800	2,200	3,300	2,300	1,800	45,300	49,000
April ...	11,500	23,000	5,100	2,100	3,100	2,400	1,900	49,100	49,000
May ...	12,000	19,000	5,100	2,100	3,100	2,300	2,200	45,800	49,000
June ...	10,700	10,000	5,100	2,000	3,000	2,400	2,400	35,600	38,000
July ...	10,000	20,000	5,400	1,900	2,800	2,000	2,400	44,500	50,000
August ...	9,600	30,000	6,700	1,800	2,700	1,700	2,400	54,900	60,000
September...	8,600	35,000	6,000	1,700	2,500	2,000	2,300	58,100	60,000
October ...	8,800	32,000	6,000	1,700	2,500	1,700	2,300	55,000	63,000
November ...	8,600	22,000	5,700	1,700	2,600	2,300	2,000	44,900	55,000
December ...	8,300	7,500	4,700	1,600	2,500	3,300	1,900	29,800	34,000
Total ...	123,700	231,500	64,800	23,100	34,800	26,300	25,000	529,200	570,000
1944—									
January ...	8,300	6,000	4,700	1,600	2,400	3,400	3,700	30,100	23,000
February ...	8,200	8,000	4,600	1,600	2,300	3,400	2,500	30,600	42,000
March ...	8,600	1,700	7,300	1,500	2,400	3,500	2,700	27,700	33,000
April ...	8,400	22,000	7,400	2,000	2,600	3,500	2,500	48,400	58,000
May ...	8,300	16,000	6,900	2,000	2,400	3,300	3,000	41,900	25,000
June ...	6,800	5,000	6,400	1,500	2,800	1,700	3,400	27,600	48,000
July ...	5,000	6,000	4,000	1,100	2,100	1,600	2,800	22,600	23,000
August ...	6,900	8,000	4,400	1,600	2,200	2,400	3,100	28,600	44,000
September...	7,100	5,000	4,400	1,600	1,200	2,000	4,200	25,500	25,000
October ...	6,600	5,000	4,800	1,400	1,500	1,900	5,500	26,700	33,000
November ...	6,300	4,000	4,100	1,300	1,500	2,200	5,200	24,600	25,000
December ...	?	?	?	?	?	?	?	?	?
11 Months' Total ...	80,500	86,700	59,000	17,200	23,400	28,900	38,600	334,300	379,000

TABLE 16.

Civilian and Industrial Lubricating Oil Consumption.

Greater Germany (excluding Protectorate).

(In Metric Tons.)

(Figures for 1939 and 1940 not available.)

Type.	1938.	1941.	1942.	1943.	1944.	Stocks of Main Products.	
						31.12.44	28.2.45
Spindle	81,766	56,515	52,621	49,620	38,550	17,061	14,997
Machine	142,510	98,286	83,800	81,453	79,000	24,321	24,317
Cylinder	34,585	29,011	27,940	26,592	23,129	5,132	4,041
Motor	155,090	45,908	36,981	37,259	40,507	8,206	6,957
Turbine	4,355	3,906	2,493	1,885	1,986		
Railway axle ...	30,480	41,964	49,472	61,872	58,385		
Black	9,875	7,618	8,666	12,999	23,472		
Cutting	10,740	22,035	22,738	22,143	19,694		
White	10,450	10,222	5,506	3,708	2,338	11,293	5,186
Transparent ...	16,480	18,406	17,338	13,570	12,996		
Insulating (cable)	4,395	5,144	4,229	2,648	2,017		
Other	7,820	5,980	5,789	3.866	2,628		
Grease	11,060	7,189	6,934	6,613	5,983		
	519,600	351,912	324,507	324,228	310,685	66,013*	55,498*

* Excluding stocks with jobbers at approximately 20,000 tons.

Source: Arbeitsgemeinschaft Schmierstoff-Verteilung G.m.b.H., Hamburg, 5th May, 1945.

TABLE 17.

Use of Generator Fuels as Gasoline Substitutes in Germany.

—	1942.			1943.			1944.			January to April, 1945.		
	Total Used.	Gasoline Equivalent Tons.	Per-centage of Total.	Total Used.	Gasoline Equivalent Tons.	Per-centage of Total.	Total Used.	Gasoline Equivalent Tons.	Per-centage of Total.	Total Used.	Gasoline Equivalent Tons.	Per-centage of Total.
Wood (c.m.)	1,200,000	90,000	72	3,000,000	225,000	70	4,400,000	330,000	66	1,700,000	127,500	69
Peat (tons)	30,000	7,500	6	50,000	12,000	4	58,000	15,000	3	12,000	3,000	1½
Charcoal and Peat Coke tons	...	...	...	1,000	600	...	7,000	4,500	1	1,000	600	...
Anthracite Coal (tons)	44,000	22,000	17	104,000	52,000	16	120,000	60,000	12	22,000	11,000	6
L.T.C Coke (tons) ...	...	...	...	30,000	13,000	4	40,000	18,000	4	6,000	2,600	1½
Brown Coal Brikette ...	18,000	6,000	5	53,000	18,000	6	200,000	70,000	14	120,000	40,000	22
Total	...	125,500	100	...	320,600	100	...	497,500	100	...	184,700	100

1 kg. gasoline = 0·8 kg. diesel oil = 0·0133 c.m. wood (about 4 kg.) = 4 kg. peat = 3 kg. brown coal = 2 kg. anthracite = 2·3 kg. L.T.C. coke = 1·6 kg. charcoal.

TABLE 18.

Over-all Attack Data of Combined Strategic Air Forces in the European Theatre of Operations.

(In Short Tons—2,000 lbs.)

—	Group I: Cities and Areas.	Group II: Industry.	Group III: Trans-portation.	Group IV: Tactical Targets.	Total.	Total Tons on Oil Targets.	Per cent. Dropped on Oil Targets.
Pre-Oil Offensive (up to May 1, 1944)—							
U.S.A.A.F.	44,141	38,165	30,452	63,796	176,554	1,752	0·10
R.A.F.	227,093	27,190	40,093	38,276	332,652	3,918	1·18
Total	271,234	65,355	70,545	102,072	509,206	5,670	1·11
Oil Offensive (May 1, 1944, to May 8, 1945)—							
U.S.A.A.F.	87,845	205,540	351,953	173,444	818,782	136,109	16·62
R.A.F.	317,767	114,654	98,985	127,029	658,435	98,697	14·99
Total	405,612	320,194	450,938	300,473	1,477,217	234,806	15·90
Totals for both Periods—							
U.S.A.A.F.	131,986	243,705	382,405	237,040	995,336	137,861	13·85
R.A.F.	544,860	141,844	139,078	165,305	991,087	102,615	10·35
Total	676,846	385,549	521,483	402,545	1,986,423	240,476	12·11
Monthly Totals during Oil Offensive U.S.A.A.F.-R.A.F.—							
Oil Offensive—							
1944—							
May...	17,546	13,600	38,159	34,438	103,743	5,571	5·37
June	32,080	20,073	36,431	53,772	142,356	17,033	11·97
July...	64,767	31,270	22,265	22,732	141,034	22,831	16·19
August	44,544	36,266	21,372	42,973	145,155	26,484	18·25
September... ...	25,239	20,322	26,335	45,679	117,575	13,585	11·55
October	47,919	26,782	28,945	20,042	123,688	13,950	11·28
November	42,750	36,325	24,121	11,335	114,531	35,558	31·05
December	21,033	21,227	59,640	7,699	109,599	15,779	14·40
1945—							
January	12,358	17,969	44,597	3,827	78,751	15,891	20·18
February	29,441	28,542	55,001	5,597	118,581	24,427	20·60
March	40,465	49,072	52,176	28,659	170,372	36,690	21·54
April	27,236	18,746	41,819	23,689	111,490	7,007	6·28
May...	234	...	77	31	342	...	...
Total	405,612	320,194	450,938	300,473	1,477,217	234,806	15·90

NOTE: Classes of targets included in the four groups above—

Group I: Cities and Areas.
- Unidentified Targets.
- Cities, Towns and Urban Areas.

Group II: Industry.
- Public Utilities.
- Government Buildings.
- General Manufacturing.
- Aircraft Factories.
- Armament and Ordnance Plants.
- Machinery and Equipment.
- Iron and Steel.
- Light Metal.
- Chemicals.
- Radio and Radar.
- Railroad Manufacturing Works.
- Rubber.
- Oil.
- Shipbuilding.

Group III: Transportation.
- Communication Facilities.

Group IV: Tactical Targets.
- Naval Installations.
- Airfields and Aircraft.
- Ground Support.

NOTE: Tables 18, 19 and 20 prepared by U.S.S.B.S.

TABLE 19.

Summary of Oil Capacity, Production and Attack Data for Greater Germany.

Type of Plant.	Hydro-genation.	Fischer Tropsch.	Refineries.	Benzol and Misc.	Total.
Number of plants attacked	16	9	40	22	87
Capacity in thousand tons per year ...	4,041	587	2,000*	1,632†	8,270
Percentage of total German capacity	49·0	7·1	24·2	19·7	100·0
Average output in 4 months, 1944 (*i.e.*, before attacks) in thousand tons	316	43	167	136	662
Percentage of total German output over period	47·8	6·7	25·2	20·3	100·0
Tons dropped by U.S.A.A.F.	50,650	7,462	35,719	5,009	98,840
Tons dropped by R.A.F.	36,298	29,176	9,379	12,148	87,001
Total tons dropped	86,948	36,638	45,098	17,157	185,841
Percentage of grand total tons dropped	46·8	19·7	24·3	9·2	100·0
Number of H.E. bombs U.S.A.A.F. ...	273,942	34,968	181,151	18,451	508,512
Number of H.E. bombs R.A.F. ...	107,002	95,685	28,052	33,203	263,942
Total	380,944	130,653	209,203	51,654	772,454
Attacks by U.S.A.A.F.	132	32	156	27	347
Attacks by R.A.F.	53	56	20	29	158
Total attacks	185	88	176	56	505
Percentage of attacks by U.S.A.A.F. by process	71·4	36·4	88·6	48·2	68·7
Percentage of attacks by R.A.F. by process	28·6	63·6	11·4	51·8	31·3
Percentage of total number of attacks by process	36·6	17·4	34·9	11·1	100·0
Average tonnage per attack (U.S.A.A.F.)	384	233	229	186	285
Average tonnage per attack (R.A.F.) ...	685	521	469	419	551
Average over-all tonnage per attack ...	470	416	256	306	368
Average weight per bomb (H.E.) pounds—					
U.S.A.A.F.	370	427	394	543	388
R.A.F.	678	610	669	731	660
Average weight all bombs H.E. (pounds)	457	561	431	664	482
Percentage of total tons dropped by U.S A.A.F.	...	...	...	...	53·2
Percentage of total tons dropped by R.A.F.	...	...	...	...	46·8

* Refineries had 3,000,000 tons crude distilling capacity and 5,000,000 tons including intermediate running capacity.

† Based on monthly production.

NOTE.—Capacity tons figured in metric tons. Bombing tons figured in short tons. All bomb tonnages from 1st May, 1944, through 8th May, 1945 (high explosive bombs only—approximately 5,400 tons of incendiaries not included). Tonnages for targets outside Greater Germany are *not* included.

TABLE 20.

Attack Data for Oil Targets in Greater Germany by Type of Target.

	Column I. Fischer Tropsch Plants.			Column II. Hydrogenation Plants.			Column III. Oil Refineries.			Column IV. Benzol Plants			Column V. Oil Storage and Depots.			Total.		
	U.S.A.A.F.	R.A.F.	Total Tons.	U.S.A.A.F.	R.A.F.	Total Tons.	U.S.A.A.F.	R.A.F.	Total Tons.	U.S.A.A.F.	R.A.F.	Total Tons.	U.S.A.A.F.	R.A.F.	Total Tons.	U.S.A.A.F.	R.A.F.	Total Tons.
PRE-OIL OFFENSIVE.																		
1940—																		
January	...	...	...	...	...	...	...	...	...	...	...	...	...	(1) 1	(1) 1	...	(1) 1	(1) 1
April	...	...	...	...	...	...	...	...	...	...	...	...	...	(3) 15	(3) 15	...	(3) 15	(3) 15
May	...	(3) 5	(3) 5	...	(5) 10	(5) 10	...	(10) 89	(10) 89	...	...	...	...	...	...	...	(18) 104	(18) 104
June	...	(12) 117	(12) 117	...	(8) 67	(8) 67	...	(28) 149	(28) 149	...	(1) 14	(1) 14	...	(6) 33	(6) 33	...	(55) 380	(55) 380
July	...	(14) 56	(14) 56	...	(11) 51	(11) 51	...	(34) 120	(34) 120	...	...	...	...	(6) 16	(6) 16	...	(65) 243	(65) 243
August	...	(12) 60	(12) 60	...	(22) 170	(22) 170	...	(14) 64	(14) 64	...	(1) 2	(1) 2	...	(11) 81	(11) 81	...	(60) 377	(60) 377
September	...	(1) 2	(1) 2	...	(11) 59	(11) 59	...	(4) 7	(4) 7	...	...	...	...	(7) 24	(7) 24	...	(23) 92	(23) 92
October	...	(7) 20	(7) 20	...	(33) 203	(33) 203	...	(24) 136	(24) 136	...	(1) 2	(1) 2	...	(3) 5	(3) 5	...	(68) 366	(68) 366
November	...	(6) 19	(6) 19	...	(18) 155	(18) 155	...	(11) 49	(11) 49	...	(1) 14	(1) 14	...	(2) 2	(2) 2	...	(38) 239	(38) 239
December	...	...	...	...	(7) 60	(7) 60	...	(1) 1	(1) 1	...	...	...	...	(2) 10	(2) 10	...	(10) 71	(10) 71
1941—																		
January	...	...	...	...	(1) 63	(1) 63	...	(2) 9	(2) 9	...	...	...	...	(3) 9	(3) 9	...	(6) 81	(6) 81
February	...	(3) 92	(3) 92	...	(1) 9	(1) 9	...	...	...	...	...	...	...	(1) 1	(1) 1	...	(5) 102	(5) 102
March	...	...	...	...	(1) 99	(1) 99	...	...	...	...	...	...	...	...	...	...	(1) 99	(1) 99
April	...	...	...	...	...	...	...	...	...	...	...	...	...	(1) 2	(1) 2	...	(1) 2	(1) 2
May	...	...	...	...	(2) 107	(2) 107	...	...	...	...	...	...	...	...	...	...	(2) 107	(2) 107
July	...	...	...	...	(1) 28	(1) 28	...	...	...	...	...	...	...	...	...	...	(1) 28	(1) 28
August	...	...	...	...	...	...	...	...	...	...	...	...	...	(1) 1	(1) 1	...	(1) 1	(1) 1
December	...	...	...	...	...	...	...	(1) 1	(1) 1	...	...	...	...	...	...	...	(1) 1	(1) 1
1943—																		
March	...	...	...	...	...	...	...	...	...	...	(2) 803	(2) 803	...	...	...	...	(2) 803	(2) 803
June	...	...	...	(1) 6	(1) 2	(2) 8	...	...	...	...	...	...	...	...	...	(1) 6	(1) 2	(2) 8
July	...	...	...	...	(2) 3	(2) 3	...	...	...	...	...	...	...	...	...	...	(2) 3	(2) 3
August	...	...	...	(2) 185	...	(2) 185	...	...	...	...	...	...	...	...	...	(2) 185	...	(2) 185
October	...	(1) 7	(1) 7	...	(1) 258	(1) 258	...	...	...	...	...	...	...	...	...	...	(2) 265	(2) 265
November	...	...	...	(1) 200	...	(1) 200	...	...	...	...	...	...	...	...	...	(1) 200	...	(1) 200
December	...	...	...	(1) 624	...	(1) 624	...	...	...	...	...	...	...	...	...	(1) 624	...	(1) 624
1944—																		
January	...	...	...	(1) 398	...	(1) 398	...	...	...	...	...	...	...	...	...	(1) 398	...	(1) 398
April	...	...	...	(2) 201	...	(2) 201	...	...	...	...	...	...	...	...	...	(2) 201	...	(2) 201
Total	...	(59) 378	(59) 378	(8) 1,614	(125) 1,344	(133) 2,958	...	(129) 625	(129) 625	...	(6) 835	(6) 835	...	(47) 200	(47) 200	(8) 1,614	(366) 3,382	(374) 4,996
Tons of Incendiary Bombs Dropped	...	16	16	704	369	1,073	...	34	34	...	3	3	...	19	19	704	441	1,145
Tons of High Explosive Bombs Dropped	...	362	362	910	975	1,885	...	591	591	...	832	832	...	181	181	910	2,941	3,851
Number of High Explosive Bombs Dropped	...	2,179	2,179	2,724	5,397	8,121	...	3,530	3,530	...	3,308	3,308	...	1,003	1,003	2,724	15,417	18,141
Average Weight per Bomb—High Explosive Pounds	...	332	332	668	361	464	...	335	335	...	503	503	...	361	361	668	382	424
Number of Missions	...	59	59	8	125	133	...	129	129	...	6	6	...	47	47	8	366	374
Number of Sorties	...	335	335	783	1,037	1,820	...	596	596	...	198	198	...	214	214	783	2,380	3,163
Average Ton per Mission (H.E.)	...	6	6	114	8	14	...	5	5	...	139	139	...	4	4	114	8	10
Average Tons per Mission (H.E. and I.B.)	...	6	6	202	11	22	...	5	5	...	139	139	...	7	7	202	9	13
OIL OFFENSIVE.																		
1944—																		
May	(1) 67	...	(1) 67	(9) 2,442	...	(9) 2,442	...	...	...	...	...	...	...	...	...	(10) 2,509	...	(10) 2,509
June	(1) 230	(3) 1,521	(4) 1,751	(4) 1,242	(7) 2,128	(11) 3,370	(21) 6,310	...	(21) 6,310	...	...	...	(3) 256	...	(3) 256	(29) 8,038	(10) 3,649	(39) 11,687
July	(1) 242	(6) 1,390	(7) 1,632	(10) 7,063	(11) 2,410	(21) 9,473	(5) 1,783	...	(5) 1,783	...	...	...	(3) 163	(2) 1,065	(5) 1,228	(19) 9,251	(19) 4,865	(38) 14,116
August	(3) 710	(6) 2,675	(9) 3,385	(15) 5,639	...	(15) 5,639	(26) 4,474	...	(26) 4,474	...	...	...	(3) 1,146	(4) 2,236	(7) 3,382	(47) 11,969	(10) 4,911	(57) 16,880
September	(3) 474	(6) 2,484	(9) 2,958	(18) 7,219	(4) 2,111	(22) 9,330	(7) 930	...	(7) 930	...	...	...	(1) 101	(1) 3	(2) 104	(29) 8,724	(11) 4,598	(40) 13,322
October	(2) 333	(5) 2,301	(7) 2,634	(14) 4,833	(3) 1,928	(17) 6,761	(12) 3,505	...	(12) 3,505	(2) 116	...	(2) 116	(4) 929	...	(4) 929	(34) 9,716	(8) 4,229	(42) 13,945
November	(3) 990	(15) 11,555	(18) 12,545	(25) 10,681	(2) 976	(27) 11,657	(19) 7,916	(3) 1,166	(22) 9,082	(8) 1,451	...	(8) 1,451	(4) 823	...	(4) 823	(59) 21,861	(20) 13,697	(79) 35,558
December	(4) 1,142	...	(4) 1,142	(15) 6,060	(5) 5,317	(20) 11,377	(11) 1,612	(2) 85	(13) 1,697	(4) 116	(4) 577	(8) 693	(5) 870	...	(5) 870	(39) 9,800	(11) 5,979	(50) 15,779
1945—																		
January	(3) 559	(6) 969	(9) 1,528	(3) 1,551	(5) 6,922	(8) 8,473	(8) 1,115	...	(8) 1,115	(2) 460	(7) 2,934	(9) 3,394	(5) 1,021	(3) 360	(8) 1,381	(21) 4,706	(21) 11,185	(42) 15,891
February	(2) 625	(6) 4,653	(8) 5,278	(8) 2,737	(6) 6,100	(14) 8,837	(17) 3,199	(5) 1,356	(22) 4,555	(4) 675	(6) 2,399	(10) 3,074	(8) 2,572	...	(8) 2,572	(39) 9,808	(23) 14,508	(62) 24,316
March	(9) 2,181	(3) 1,845	(12) 4,026	(11) 3,297	(7) 6,903	(18) 10,200	(30) 6,028	(8) 5,890	(38) 11,918	(7) 2,208	(11) 5,682	(18) 7,890	(7) 1,250	(2) 1,011	(9) 2,261	(64) 14,964	(31) 21,331	(95) 36,295
April	...	...	...	...	(3) 3,249	(3) 3,249	...	(2) 959	(2) 959	...	(1) 556	(1) 556	(6) 1,434	(2) 809	(8) 2,243	(6) 1,434	(8) 5,573	(14) 7,007
Total	(32) 7,553	(56) 29,393	(88) 36,946	(132) 52,764	(53) 38,044	(185) 90,808	(156) 36,872	(20) 9,456	(176) 46,328	(27) 5,026	(29) 12,148	(56) 17,174	(49) 10,565	(14) 5,484	(63) 16,049	(396) 112,780	(172) 94,525	(568) 207,305
Tons of Incendiary Bombs Dropped	91	217	308	2,114	1,746	3,860	1,153	77	1,230	17	...	17	196	5	201	3,571	2,045	5,616
Tons of High Explosive Bombs Dropped	7,462	29,176	36,638	50,650	36,298	86,948	35,719	9,379	45,098	5,009	12,148	17,157	10,369	5,479	15,848	109,209	92,480	201,689
Number of High Explosive Bombs Dropped	34,968	95,685	130,653	273,942	107,002	380,944	181,151	28,052	209,203	18,451	33,203	51,654	52,305	14,984	67,289	560,817	278,926	839,743
Average Weight per Bomb—High Explosive Pounds	427	610	561	370	678	457	394	669	431	543	731	664	397	731	471	390	664	480
Number of Missions	32	56	88	132	53	185	156	20	176	27	29	56	49	14	63	396	172	568
Number of Sorties	3,017	6,360	9,377	21,781	8,954	30,735	14,549	2,758	17,307	1,778	2,515	4,293	4,055	1,604	5,659	45,180	22,191	67,371
Average Tons per Mission (H.E.)	233	521	416	384	685	470	229	469	256	186	419	306	212	391	252	276	538	355
Average Tons per Mission (H.E. and I.B.)	236	525	420	400	718	491	236	473	263	186	419	306	363	562	407	285	550	365
GRAND TOTAL	7,553	29,771	37,324	54,378	39,388	93,766	36,872	10,081	46,953	5,026	12,983	18,009	10,565	5,684	16,249	114,394	97,907	212,301
Tons of Incendiary Bombs Dropped	91	233	324	2,818	2,115	4,933	1,153	111	1,264	17	3	20	196	24	220	4,275	2,486	6,761
Tons of High Explosive Bombs Dropped	7,462	29,538	37,000	51,560	37,273	88,833	35,719	9,970	45,689	5,009	12,980	17,989	10,369	5,660	16,029	110,119	95,421	205,540
Number of High Explosive Bombs Dropped	34,968	97,864	132,832	276,666	112,399	389,065	181,151	31,582	212,733	18,451	36,511	54,962	52,305	15,987	68,292	563,541	294,343	857,884
Average Weight per Bomb—High Explosive Pounds	427	604	557	373	663	457	394	631	430	543	711	654	396	708	470	390	648	480
Number of Missions	32	115	147	140	178	318	156	149	305	27	35	62	49	61	110	404	538	942
Number of Sorties	3,017	6,695	9,712	22,564	9,991	32,555	14,549	3,354	17,903	1,778	2,713	4,491	4,055	1,818	5,873	45,963	24,571	70,534
Average Tons per Mission (H.E.)	233	257	252	368	209	279	229	67	150	186	371	290	212	93	146	273	177	218
Average Tons per Mission (H.E. and I.B.)	236	259	254	388	221	295	236	68	154	186	371	290	363	135	236	283	182	225

NOTES.

Figures in ( ) denote number of missions.
Tons are short tons (2,000 lbs.).
105 tons fragmentation bombs not included in these totals.

Targets outside Greater Germany are not included.
10,368 tons included here not shown in Tabulating Section data, but shown in Group Mission Reports.

TABLE 21.

Chronological Summary of Strategic Attacks on Oil Targets from the Start of the Oil Offensive.†

(NOTE.—R.A.F. Command Bomb Loads shown in Long Tons.)

Date.	U.S. 8th A.F.	Short Tons.	U.S. 15th A.F.	Short Tons.	R.A.F. Bomber Command.	Long Tons.
1944—						
Apr. 3 ...	...	...	Budapest-Magyar ...	690*		
4 ...	...	...	Bucharest-Prahova ...	863*		
5 ...	...	...	Astra Romana ...	588*		
25 ...	...	...	Astra Romana ...	545*		
	...	...	...	*		
May 6 ...	...	...	Steaua Romana ...	31		
12 ...	Bruex	309				
	Lutzkendorf	169				
	Leuna	451				
	Zeitz	255				
	Boehlen	220				
18 ...	...	...	Romano Americana ...	105		
			Concordia Vega ...	124		
			Dacia	186		
			Redeventa Xenia ...	84		
28 ...	Lutzkendorf	155				
	Magdeburg	114				
	Leuna	146				
	Ruhland	70				
	Zeitz	447				
29 ...	Poelitz...	547				
31 ..	...	...	Romano Americana ...	257		
			Dacia	130		
			Unirea Sperantza ...	248		
			Redeventa Xenia ...	130		
			Concordia Vega ...	245		
	Total, May ...	2,883	...	1,540		
June 6 ...	...	...	Romano Americana ...	308		
			Redeventa Xenia ...	158		
			Dacia	169		
			Astra Romana ...	58		
9 ...	...	...	Trieste (Aquila) ...	123		
10 ...	...	...	Trieste (Aquila) ...	277		
			Romano Americana ...	18		
11 ...	...	...	Bosanski Brod ...	91		
			Smederevo	252		
12/13 ...	...	...	...	...	Nordstern	1,444
14 ...	Emmerich	176	Budapest-Csepel ...	} 31		
			Budapest-Fanto ...			
			Petfurdo	288		
			Szoeny...	210		
			Osijek	174		
			Caprag...	166		
14/15 ...	...	...	...	...	Scholven	49
	Misburg	500				
15/16 ...	...	...	...	...	Scholven	44
16 ...	...	...	Vienna-Floridsdorf ...	197		
			Vienna-Kagran ...	195		
			Bratislava	242		
			Vienna-Lobau ...	203		
			Vienna-Schwechat ...	176		
16/17 ...	...	...	...	..	Sterkrade-Holten...	1,275
17/18 ...	...	...	...		Scholven	7
18 ...	Harburg (Ebano) ...	48		...		

* These tonnages were dropped in attacks on neighbouring transportation targets, resulting in incidental damage to the oil targets named.

† Attacks by the Tactical Air Forces not included in this Table.

Date.	U.S. 8th A.F.	Short Tons.	U.S. 15th A.F.	Short Tons.	R.A.F. Bomber Command.	Long Tons.
1944—						
June 18 ...	Hamburg (Eurotank)	146				
	Hamburg-Neuhof (Schindler)	80				
	Hamburg - Grasbrook (Rhenania)	97				
	Ostermoor	210				
	Misburg	232				
	Oslebshausen	166				
20 ...	Hamburg (Eurotank)	315				
	Hamburg - Wilhelmsburg (Schliemann)	161				
	Hamburg - Grasbrook (Rhenania)	145				
	Hamburg - Grasbrook (Albrecht and Schliemann)	148				
	Hamburg-Neuhof (Schindler)	72				
	Hamburg - Wilhelmsburg (D.P.A.G.)	156				
	Harburg (Ebano) ...	179				
	Harburg (Rhenania)...	155				
21 ...	Ruhland	?				
21/22 ...	...	...	...	...	Scholven	570
					Wesseling	578
23 ...	...	...	Romano Americana ...	168		
			Dacia	104		
24 ...	...	...	Romano Americana ...	328		
	Oslebshausen	526				
25/26 ...	...	...	...	...	Homberg	44
26 ...	...	...	Budapest-Csepel ...	?		
			Vienna-Schwechat ...	156		
			Vienna-Lobau ...	222		
			Korneuburg	236		
			Vienna-Floridsdorf ...	245		
			Moosbierbaum ...	277		
			Drohobycz	140		
27 ...	...	...	Trieste (Aquila) ...	185		
28 ...	...	...	Titan	147		
			Prahova	109		
28/29 ...	...	...	...	...	Scholven	10
	Boehlen	177				
30/31 ...	...	...	...	...	Homberg	52
	Total, June ...	3,689	...	5,653	...	4,073
July 1/2 ...	...	...	...	...	Scholven	5
					Homberg	4
2 ...	...	...	Budapest-Csepel ...	122		
			Almasfuzito	153		
3 ...	...	...	Prahova	57		
			Titan	84		
3/4 ...	...	...	...	...	Scholven	4
		...			Homberg	1
4 ...	...	...	Brasov...	430		
4/5 ...	...	...	...	...	Scholven	3
5/6 ...	...	...	...	...	Scholven	41
6/7 ...	...	...	...	...	Scholven	48
7 ...	Lutzkendorf	229				
	Boehlen	148				
	Leuna	106				
			Blechhammer North...	429		
			Blechhammer South...	479		
			Deschowitz	221		
7/8 ...	...	...	...	...	Scholven	12
8 ...	...	...	Vienna-Vosendorf ...	?		
			Vienna-Floridsdorf ...	?		
			Korneuburg	?		
8/9 ...	...	...	...	..	Scholven	9
9 ...	...	...	Bosanki Brod... ...	194		
			Redeventa Xenia ...	362		
			Concordia Vega ...	238		

Date.	U.S. 8th A.F.	Short Tons.	U.S. 15th A.F.	Short Tons.	R.A.F. Bomber Command.	Long Tons.
1944—						
July 9/10 ...	...	...	...	...	Scholven	9
11/12 ...	...	...	...	...	Homberg	5
12/13 ...	...	...	...	...	Homberg	8
13/14 ...	...	...	...	...	Scholven	5
					Homberg	4
14 ...	...	...	Budapest–Fanto ...	165		
			Budapest–Csepel ...	159		
			Budapest–Magyar ...	159		
			Petfurdo	200		
			Bosanki Brod... ...	182		
15 ...	...	...	Romano Americana ...	757		
			Dacia	234		
			Creditul Minier ...	200		
			Standard Petrol Block	147		
			Unirea Sperantza ...	89		
16/17 ...	...	...	...	...	Homberg	56
17 ...	...	...	Smederevo	196		
18 ...	Heide Hemmingstedt	133				
18/19 ...	...	...	...	...	Wesseling	711
					Scholven	787
20 ...	...	...	Fiume (Romsa) ...	263		
	Lutzkendorf	129				
	Leuna	374				
20/21 ...	...	...	...	...	Bottrop–Welheim...	587
					Homberg	750
21 ...	...	...	Breux	338		
22 ...	...	...	Pardubice	101		
			Romano Americana ...	1,128		
23 ...	...	...	Kucova	119		
25/26 ...	...	...	...	...	Wanne Eickel ...	420
26 ...	...	...	Kucova	63		
27 ...	...	...	Romano Americana ...	131		
28 ...	...	...	Prahova	138		
	Leuna	1,601				
28 ...	...		Standard Petrol Block	255		
			Astra Romana ...	588		
29 ...	Oslebshausen... ...	1,249				
	Leuna	1,410				
30 ...	...	...	Lispe (Casinghead) ...	192		
			Redeventa Xenia ...	436		
			Creditul Minier ...	48		
			Prahova	156		
	Total, July ...	5,379	...	9,313	...	3,419
Aug. 3 ...	Merkwiller	261				
	Courchelettes... ...	84				
	Harnes	147				
4 ...	Harburg (Rhenania and Ebano)	464				
	Oslebshausen	130				
	Heide–Hemmingstedt	115				
5 ...	Dollbergen	274				
	Nienburg (Depot) ...	665				
5/6 ...	...	...	...	...	Wanne Eickel ...	27
6 ...	Hamburg - Wilhelmsburg (D.P.A.G. and Rhenania)	156				
	Harburg (Rhenania)...	162				
	Harburg (Ebano) ...	88				
	Schulau	203				
	Heide Hemmingstedt	63				
	Hamburg - Wilhelmsburg (Schliemann)	90				
6/7 ...	...	...	...	...	Castrop–Rauxel ...	53
7 ...	...	...	Blechhammer South	622		
			Blechhammer North	196		
	Trzebinia	110				
9 ...	...	...	Almasfuzito	190		
10 ...	...	..	Romano Americana ...	161		
			Concordia Vega ...	134		
			Unirea Sperantza ... Standard Petrol Block }	172		
			Redeventa Xenia .	172		
			Astra Romana ...	165		
			Steaua Romana ...	85		

Date.	U.S. 8th A.F.	Short Tons.	U.S. 15th A.F.	Short Tons.	R.A.F. Bomber Command.	Long Tons.
1944—						
Aug. 11 ...	Strasbourg (Depot) ...	188				
14 ...	Ludwigshafen ...	307				
14/15 ...	...	...	...	...	Sterkrade-Holten ...	1
15/16 ...	...	...	...	...	Dortmund-Wambelerholz	2
					Kamen	3
16 ...	Boehlen	206			Sterkrade-Holten ...	3
	Zeitz	226				
	Rositz	250				
	Magdeburg	225				
16/17 ...	...	...	...	...	Dortmund-Wambelerholz	2
					Kamen	2
					Sterkrade-Holten ...	2
Aug. 17 ...	...	...	Standard Petrol Block / Unirea Sperantza ...	} 60		
			Redeventa Xenia ...	4		
			Astra Romana ...	192		
			Romano Americana ...	280		
17/18 ...	...	...	...	...	Sterkrade-Holten ...	2
					Kamen	2
					Dortmund-Wambelerholz	1
18 ...	...	...	Steaua Romana ...	86		
			Dacia	103		
			Romano Americana ...	468		
			Creditul Minier ...	24		
			Astra Romana ...	17		
18/19 ...	...	...	...	...	Sterkrade-Holten ...	770
					Wanne Eickel ...	1
19 ...	...	...	Redeventa Xenia ...	65		
			Dacia	78		
20 ...	...	...	Czechowice	209		
			Oswiecim	337		
			Dubova	177		
23/24 ...	...	...	...	...	Homberg	3
					Castrop Rauxel ...	2
24 ...	Misburg	218				
	Leuna	436				
	Bruex	311				
	Ruhland	293				
	Freital...	162				
25 ...	Poelitz... ...	385				
25/26 ...	...	...	...	...	Homburg	3
					Castrop Rauxel ...	2
26 ...	Ludwigshafen ...	106				
	Scholven	159				
	Nordstern	94				
	Salzbergen	216				
	Emmerich	103				
	Dulmen (Depot) ...	219				
26/27 ...	...	...	...	...	Dortmund-Wambelerholz	1
27 ...	...	...	...	...	Homberg	775
	Total, Aug. ...	7,116	...	3,997	...	1,657
Sept. 3 ...	Ludwigshafen ...	969				
5 ...	Ludwigshafen ...	723				
10 ...	...	...	Vienna-Lobau ...	175		
			Vienna-Schwechat ...	226		
11 ...	Leuna	279				
	Boehlen	173				
	Magdeburg	78				
	Ruhland	53				
	Misburg	243				
	Lutzkendorf	233				
	Bruex	95				
					Kamen	586
					Nordstern	398
					Castrop Rauxel ...	439

Date.	U.S. 8th A.F.	Short Tons.	U.S. 15th A.F.	Short Tons.	R.A.F. Bomber Command.	Long Tons.
1944—						
Sept. 12 ...	Boehlen	83				
	Magdeburg	330				
	Misburg	90				
	Heide Hemmingstedt	165				
	Bruex	190				
	Ruhland	143				
					Scholven	514
					Dortmund-Wambelerholz	403
					Wanne Eickel ...	438
13 ...	...	...	Blechhammer North...	288		
			Deschowitz	274		
			Oswiecim	286		
	Ludwigshafen ...	212				
	Leuna	335				
	Lutzkendorf	188				
16/17 ...	...	...	...	...	Nordstern	366
					Dortmund-Wambelerholz	3
17 ...	...	...	Budapest-Csepel ...	135		
			Budapest-Magyar ...	119		
21 ...	Ludwigshafen ...	412				
23 ...	...	...	Bruex	376		
25 ...	Ludwigshafen ...	1,134				
26/27 ...	...	...	...	...	Homberg	5
27 ...	Ludwigshafen ...	585				
					Bottrop-Welheim...	490
					Sterkrade-Holten...	286
28 ...	Magdeburg	57				
	Leuna	725				
30 ...	...	...	...	...	Bottrop-Welheim...	5
					Sterkrade-Holten...	72
30/31 ...	...	...	...	...	Sterkrade-Holten...	2
	Total, Sept. ...	7,495	...	1,829	...	4,007
Oct. 1/2 ...	...	...	...	...	Dortmund-Wambelerholz	5
2/3 ...	...	...	...	...	Dortmund-Wambelerholz	2
3 ...	Wesseling	218				
3/4 ...	...	...	...	...	Kamen	2
5/6 ...	...	...	...	...	Dortmund-Wambelerholz	3
6 ...	Harburg (Ebano and Rhenania)	320				
					Scholven	522
					Sterkrade-Holten...	517
7 ...	...	...	Vienna-Schwechat ...	44		
			Vienna-Lobau ...	297		
	Poelitz	348				
	Ruhland	148				
	Boehlen	205				
	Lutzkendorf	206				
	Leuna	309				
	Magdeburg	65				
11 ...	Wesseling	169				
12 ...	...	...			Wanne Eickel ...	569
13 ...	...	...	Blechhammer South...	665		
			Vienna-Floridsdorf ...	216		
14 ...	...	...	Blechhammer North...	243		
			Deschowitz	183		
15 ...	Reisholz	160				
	Monheim	155				
16 ...	...	...	Bruex	79		
			Linz Benzol	161		
17 ...	...	...	Blechhammer South...	280		
20 ...	...	...	Bruex	347		
25 ...	Harburg (Rhenania)...	789				
	Harburg (Ebano) ...	494				
	Scholven	246				
	Nordstern	60				
					Homberg	972
26 ...	Bottrop-Welheim ...	180				

Date.	U.S. 8th A.F.	Short Tons.	U.S. 15th A.F.	Short Tons.	R.A.F. Bomber Command.	Long Tons.
1944—						
Oct. 30 ...	Harburg (Rhenania)...	193				
	Harburg (Ebano) ...	197				
					Wesseling	527
31 ...	...	...	...	...	Bottrop-Welheim...	531
	Total, Oct. ...	4,462	...	2,515	...	3,650
Nov. 1 ...	Nordstern	277				
	Scholven	406	•			
					Homberg	955
2 ...	Leuna	1,370				
	Castrop Rauxel ...	403				
	Sterkrade Holten ...	251				
					Homberg	952
3 ...	...	...	Moosbierbaum ...	125		
4 ...	Bottrop Welheim ...	220				
	Harburg (Rhenania)...	450				
	Harburg (Ebano) ..	701				
	Nordstern	368				
	Misburg	591				
	Neunkirchen (Depot)	426				
			Linz Benzol	325		
5 ...	Ludwigshafen ...	507				
			Vienna-Floridsdorf ...	1,100		
6 ...	Sterkrade Holten ...	434				
	Bottrop Welheim ...	198				
	Harburg (Rhenania)...	423				
	Harburg (Ebano) ...	413				
	Meiderich Tar ...	134				
			Moosbierbaum ...	403		
7 ...	...	...	Vienna-Floridsdorf ...	27		
			Moosbierbaum ...	11		
8 ...	Leuna	478				
					Homburg	732
9 ...	...	...	...	...	Wanne Eickel ...	1,283
			Regensburg (Depot)...	195		
11 ...	Scholven	237				
	Bottrop Welheim ...	344				
					Castrop Rauxel ...	593
11/12 ...	...	...	...	...	Dortmud-Wambelerholz	1,127
					Kamen	26
					Harburg (Rhenania and Ebano)	1,099
13 ..	...	...	Blechhammer South...	35		
15 ...	...	...	Linz Benzol	101		
					Dortmund-Wambelerholz	902
15/16 ...	...	...	...	...	Wanne Eickel ...	9
					Scholven	8
17 ...	...	...	Blechhammer South...	199		
			Vienna-Floridsdorf ..	402		
18 ...	...	...	Vienna-Floridsdorf ...	230		
			Korneuburg	180		
18/19 ...	...	...	...	...	Wanne-Eickel ...	1,516
			Vienna-Schwechat ...	112		
			Vienna-Vosendorf ...	90		
			Vienna-Lobau ...	214		
20 ...	..	...	Blechhammer South...	314		
	Scholven	141				
					Homberg	869
20/21 ...	...	...	...	...	Homberg	10
21 ...	Leuna	476				
	Harburg (Rhenania and Ebano)	474				
	Hamburg - Wilhelmsburg (D.P.A.G. and Rhenania)	479				
					Homberg	806
21/22 ...	...	...	...	...	Castrop Rauxel ...	960
					Sterkrade Holten ...	869
23 ...	Nordstern Benzol ...	316				
					Nordstern	866

Date.	U.S. 8th A.F.	Short Tons.	U.S. 15th A.F.	Short Tons.	R.A.F. Bomber Command.	Long Tons.
1944—						
Nov. 25 ...	...	...	Linz Benzol	61		
	Leuna	1,391				
26 ...	Misburg	863				
29 ...	Misburg	1,152				
					Meiderich Tar ...	43
30 ...	...	...	Linz Benzol	44		
	Boehlen	166				
	Zeitz	367				
	Leuna	1,015				
	Lutzkendorf	413				
					Prosper	312
					Osterfeld	312
					Meiderich Tar ...	51
	Total, Nov. ...	15,884	...	4,168	...	14,312
Dec. 1/2 ...	...	...	...	...	Bruckhausen ...	7
2 ...	...	...	Blechhammer North...	258		
			Blechhammer South...	161		
			Deschowitz	133		
			Vienna–Floridsdorf ...	168		
					Hansa	485
Dec. 5/6 ...	...		...	...	Bruckhausen ...	3
6 ...	Leuna	1,076				
6/7 ...	...	...	...	...	Leuna	1,847
8 ...	...	...	...	...	Meiderich Tar ...	32
9/10 ...	...	...	...	...	Meiderich Tar ...	4
11 ...	...	...	Moosbierbaum ...	404		
					Osterfeld	284
					Bruckhausen ...	61
					Meiderich Benzol...	40
11/12 ...	Leuna	988			Bruckhausen ...	4
			Blechhammer South...	98		
			Moravska Ostrava ...	16		
15/16 ...	...	...	...	...	Bruckhausen ...	2
16 ...	...	...	Bruex	675		
			Linz Benzol	43		
17 ...	...	...	Blechhammer North...	355		
			Blechhammer South...	172		
			Deschowitz	363		
17/18 ...	...	...	Moravska Ostrava ...	45		
					Salzgitter	5
18 ...	...	...	Blechhammer South...	110		
			Blechhammer North...	156		
			Deschowitz	348		
			Oswiecim	106		
			Vienna–Floridsdorf ...	194		
			Moravska Ostrava ...	42		
19 ...	...	...	Blechhammer South...	55		
			Blechhammer North...	248		
			Moravska Ostrava ...	10		
20 ...	...	...	Bruex	114		
			Regensburg (Depot)...	365		
21/22 ...	...	...	...	...	Poelitz	694
25 ...	...	...	Bruex	351		
26 ...	...	...	Blechhammer South...	245		
			Deschowitz	237		
			Oswiecim	151		
27 ...	...	...	Vienna-Vosendorf ...	92		
28 ...	...	...	Regensburg (Depot)...	229		
			Roudnice (Depot) ...	50		
			Pardubice	122		
			Kolin	110		
29/30 ...	...	...	...	...	Scholven	1,625
30/31 ...	...	...	...	...	Carolinengluck ...	16
31 ...	Harburg (Rhenania)...	205				
	Hamburg – Grasbrook (Rhenania)	206				
	Hamburg – Wilhelmsburg (D.P.A.G. and Rhenania)	172				
	Hamburg – Wilhelmsburg (Schliemann)	57				
	Misburg	233				
	Total, Dec. ...	2,937	...	6,226	...	5,109

Date.	U.S. 8th A.F.	Short Tons.	U.S. 15th A.F.	Short Tons.	R.A.F. Bomber Command.	Long Tons.
1945—						
Jan. 1 ...	Dollbergen	128				
	Magdeburg	24				
	Ehmen (Depot) ...	43				
1/2 ...	...	...	...	...	Minster Stein ...	375
2/3 ...	...	...	...	...	Castrop Rauxel Tar	13
3 ...	...	...	...	...	Castrop Rauxel Tar	214
					Hansa	235
5/6 ...	...	...	...	...	Castrop Rauxel Tar	5
6/7 ...	...	...	...	...	Castrop Rauxel Tar	3
12/13 ...	...	...	...	...	Carolinengluck ...	17
					Ewald-Fortsetzung	12
13/14 ...	...	...	...	...	Poelitz	813
14 ...	Magdeburg	223				
	Heide Hemmingstedt	251				
	Derben (Depot) ...	551				
	Ehmen (Depot) ...	237				
	Salzgitter	475				
14/15 ...	...	...	...	...	Leuna	2,213
					Dulmen (Depot) ...	304
15 ...	...	...	...	...	Ewald-Fortsetzung	376
					Robert Muser ...	282
16 ...	Ruhland	150				
	Magdeburg	113				
16/17 ...	...	...	...	...	Bruex	857
					Zeitz	1,329
					Wanne Eickel ...	572
17 ...	Harburg (Rhenania and Ebano)	187				
	Hamburg - Grasbrook (Rhenania)	106				
	Hamburg - Grasbrook (Albrecht and Schliemann)	100				
17/18 ...	...	...	...	...	Ruthen (Depot) ...	9
18/19 ...	...	...	...	...	Sterkrade-Holten ...	63
					Ruthen	8
20 ...	Sterkrade Holten ...	93				
21 ...	...	...	Vienna-Lobau ...	207		
			Vienna-Schwechat ..	192		
			Regensburg (Depot) ...	228		
			Fiume (Romsa) ...	40		
22 ...	Sterkrade Holten ...	402				
22/23 ...	...	...	...	...	Bruckhausen ...	1,297
26/27 ...	...	...	...	...	Castrop Rauxel Tar	14
28 ...	Kaiserstuhl	300				
	Gneisenau	154				
31 ...	...	...	Moosbierbaum ...	1,356		
31/1 ...	...	...	...	...	Hansa	11
					Bruckhausen ...	8
	Total, Jan. ...	3,537	...	2,023	...	9,030
Feb. 1 ...	...	...	Moosbierbaum ...	165		
1/2 ...	...	...	...	...	Bruckhausen ...	6
2/3 ...	...	...	...	...	Wanne Eickel ...	872
3 ...	Magdeburg	269				
3/4 ...	...	...	...	...	Prosper	1,007
					Hansa	591
4/5 ...	...	...	...	...	Osterfeld	321
					Nordstern Benzol ...	348
5 ...	...	...	Regensburg (Depot) ...	1,123		
7 ...	...	...	...		Wanne Eickel ...	340
			Vienna-Lobau ...	214		
			Vienna-Schwechat ...	210		
			Vienna-Floridsdorf ...	214		
			Vienna-Kagran ...	10		
			Kerneuburg	65		
			Moosbierbaum ...	528		
8/9 ...	...	...	...	...	Poelitz	1,659
					Wanne Eickel ...	654
9 ...	Lutzkendorf	578				
	Magdeburg	23				
	Dulmen (Depot) ...	316				
			Moosbierbaum ...	57		
10 ...	Dulmen (Depot) ...	414				
11 ...	Dulmen (Depot) ...	336				

Date.	U.S. 8th A.F.	Short Tons.	U.S. 15th A.F.	Short Tons.	R.A.F. Bomber Command.	Long Tons.
1945—						
Feb. 12/13 ...	...	...	...	...	Misburg	20
13/14 ...	...	...	...	...	Boehlen	789
					Misburg	13
14 ...	Bruex	63				
	Dulmen (Depot) ...	104				
			Vienna-Schwechat ...	216		
			Vienna-Lobau ...	303		
			Moosbierbaum ...	4		
			Vienna-Floridsdorf ...	106		
14/15 ...	...	...	...	...	Rositz	831
15 ...	Magdeburg ...	899				
			Korneuburg	100		
			Fiume (Romsa) ...	197		
16 ...	Nordstern Benzol ...					
	Graf Bismarck ...	308				
	Hugo II					
	Minster Stein... ...					
	Kaiserstuhl	330				
	Dortmund-Harpenerweg					
	Castrop Rauxel Tar ...	243				
	Robert Muser... ...					
17 ...	...	...	Linz Benzol	117		
18 ...	...	...	Linz Benzol	417		
19 ...	Dortmund - Wambelerholz	217				
	Carolinengluck ...	294				
	Alma Pluto	107				
	Scholven Benzol ...	104				
19/20 ...	...	...	...	...	Boehlen	968
20 ...	...	..	Vienna-Lobau ..	148		
			Vienna-Schwechat ..	115		
20/21 ...	...	...	...	..	Reisholz	373
					Monheim	306
Feb. 22 ...	...	...	...	...	Scholven Benzol ...	375
					Osterfeld	333
23 ...	...	...	...	...	Alma Pluto ...	580
24 ...	Harburg (Ebano and Rhenania)	205				
	Hamburg - Grasbrook (Rhenania)					
	Hamburg - Grasbrook (Albrecht and Schliemann)					
	Hamburg - Wilhelmsburg (Schliemann)	810				
	Hamburg - Wilhelmsburg (D.P.A.G. and Rhenanai)					
	Hamburg - Neuhof (Schindler)					
	Misburg	280				
					Kamen	1,035
25 ...	Neuburg (Depot) ...	261				
					Kamen	651
			Linz Benzol	53		
26 ...	...	...	...	...	Dortmund - Wambelerholz	652
27 ...	...	...	...	...	Alma Pluto ...	641
28 ...	...	...	...	...	Nordstern Benzol ...	697
	Total, Feb. ...	6,161	...	4,362	...	14,062
1945—						
Mar. 1 ...	...	...	...	...	Kamen	660
			Moosbierbaum ...	997		
2 ...	Boehlen	138				
	Rositz	81				
	Ruhland	58				
	Magdeburg	85				
3 ...	Misburg	69				
	Magdeburg	479				
	Ruhland	60				
	Dollbergen	126				
	Dedenhausen	160				
	Nienhagen	195				

Date.	U.S. 8th A.F.	Short Tons.	U.S. 15th A.F.	Short Tons.	R.A.F. Bomber Command.	Long Tons.
1945—						
Mar. 3/4 ...	...	...	...	...	Kamen	737
5 ...	...	...	...	...	Consolidation I/IV	793
	Harburg (Rhenania and Ebano)	324				
5/6 ...	...	...	...	...	Boehlen	872
					Moelbis-Espenhain	
6 ...	...	...	...	...	Salzbergen	548
7 ...	Horder Verein ...	69				
	Dortmund-Harpenerweg	188				
	Emscher-Lippe ...	584				
	Castrop Rauxel Tar ...	267				
7/8 ...	...	...	...	...	Harburg (Rhenania and Ebano)	1,039
					Heide-Hemmingstedt	730
8 ...	Robert Muser... ...	335				
	Bruchstrasse	204				
	Gneisenau	381				
	Emil	339				
	Mathias Stinnes III/IV	106				
	Scholven	224				
	Augustus Viktoria ...	328				
9 ...	...	...	...	...	Emscher Lippe ...	782
10 ...	...	...	...	...	Scholven	755
11 ...	Hamburg-Wilhelmsburg (D.P.A.G. and Rhenania)	1,124				
	Hamburg-Wilhelmsburg (Schliemann)					
	Hamburg-Neuhof (Schindler)					
12 ...	...	...	Vienna-Floridsdorf ...	1,667		
13/14 ...	...	...	...	...	Erin...	464
					Dahlbusch	450
14 ...	...	...	...	...	Henrichshutte ...	381
					Emscher Lippe ...	349
	Nienhagen	197				
	Misburg	193				
			Szoeny	279		
			Almasfuzito	144		
14/15 ...	...	...	...	...	Lutzkendorf ...	936
15 ...	...	...	...	...	Mathias Stinnes III/IV	279
					Castrop Rauxel Tar	266
			Ruhland	315		
			Kolin	298		
			Moosbierbaum ...	286		
			Vienna-Schwechat ...	103		
			Vienna-Floridsdorf ...	80		
15/16 ...	...	...	...	...	Misburg	1,034
16 ...	...	...	Vienna-Floridsdorf ...	245		
			Vienna-Schwechat ...	238		
			Korneuburg	34		
			Moosbierbaum ...	151		
17 ...	Ruhland	594				
	Boehlen	378				
	Moelbis-Espenhain ...	351				
					Auguste Viktoria ...	429
					Gneisenau	368
18 ...	...	...	...	...	Bruchstrasse ...	251
					Henrichshutte ...	261
19 ...	...	...	...	...	Alma Pluto ...	370
20 ...	...	...	Korneuburg	172		
			Vienna-Kagran ...	100		
	Heide Hemmingstedt	308				
20/21 ...	...	...	...	...	Boehlen	946
					Heide Hemmingstedt	711
21 ...	...	...	...	...	Oslebshausen ...	643
21 ...	...	...	Vienna-Vosendorf ...	71		
			Vienna-Kagran ...	21		
			Vienna-Floridsdorf ...	1[illegible]		

Date.	U.S. 8th A.F.	Short Tons.	U.S. 15th A.F.	Short Runs.	R.A.F. Bomber Command.	[illegible]g Tons.
1945—						
Mar. 21/22...	...	...	...	...	Hamburg-Wilhelmsburg (D.P.A.G. and Rhenania)	671
					Bruchstrasse ...	622
22 ...	...	...	Ruhland	332		
			Kralupy	288		
			Vienna-Floridsdorf ...	50		
			Kralupy	30		
23 ...	...	...	Ruhland	427		
			Vienna-Kagran ...	156		
24 ...	...	...	...	...	Dortmund - Harpenerweg	427
					Mathias Stinnes III/IV	426
25 ...	Buchen-Nussau (Depot)	144				
	Hitzacker (Depot) ...	327				
	Ehmen (Depot) ...	157				
26 ...	Zeitz	30				
27 ...	...	...	...	...	Farge (Depot) ...	539
					Konigsborn... ...	390
					Sachsen	363
29 ...	...	...	...	...	Salzgitter	446
30 ...	Hamburg (Eurotank)	467				
31 ...	Bad Berka (Depot) ...	81				
	Zeitz	399				
	Total, Mar. ..	9,550	...	6,628	...	18,938
Apr. 4/5 ...	...	...	...	...	Leuna	991
					Lutzkendorf ...	968
					Harburg (Rhenania)	1,028
7 ...	Hitzacker (Depot) ...	284	...	...		
	Buchen-Nussau (Depot)	108				
7/8 ...	...	...	...	...	Moelbis-Espenhain	505
8 ...	Derben (Depot) ...	78				
8/9 ...	...	...	...	...	Lutzkendorf ...	957
9 ...	Neuburg (Depot) ...	242				
					Hamburg (Eurotank)	235
11 ...	Freiham (Depot) ...	717				
	Regensburg (Depot) ...	205				
			Ghedi	124		
14 ...	...	...	...	...	Riesa (Depot) ...	11
17 ...	Roudnice (Depot) ...	315				
20 ...	...	...	...	...	Regensburg (Depot)	341
25/26 ...	...	...	...	...	Vallo	315
	Total, Apr. ...	1,949	...	124	...	5,351

TABLE 22.

Weight of H.E. Bombs dropped on Oil Production Facilities and the Effects on Capacity and Production.

(Oil in Metric Tons; Bombs in Short Tons.)

—	Hydrogenation.	Fischer.	Crude.	Misc.	Total.
1. Capacity, in tons per month per cent. of total	316,000 47·7	43,000 6·5	167,000 25·2	136,000* 20·6	662,000 100·0
2. Production loss, in tons per cent. of total	3,113,000 65·3	358,000 7·5	800,000 16·8	495,000* 10·4	4,766,000 100·0
3. H.E. bombs dropped, in tons per cent. of total	87,000 46·8	37,000 19·9	45,000 24·2	17,000† 9·1	186,000 100·0
4. Tons of bombs per ton per month capacity	0·28	0·86	0·27	0·13	0·28
5. Tons of production loss per ton of bombs dropped	36	10	17·7	29	26

* Includes coal tar, benzol, alcohol. † Bombs dropped on benzol plants only.

Line 1.—Average monthly output in the first four months of 1944 before bombing of oil targets began.

Line 2.—Based on capacity in line 1 from 1st May, 1944, to 1st May, 1945. These figures include plants knocked out and then captured, which would otherwise possibly have required more bombing to keep them inactive.

Line 3.—From 1st May, 1944, to 1st May, 1945. Includes all types of bombs used.

Line 4.—Equals line 3 divided by line 1.

Line 5.—Equals line 2 divided by line 3.

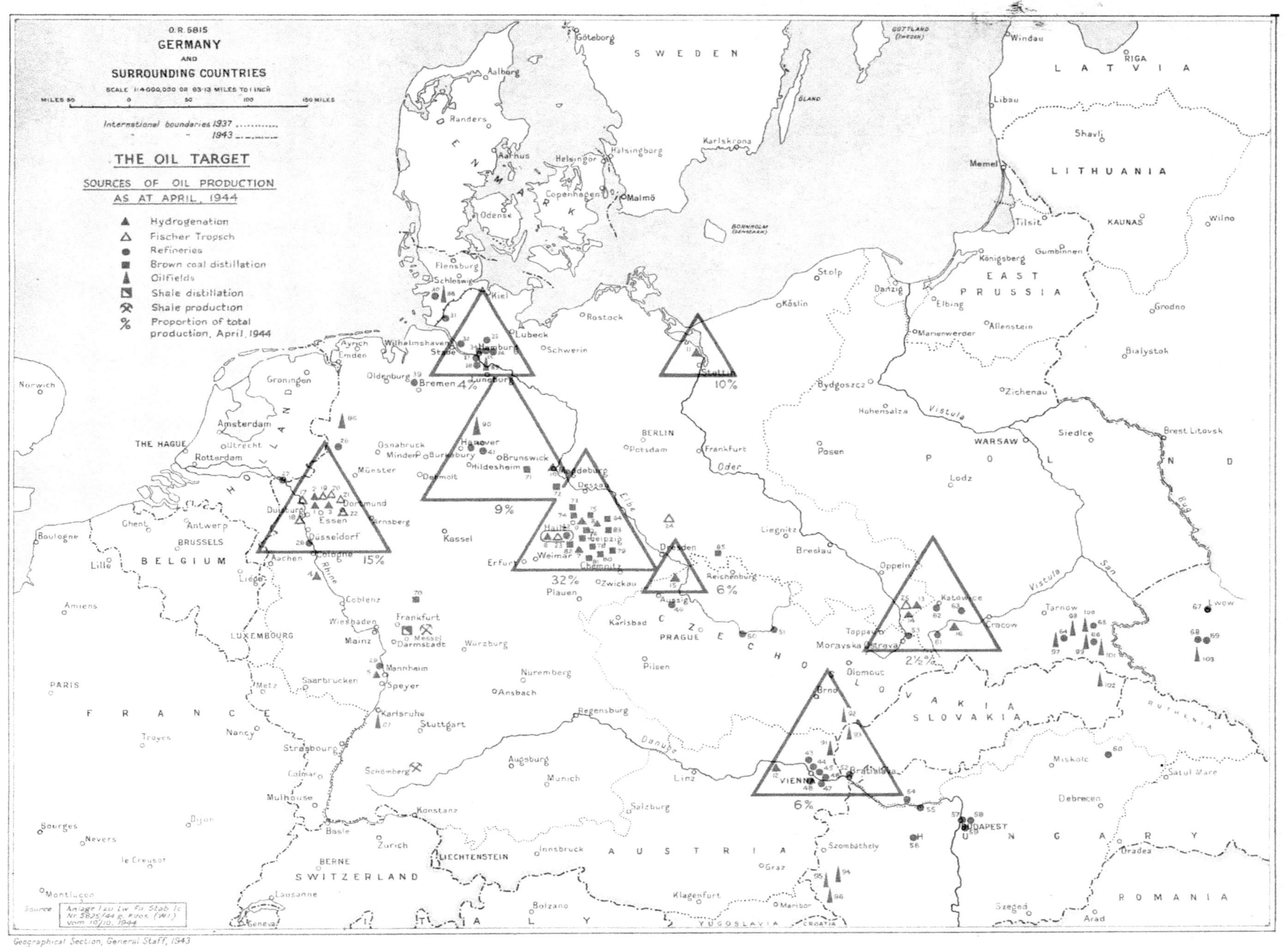

KEY
OVERLEAF.

INDEX.

A.

A. 2. United States Air Force. 122.
Abbreviations. 155.
A.C.I.U. (Central Photographic Interpretation Unit). 94.
A.C.I.U., " D " Section, Industrial. 119, 121.
Adam. 84.
Admiralty, British. 11.
Administration, German, the Organisation of. 96, 97, 98.
Administrative Organisation of the Oil Industry, changes in. 42.
Aerial Reconnaissance Units. 94, 120, 121.
Ahrens. 6, 23, 138.
Air attack data for oil targets in Greater Germany by type of target. Table 20, facing **172.**
Air attack data of strategic Air Forces in the European Theatre of Operations. Table 18. **171.**
Air attack data, oil production and capacity for Greater Germany, summary. Table 19. **172.**
Air attacks directed against—
  Aircraft production. 54, 55.
  Austrian and Hungarian refineries. 60.
  Aviation spirit production. 57.
  Ball bearing industry at Schweinfurt. 54, 55.
  Benzol plants. 61, 68.
  Bohlen synthetic oil plant. 55.
  Bruex synthetic oil plant. 59.
  Communications and tactical targets. 66.
  Communications in Eastern Europe. 54, 55.
  Gelsenkirchen. 51.
  Hungarian oil. 77, 78.
  Leuna, synthetic oil plant. 51, 55, 64, 126, 129, 130. Plate 6.
  Luetzkendorf synthetic oil plant. 55.
  Magdeburg synthetic oil plant. 55.
  Oil targets. 55, 71, 60, 66.
  Oil targets, aiming technique. 151.
  Oil targets, Chronological Summary. **173–183.**
  Oil targets, effect of the weather. 151.
  Oil targets, precision ground control (" OBOE "). 151.
  Oil targets, priority of. 89.
  Oil targets, Speer's Reports to Hitler. **123–136.**
  Oil targets, tonnage of bombs dropped. 151.
  Oil targets, weapon effectiveness. 151.
  Pipeline terminals. 60.
  Ploesti (Allied). 55, 57, 60.
  Ploesti by the Russian Air Force. 18.
  Ploesti, low level. 53.
  Poelitz, synthetic oil plant. 55.
  Rail transport. 60, 78.
  Refineries in Germany. 129, 131.
  Refineries in Rumania, results of. **137.**
  Refineries in Rumania, types of bombs dropped. 137.
  Road transport. 73.
  Ruhland synthetic oil plant. 55.
  Storage depots. 68, 69, 70, 71.
  Storage installations. 51.
  Strategic depots in France and Germany. 70–71.
  Synthetic oil plants. 51, 55–71, 74, 126–131, 134.
  Transportation targets. 57.
  Underground storage depot at Montbartier. 61, 70.
  W.I.F.O. depot at Nienburg. 61, 70, Plate 4.
  Zeitz synthetic oil plant. 55.
Air attacks—
  Loss of production due to. 51.
  Lack of protection at plants. 56.
Aircraft production.—
  Jet. 82.
  New programme. 82.
Air crew, replacements in the **Luftwaffe,** effects of fuel shortage upon. **143–144.**
Air fields at Foggia, the occupation of. 54.
Air Force, Italian, fuel requirements of. 39, 40.
Air Ministry—
  (Br.) A.I.3.e. 117, 121.
  (Br.) 88, 89.
  (Br.), A.I.3 (C). 92, 117.
Air reconnaissance planes, the shooting down of. 58.
Alamein, Allied victory at. 40.
Albania, activities of the **Kontinentale Oel A.G.** in. 106.
Albanien Oel G.m.b.H. 106.
Alcohol. 9, 10.
Alcohol, consumption of, 1935–38. Table 11. **164.**
Allied Central Interpretation Unit (A.C.I.U.). 119, 121.
Allied strategic bombing policy, some German views upon. **147–148.**
Allocations to the three Services by the O.K.W. 80.
Alsace Area—
  Annual refining capacity in. 44.
  Crude oil available in. 44.
American delegation. 93.
Ammoniak Werke Merseburg. 4, 9, Plate 1, Plate 6.
Analyses of captured enemy oil samples. 88, 91.
Anglo-American Oil Company. 107.
Anglo-American Oil Conversations. 91.
Anglo-Danubian Transport Company. 16.
Anglo-Iranian Oil Company. 107.
Annaburg, strategic storage at. 71.
Antonescu. 26.
Arbeitsgemeinschaft—
  Benzolerzeuger. 96.
  Erdoelgewinnung und Vorarbeitung. 96.
  Fuer Petroleumverteilung. 97.
  Fuer Test-Benzin. 97.
  Hydrierung, Synthese und Schwelerei. 96.
  Mineraloelverteilung. 97.
  Schmierstoff Verteilung G.m.b.H. 2.
  Steinkohleteererzeugnisse. 96.
  Fuer White Spirit. 97.
Ardennes Counter-Offensive—
  Combating the. 64, 66, 67.
  The effect of oil shortage upon the outcome of. 75, 78.
Argenteuil, stocks held at. 69.
Armament Programme, the development of. 72.
Armed Forces—
  British calculation of consumption. 91, 92.
  Estimated war-time consumption requirements. 10.
Armoured and armoured infantry divisions, fuel supplies for. 73.
Armoured divisions—
  Economy of fuels in. 76.
  Operating on diesel oil and producer gas. 75, 76.
Army—
  Consumption in 1941. 32.
  Consumption in the Western Campaigns. 21, 23.
  Distribution organisation for liquid fuels. 69.
  Effect of fuel shortages upon. 72.
  Italian requirements. 39, 40.
  Oil supplies from Rumania. 53.
  Oil supplies from Rumania, 1938–1943. Table 7. **163.**
  Requirements of liquid fuel. 2, 3, 49, 72, 135, 136.
  Restricted training of. 77.
  Stocks. 75.
  Stocks of diesel oil. 36.
  Stocks of motor gasoline. 36, 49, 65.
Astley-Bell, L. A. 116.
Asiatic Petroleum Company, Limited. 14, 107.
Astra Romana. 14.
Astra Romana, contract with **Kontinentale Oel A.G.** 103.
Auld, Lieut.-Colonel S. J. M. 88, 91, 116.
Auschwitz—
  Capacity of. Table 3. 160.
  (Oswiecim) plant at. 9, 49, 91.
  Production from. 46.

Austria—
Annexation of. 89.
Area annual refining capacity in. 44.
Area crude oil available in. 44.
Crude oil production, 1940–44. Table 2. **160.**
Deggendorf. 44.
Increase in crude oil production. 42, 44.
Lobau. 44.
Moosbierbaum. 44.
Oilfields. 8, 36, 77.
Planned and actual production in 1942 and 1943. 36.
Refineries, attacks upon. 60.
Aviation diesel oil, estimated war-time consumption requirements, civil and military. 10.
Aviation lubricants—
Estimated war-time consumption requirements, civil and military. 10.
Production of. 138.
Aviation spirit—
Allocations for flying training. 80.
And motor gasoline consumption in 1938. 10.
British calculations of consumption. 92.
Captured in France and Holland. 23.
Components, shortage of. 68.
Components, storage of. 71.
Demands made to the Zentrale Planung. 80.
Estimated war-time consumption requirements, civil and military. 10.
German exports to Italy. 49.
German production capacity. 79.
German use of aromatics. 91.
High octane. 138, 142.
High octane, the production of. 9.
Luftwaffe consumption. 31, 32, 49, 78, 126, 128, 135, 153.
Luftwaffe consumption, 1940–44. 43.
"Aviation Spirit Plan," **Goering.** 33, 35, 36, 46.
Planned production, 1942–43. 4, 33.
Planned stocks, 1943. 5.
Position, improvement in. 49.
Production. 32, 43, 49, 58, 63, 65, 66, 67, 82, 83, 123–136.
Production, air attacks against. 57.
Production from underground plants. 152–153.
Production in Greater Germany, 1940–44 (all processes). Table 1. 159.
Production, September 1944. 133.
Reduction in supplies for training. 79.
Shortage of. 85.
Shortage, effect upon Luftwaffe aircrew replacements. **143–144.**
Stocks of. 22, 23, 36, 43, 50, 83.
Stocks, dwindling of. 141.
Stocks, exhaustion of. 37.
Stocks held by the Luftwaffe. 82.
Storage of. 79.
Azienda Generale Italiana Petrolii. 106.

B.

B. 4. Fuel. 41.
Balaton Lake. 78.
Balkan Campaign, estimated oil consumption in. 31.
Baltic area, activities of the **Kontinentale Oel A.G.** in. 105.
Baltische Oel G.m.b.H. 45, 105.
Baranovo Bridgehead, the Russian breakout from. 77.
Bari refinery. 38, 39.
Bastico. 41.
Bastogne and Malmedy, the defence of. 76.
Bataafsche Petroleum Mij. 105.
Battle of Crete. 31.
Battle of Stalingrad. 35.
Battle of the Falaise Gap. 75.
Bauxite. 77.
Bayerlein. 76.
Beauftragter fuer—
Die Erdoelgewinnung. 2. 96.
Sonderfragen der Chemischen Erzeugung. 2, 96.
Bentz, Dr. 2, 21, 34 and 96.
Benzol—
Plants, bombing attacks upon. 61, 68, 85.
Plants, bombing of, chronological survey of. **173–183,** Table 21.
Production of. 9, 10, 61.
Verband. 103.
Berthoud. 15, 18, 35, 116.
Blechhammer synthetic oil plant. 36, 58, 63, 126, 132, 134, 136.
Capacity of. Table 3. 160.
Production from 1940–44. 46.
Blending of fuels for the Luftwaffe. 69.
Blessing, Karl. 103.
Blockade, the. 11, 28, 38.
Breaches of. 11.
Blumentritt.
Board of Trade. 87.
Behlen synthetic oil plant—
Attacks upon. 55.
Capacity of. Table 3. 160.
Production from 1940–44. 46.
Bomber Command—
Air Officer Commanding-in-chief. 52.
R.A.F. 63, 64, 67, 70, 94, 121.
R.A.F., weight of bombs dropped on oil plants by 151, **173–183.**
Bomber output, shortage of in Germany. 143.
Bombing—
Allied, accuracy of. 151.
Allied, aiming technique. 151.
Allied, effect of weather upon. 151.
Allied policy, some German views upon. **147–148.**
Allied, tonnage of bombs dropped. 151.
Attack data, oil capacity and production for Greater Germany. Table 19, 172.
Attack data for oil targets in Greater Germany by type of target. Table 20, facing 172.
Objectives, Intermediate. 52.
Objectives, Primary. 53.
Objectives, Secondary. 53.
Of aircraft production. 54, 55.
Of Austrian and Hungarian refineries. 60.
Of aviation spirit production. 57.
Of ball bearing industry at Schweinfurt. 54, 55.
Of benzol plants. 61, 68.
Of Bohlen. 55.
Of Bruex. 127, 129, 130, 134.
Of communications in Eastern Europe. 54, 55.
Of Gelsenkirchen. 51.
Of Leuna. 51, 55, 64, 126, 129, 130, Plate 6.
Of Luetzkendorf. 55.
Of Magdeburg. 55.
Of oil plants, Speers reports to Hilter. **123–136.**
Of oil targets, chronological summary of. **173–183.**
Of pipeline terminals. 60.
Of Ploesti. 53, 55, 57, 60.
Of Poelitz. 55.
Of rail transport. 60, 78.
Of refineries in Rumania. 137.
Of refineries in Rumania, types of bombs dropped. 137.
Of road movements. 73.
Of Ruhland. 55.
Of storage in France. 70.
Of storage installations. 51, 66, 68, 69, 70, 71.
Of synthetic oil plants. 55–71, 74, 126–131, 134.
Of the underground depot at Montbartier. 61, 70.
Of the underground WIFO depot at Nienburg. 61, 70, Plate 4.
Of transportation targets. 57.
Of Zeitz. 55.
Operations and Intelligence Directorates of the Air Ministry. 59.
Overall attack data, of Combined Strategic Air Forces in the European Theatre of Operations. Table 18. 171.
Priorities. 52.
Strategic offensive against oil. 55–71.
Bomb tonnage dropped on oil targets. 57, 66.
Bombs, weight dropped on oil production facilities and effects upon capacity and production. Table 22, **184.**

Bordeaux depots, attacks upon. 70.
Bottrop synthetic oil plant, production from 1940–44. 46.
Boyle, Air Commodore A. R. 115.
Brauchitsch, von. 21, 36.
Braunkohlen Benzin A.G. 2, 103.
Brayne, A. F. L. 115, 116.
Bridgeman, Hon. M. R. 15, 108, 109, 115, 116.
Browett, Sir Leonard. 115.
Bruce, R. H. W. 91, 116.
Bruex—
Synthetic oil plant. 9, 29, Plate 2, 37, 46, 59, 67, 103, 123, 132, 136.
The capacity of. Table 3, 160.
Buchen depots, attacks upon. 70, 71.
Budafapuszta, oil field. 27.
Budapest, the fall of. 78.
Budczies. 84.
Buhle. 32, 75, 82.
Bulgaria, activities of the Kontinentale Oel A.G. in. 105.
Bulgaria, imports from Rumania. 38–43, Table 7, 163.
Bulgarische Mineraloel, A.G. 105.
Buna—
(Methanol), shortage of. 59.
Production. 129.
Stocks of. 129.
Bunker fuel, consumption of. 84.
Bunkers, shortage of. 85.

C.

C. 3. fighter fuels. 71.
Cadman, Lord. 115.
Calinescu, the assassination of. 17.
Campaign—
In North Africa. 72.
Russian. 32, 79.
Campbell, I. M. R., Captain, R.N. 115.
Cannan, G. W., Engineer-Commander, R.N. 116.
Capacities of—
Refineries in Greater Germany before June 1944. Table 4, **161–162.**
Synthetic oil plants. Table 3, **160.**
Capel-Dunn, D., Colonel. 116.
Casablanca Directive, The. 52.
Castrop Rauxel, the capacity of. Table 3, **160.**
Catalytic cracking plants underground (" MEISEN "). 153.
Caucasus—
Demolition experts sent to. 35.
Oil, German need to obtain. 33.
Cavendish-Bentinck, V. F. W. 115.
Central Office for Mineral oil production. 3.
Central Photographic Interpretation Unit (A.C.I.U.). 94.
Charley, J. P., Captain, R.N. 116.
Chatfield, Lord. 87.
Chief of Air Staff, R.A.F. 62, 87.
Ciano, Count. 38, 40.
Clifford, E. G. A., Commander, R.N. 115.
Combined Bomber Offensive Plan. 52.
Combined Chiefs of Staff. 52, 55, 62.
Combined Strategic Targets Committee. 62, 86, 94, 88, **117–122.**
Accuracy of damage assessments. 122.
Allotment of target priorities. 118, 119, 120.
Assessment of German oil production. 118, 119, 120.
Communication of recommendations. 121–122.
Research into Germany's oil distribution system. 121–122.
The constitution of. 117.
Working methods and sources of information. **118–121.**
Working Committee of. 62, **117–122.**
Working Committee terms of reference. 117.
Commanding General, United States Army Air Force. 62.
Commanding General, United States 8th Army Air Force. 52.
Commissioner for the Four-Year Plan. 96.
Committee of Imperial Defence. 87.
Committee on the Prevention of Oil from reaching Germany. 14, 87.
Committees, the composition of various. **115–116.**
Communications, dislocation of. 75, 78.
Conservation of fuel in Italy. 74.
Constanza. 28.
Consumption—
Accuracy of statistical estimates of. 95.
Alcohol as a motor fuel. Table 11, **164.**
Army. 32.
Aviation fuel, Allied estimates of. 92.
Aviation spirit, Luftwaffe. 31, 49, 50, 126, 135, 153.
Aviation spirit, military and civil in 1940–44. 43.
Aviation spirit, necessity to reduce. 128.
Bunker fuel. 84.
Civil, approximate allocation of main products in 1938. Table 10, 164.
Civil, automotive fuels, 1940–44. Table 13, **166.**
Civil, controlling organisation. 96.
Civil, restrictions in. 19, 98.
Civilian, drastic reduction in. 37.
Civilian in 1940. 23.
Cuts in Germany. 58.
Diesel oil, 1938–44. 19.
Diesel oil by the army. 49, 50, 136.
Diesel oil by the German armies in Russia. 84.
Diesel oil, civilian, 1941–44. Table 15, 168.
Diesel oil, military and civil. 43.
Economies in. 79.
Estimated in the Balkan Campaign. 31.
Estimated war-time requirements, military. 10.
Estimated war-time requirements, civil. 10.
Fuel oil, 1938–44. 19.
German army, Allied estimates of. 91–92.
German navy, Allied estimates of. 92.
Germany, Allied estimates of. 89, 91.
High temperature tars. 85.
Industrial, in 1940. 23.
In North Africa. 40.
Italy, air force. 39, 40.
Italy, army. 39, 40.
Italy, economies in. 39.
Italy, naval. 39.
Liquid fuels. 58.
Lubricating oil in Greater Germany, 1938–44, industrial and civil. Table 16, **169.**
Luftwaffe. 32.
Motor gasoline by the army. 49, 50, 135, 152.
Motor gasoline by the National Economy. 135.
Motor gasoline, civilian, 1941–44. Table 14, 167.
Motor gasoline, military and civil in 1940–44. 43.
Motor spirit, 1938–44. 19.
Navy. 32.
Occupied Territories. 23.
Of oil in the conquest of Poland. 20.
Of oil in the Western Campaigns. 20, 21, 22.
Principal oil products in Germany, military and civilian, 1936–44. Table 12, **165–166.**
Rumania, German endeavours to reduce. 25.
Treibgas. 125.
Conversion—
Of bomber units of the Luftwaffe to jet-propelled types. 144.
Of diesel vessels to steam propulsion. 85.
Of iso-butylene to iso-octane. 49.
Of military vehicles to producer-gas. 125.
Of motorised supply columns to horse-drawn transport. 75.
To substitute fuels. 19, 47.
Co-ordination of Intelligence. 90.
Cracking plants, underground, " TAUBE " and " JAKOB." 152.
Cramer. 41.
Cripps, Sir Stafford. 35.
Crude oil available in the—
Alsace area. 44.
Austrian area. 44.
Dortmund area. 44.
Hamburg area. 44.

Crude oil available in the (*continued*)—
Hannover area. 44.
Hungarian area. 44.
Rumanian area. 44.
Crude oil production—
Austria. 8, 42.
Austria, 1940–44. Table 2, 160.
Germany. 1, 8, 20.
Germany, 1940–44. Table 2, 160.
Germany, Allied calculations of. 91.
Devoli area. 38.
Heide field. 145.
Hungary, 1936–44. Table 8, 164.
Lispe oilfield. 78.
Plans in connection with. 21.
Poland. 20.
Crude oil, underground distillation units ("OFEN" and "ROESTE"). 152.
Currency questions, German organisation controlling. 96.
Czechoslovakia, the annexation of. 89.

D.

"Dachs" underground lubricating oil plants. 153.
Dacia Romana. 14.
Dalton, the Right Hon. Hugh. 19.
Damaged plants, attempts to put back into operation. 58, 59.
Danube—
Capacity of. 89.
Freezing of the. 16, 19, 39.
Mining of. 54, 60.
Plans to block. 17, 18.
River pilots, withdrawal from German service. 17.
Shipping, Allied denial to Germany of. 16.
Shipping, control of by the Goeland Transport and Trading Company. 108.
Waterways, German domination over. 31.
Davidson, F. H. N., Major-General. 115
Davidson-Houston, V., Major, R.E. 17.
Davies, Captain J. S. 121.
Davis, W. R., of Davis & Co., Inc., New York. 84.
Decoy plants, the erection of. 149.
Deggendorf. 106, 44, 71.
Defence Committee. 88.
Delegation, American. 93.
Denmark, the invasion of. 20–21.
Department of Overseas Trade. 87.
Depot at Freiham, the capture of. 83.
Depots—
Air attacks upon Montbartier. 61.
Air attacks upon major. 68, 69.
Air attacks upon Nienburg. 61, 70, Plate 4?
At Argenteuil. 69.
At Ehmen, Buchen, Dulmen, Neuburg, Ebenhausen, Hamburg, Dresden and Vienna Lobau. 70, 71.
In France and the Lowlands. 69.
At Gennevilliers and Rouen and Montbartier. 70.
At Stassfurt. 69.
Attacks upon storage at Derben, Hitzacker, Farge, Freiham, Ebrach, Regensburg and Annaburg. 71.
Capacities of. 101.
Destruction of. 75.
Extension of. 100.
Underground storage. 59, 69.
W.I.F.O. **99–103.**
W.I.F.O., list of. 101, 102.
Derben, capture of the strategic depot at. 71.
Deschowitz synthetic oil plant. 47.
Capacity of. Table 3. 160.
Deutsche, Amerikanische Petroleum G.m.b.H. 28.
Deutsche Erdoel A.G. 85, 103.
Deutsche Erdoel A.G., contract with German Navy for Heide-Meldorf oil. **145–146.**
Deutsche-Oelschiefer Forschungs-Ges. 113.
Deutsche Vacuum A.G. 44.
Dewdney, D. A. C., Wing-Commander. 116.
D.H.D. process. 132.
Diesel fuel–
Allocation to the Civil Economy, 1941–44. Table 15, 163.
Civil consumption in 1938–44. 19.
Consumption. 124.
Consumption, 1938. 10.
Consumption by the German Army. 136.
Consumption by the Germans in Russia. 84.
Consumption, civilian. 49, 50.
Consumption, military and civilian, 1940–44. 43
Consumption, requirements, civil. 10.
Consumption, requirements, military. 10.
Consumption, sea and land forces. 49, 50.
Driven vehicles. 74.
German Navy, requirements of. 85.
Imports. 49, 50.
Naval stocks of. 6.
Planned production, 1942–43. 4
Planned stocks in 1943. 5.
Production in Greater Germany, 1940–44 (all processes). Table 1, 159.
Production. 31, 49, 50, 124, 127, 130–135.
Production, military and civilian, 1940–44. 43.
Purchase of Mexican in 1939. 84.
Shortage of. 59, 85.
Stocks. 22–23, 31, 36, 49, 50.
Stocks held in commercial depots. 84.
Stocks, military and civilian, 1940–44. 42.
U-Boat quality, stocks in naval storage. 86.
Use of, in armoured vehicles. 75.
Dietrich. 76.
Dihlmann Walther. 103, 106.
Directive for the control of Strategic Bomber Forces in Europe. 62.
Directive, the Casablanca. 52.
Distillation units—
Dispersed and concealed. 67.
(Underground) "OFEN." 152.
(Underground) "ROESTE." 152.
Distribution of oil products, interruption in. 54.
Distribution organisation for liquid fuels to the Armed Forces. 69, 72.
Donn. 106.
Doenitz. 45, 77, 86.
Dortmund Area—
Annual refining capacity at. 44.
Crude oil available in. 44.
Dortmund Fischer Tropsch plant, capacity of. Table 3, 160.
Dresden, attacks upon strategic storage at. 70.
D. Section (Industrial), A.C.I.U. 119.
Dulmen, attacks upon strategic storage at. 70, 71.

E.

Ebenhausen, attacks upon storage at. 70, 71.
Ebrach, attacks upon storage at. 71.
Economic Advisory Branch, Foreign Office and Ministry of Economic Warfare. 117, 121.
Economic Objectives Unit of the United States Economic Warfare Department. 94.
Economic Warfare Division, United States Embassy. 117, 119.
Economies in oil consumption. 76, 79.
Edwards, C. L., Major, R.M. 116.
Ehmen, attacks upon storage. 70, 71.
El Alamein, Allied victory. 40.
Elwerath. 103.
Enemy Oils and Fuels Committee, Washington. 88, 90, 91, 93, 118.
Enemy Oil Intelligence Committee, Ministry of Economic Warfare. 94.
Equipment—
At oil plants indiscriminate seizure of. 154.
At oil plants, protection of. **149–150.**
Bridge building. 76.
Dismantled from French refineries. 35, 44, 106.
Horse-drawn, use of by the Germans in Italy. 74.
Refining, plans to safeguard. 53.

Estonia—
German Army retreat from. 85.
Shale oil deposits. 45.
Shale oil industry. 105.
Shale plants. 58.
Ethyl Chloride, shortage of. 140.
Ethylene Dibromide. 139.
Production. 140.
Production from Marseilles plant. 141.
Production from Tornesch plant. 141.
Stocks, in W.I.F.O. depots. 141.
Ethyl fluid—
Production. **139–142.**
Production, reasons for not attacking. 139.
Stocks. 139.
Vunerability as a target system. **139–142.**
Eurotank, Hamburg. 84.
Exports Credits Guarantee Department. 87.
Exports—
Hungary to Italy. 39.
Rumanian. 8, 10, 23, 25.
Rumanian, 1938–43, by-products. Table 6, 163.
Rumanian, 1938–43, by countries. Table 7, 163.
Rumanian, decline in. 28.
Rumanian fuel oil, reduction in. 85.
Rumanian, to Germany. 49.
Rumanian, to Italy. 39, 49.
Russian. 8, 10, 23.
To Italy, aviation spirit. 49.

F.

Falmouth Report, The. 89.
Farge, attacks upon depôt at. 71.
Faulkner, Sir Alfred. 115.
F.I.D.O. 64.
Fighter aircraft. 58.
Fighter fuels, C.3. 71.
Fighter programme. 80.
Fighter protection at synthetic oil plants. 58, 64, 128.
Fighters, total of, on all fronts. 128.
Fischer, Dr. E. R. 2, 21, 28, 65, 103.
Fischer, Dr. E. R., views upon Allied strategic bombing policy. **148.**
Fischer Tropsch—
Planned production. 5.
Plants. 46, 47.
Plants, capacity of. Table 3, 160.
Plants, underground, "KARPFEN." 153.
Plants, Working Committee (Oil) estimates of output. 122.
Flak protection at synthetic oil plants. 58, 126, 149, 150.
Flights, the curtailment of. 125.
Flying, training, reductions in. 72, 79, 80, 81.
Foggia, airfields, occupation of. 54.
Foreign Office. 16, 38, 87.
Foreign Office and M.E.W. 94.
Foster, Col. W. L. 35.
Four-Year Plan. 3–5, 31, 35.
Commissioner of the. 96.
New. 4, 90.
Office of the. 24, 96.
Revision of the. 4, 33.
Frommern, shale oil production at. 45, 113.
France—
Activities of the Kontinentale Oel A.G. 106.
Imports from Rumania. Table 7, 163.
Oil stocks. 50.
Oil stocks set aside in 1943. 81.
Removal of equipment from refineries. 106.
Storage depôts in. 69, 70.
Freiham—
Attacks upon depôt at. 71.
The capture of. 83.
Friedrichswerk oilfields. 145.
Froese, Tetra-ethyl-lead plant at. 6, 139, 141.
Production in 1940–44. 140.
Fuel—
Allocations to the three Services by the O.K.W. 80.
B.4. 41.
Fuel (*continued*)—
Capacity at W.I.F.O. depôts. 101.
Conservation of, in Italy. 74.
Consumption and production in 1944, discrepancy between. 72.
Cuts in Hungary. 58.
Depletion of supplies on the Western Front. 75.
Diesel, consumption by the German Army. 136.
Diesel, consumption, civilian. 136.
Diesel, production of. 127, 130–135.
Economy of. 76.
For bunkers, consumption of. 84.
Heavy expenditure on the Russian Front and in North Africa. 79.
High octane, fighter, the deficiency of. 32.
Oil, civilian consumption of, in 1938. 19.
Oil, difficulties in the Italian Navy. 39.
Oil, for the German Navy, shortage of. 37, 84, 85, 86.
Oil, heavy, steps taken to increase supplies of. 84.
Oil, production in Greater Germany, 1940–44, all processes. Table 1, 159.
Oil, purchases from Rumania, Hungary and Russia. 84.
Oil, shortage, effect upon industrial output. 86.
Oil, shortage, effect upon the German Navy. 84, 85, 86.
Oil, shortage, effect upon the Merchant Marine. 85.
Production, decline in. 82.
Requirements for submarines. 84.
Requirements, military. 72.
Shortage in the North African Campaign. 41.
Shortages and effect upon Panzer replacements. 73.
Shortages, crippling effect upon the Luftwaffe. 79.
Shortages, grounding of combat planes owing to. 82.
Shortages, in Africa. 74.
Shortages, military effects of. 72, 73, 76, 77, 78.
Stocks of. 50.
Supplies for armoured and armoured infantry divisions. 73.
Supplies to the German Navy. 84.
Fuels and lubricants, production from—
Alcohol. 46.
Benzol. 46.
Coal Tar. 46.
Crude Oil. 46.
Fischer Tropsch process. 46.
Hydrogenation process. 46.
Fuels—
Consumption in 1938. 10.
Consumption requirements, military and civilian. 10.
Derived from tar. 10.
High octane. 49.
Increase in the demands of the Luftwaffe. 72.
Restriction in the use of, for military training. 72, 77.
Savings in, as a result of conversion to producer gas. 48.
Substitute, the use of. 6.
Furnace oil, Naval stocks of. 6.
Funk, W., Reichsminister. 103.

G.

Galland. 64, 133.
Views upon Allied strategic bombing policy. 147.
Gapel-Doeberitz—
Production from in 1940–44. 140.
Tetra-ethyl lead plant. 6, 139, 140, 141.
Geilenberg—
Oil plant dispersal programme. 27, 30, 45, 65, 67, 152–154.
Plants. 68.
Plants, actual output from. Table 9, 164.
Plants, "OFEN." 78.
Gelsenberg synthetic oil plant. 124.
Attack upon. 134.
Capacity of. Table 3, 160.
Production from. 46.
General Staff of Supreme Headquarters, A.E.F. 56.
Generator, A.G. 48.

Generator fuels, the use of in Germany, 1942–45. Table 17. 170.
Generators, wood gas, the use of. 37.
Gennevilliers, attacks upon strategic depot at. 70.
Germany, capitulation of. 69.
Giurgiu. 18, 28.
" Giurgiu Incident." 108.
Godber, Sir Frederick F. 115.
Goeland Transport and Trading Company, Limited. 16, 17, 28.
  Control of Danube vessels. 108.
  Chartering of Greek sea-going vessels. 108.
  Purchase of vessels of the French fleet. 108.
  Schemes to cause congestion on the Danube. 108.
  The operations of. **108–109.**
Goering. 4, 21, 31, 33, 34, 36, 37, 42, 46, 63, 81, 82, 133, 149.
  Views upon Allied strategic bombing policy. 147.
Greece—
  Activities of the Kontinentale Oel A.G. in. 105.
  Imports from Rumania, 1938–43. Table 7, 163.
  The attack on. 26.
Grossbunkergemeinschaft. 97.
Guderian. 77.
Gulf Oil Company (U.S.A.). 12.

H.

Haddon, T., Lieut.-Colonel. 116.
Halder. 36.
Hall Patch, Mr. E. L., C.M.G. 15.
Hall, Professor N. F. 115.
Hamburg—
  Area, crude oil available in. 44.
  Attacks upon strategic depots at. 70.
  Eurotank refinery. 84.
Hankey—
  Committee. 14, 15, 87, 88, 89, 90, 107.
  Committee reports on trends of German production and consumption. 90.
  Lord. 14, 15, 87.
Hanover oilfields. 8.
  Crude oil available in. 44.
Harris Buxland, Mr. W. 109.
Hartley Committee. 34, 87, 88, 90, 93, 118.
  And Enemy Oil Committee, joint recommendations. 93.
  Calculations of German stocks. 93.
Hartley, Sir Harold. 87, 115.
" He. 162," adoption for large-scale production. 82.
Heide-Meldorf oil—
  Contracts between Deutsche Erdoel and the German Navy. 145–146.
  Chalk deposits, agreement with D.E.A. and the German Navy. 146.
  Fields, drilling in. 145.
  Fields, mining in. 145.
Hemmingstedt—
  Pipeline to Gruenenthal, built by the German Navy. 145.
  Refinery at. 145.
Hertslet. 84.
Hettlage. 4, 36.
  Views upon Allied strategic bombing policy. 148.
Heydebreck—
  (Blechhammer S.) 9, 36, 58, 91, 126.
  Capacity of. Table 3, 160.
  Production from. 46, 49.
High octane aviation fuels. 138.
High temperature tars, the consumption of. 85.
Hitler. 6, 8, 21, 26, 36, 37, 45, 47, 48, 57, 58, 59, 61, 64, 65, 72, 75, 77, 81, 82.
  Reports from Speer on the effects of Allied air attacks upon oil. 123–136.
Hitzacker, attacks upon strategic storage at. 71.
Hoelle, drilling in. 145.
Holland—
  Allied negotiations with. 12.
  Oil stocks in. 12.
Homberg—
  Capacity of. Table 3, 160.
  Refineries, attacks upon. 57.
Holstein. See under HEIDE-MELDORF.
Houseman, A.E., Wing-Commander. 116.
Hungary—
  Crude oil available in. 44.
  Crude oil production, importance of, in connection with supplies to the German Army. 134, 135.
  Exports of heavy fuel to Germany. 84.
  Exports to Italy. 39.
  Fuel cuts in. 58.
  German desire to safeguard oil. 77.
  German domination over. 31.
  German imports from. 27, 28.
  German organisation controlling imports from. 96.
  Imports from Rumania, 1938–43. Table 7, 163.
  Lispe oilfields. 27, 77, 78.
  Offensive in. 77.
  Problems of oil transport. 28.
  Production, crude oil, 1936–44. Table 8, 164.
  Production in 1937–44. 27.
  Railways. 54.
  Refineries, attacks upon. 59, 60.
  Refineries, lubricating oil facilities in. 27.
  Refining capacity, damage to. 28.
  The fall of Budapest. 78.
  Trade Agreement with Italy. 39.
Hydrogenation plants. 46, 47, 58, 59.
  Allied estimates of capacity of. 91.
  Attacks upon. 65, 77.
  Capacities of. Table 3, 160.
  Output of aviation and motor gasoline. 65.
  Production from. 65.
  Selected by the Germans to withstand Allied air attacks. 65.
  Smoke protection for. 149.
  Underground, " Schwalben." 152.

I.

I.G. Farbenindustrie, A.G. 2, 3, 4, 9, 42, 44, 103.
Imports—
  German, diesel oil. 50.
  German, organisation controlling. 96.
  German, from Rumania, 1938–43. 8, 10, 23, 25, 49, Table 7, 163.
  German, from Russia. 8, 10, 23.
  Into Holland, embargo upon. 12.
  Into neutral countries, the limiting of. 11.
  Italy, from Hungary. 39.
  Italy, from Rumania. 25, 39, 49.
  Italy, from Rumania, 1938–43. 25, Table 7, 163.
  Italy, aviation spirit from Germany. 49.
Industrial Intelligence Centre. 87, 88, 89, 93.
  In foreign countries. 87.
Intelligence—
  Assessment of the German oil position. **87–95.**
  Branch, Air Ministry. 92.
  The co-ordination of. 90, 93.
  Departments of the Services. 94.
  Division, Navy. 92.
  Organisation (British), results achieved by. 95.
  Sifting and dissemination of. 94.
Interpretation of reconnaissance photographs. 94.
Interrogation of prisoners of war. 94.
Iron Gates—
  Attempt to block. **110.**
  Pilots. 108.
  Plans to block. 16–18, 28.
Iso-butylene, conversion to iso-octane. 49.
Iso-octane. 83.
Italy—
  Activities of the Kontinentale Oel A.G. in. 106.
  Allocations from Germany. 40.
  Civil consumption, economies in. 39.
  Crippling shortages of fuel. 40.
  Gasoline requirements, Air Force. 39, 40.
  Gasoline requirements, Army. 39, 40.
  German use of horse-drawn equipment in. 74.

Italy (*continued*)—
Imports from Rumania. 25, 49, 39.
Imports from Rumania, 1938–43. 25, **Table 7.** 163.
Import of aviation spirit from Germany. 49.
Naval fuel difficulties. 39.
Navy, consumption. 39.
Navy, oil stocks. 38.
Navy, shortage of oil in. 85.
Oil requirements of. 38.
Oil resources. 38.
Oil stocks. 38.
Question of limiting supplies to. 38.
Refineries at Bari. 38.
Refineries at Leghorn. 38.
Refineries, production from. 40.
Shortage of fuels. 106.
Trade agreement with Hungary. 39.
Trade agreement with Rumania. 39.

J.

"JAKOB" underground cracking plants. 152.
Janszen. 84.
Jet—
Aircraft. 67, 82.
Aircraft production programme. 141.
Fuels, J. 2. 71, 131.
Fuels, J. 2, output from underground catalytic cracking plants. 153, 154.
Fuels, J. 2, requirements of the Luftwaffe. 78.
Fuels, J. 2, requirements of the Wehrmacht. 136.
Jodl. 24, 26, 31, 34, 77.
John. 70, 72, 73.
Joint Intelligence Sub-Committee. 67, 88.
Joint Oil Targets Committee. 59, 62, 67, 68, 117.
Joint Photographic Reconnaissance Committee. 120.

K.

Kamen-Dortmund—
Synthetic oil plant at. 47.
Synthetic oil plant, capacity of. **Table 3,** 160.
Karinhall Plan. 4, 35.
"**Karpfen**" underground Fischer-Tropsch plants. 153.
Kehrl. 30, 42.
Keitel. 21, 33, 39, 64, 72, 75, 80, 123.
Kent, Flight-Lieutenant P. E. 120.
Keppler. 4.
Kerosene production in Greater Germany, 1940–44, all processes. Table I, 159.
Kilbey, Mr. S., 90.
Kisch, Sir Cecil. 115.
Kohle Oel Union. 113.
Koller. 31, 34, 45, 71, 77, 78–79.
Kontinentale Oel A.G. 14, 45, **103–106.**
Activities of the. 26, 103.
Activities in Albania. 106.
Activities in Baltic area. 105.
Activities in Bulgaria. 105.
Activities in France. 106.
Activities in Greece. 105.
Activities in Italy. 106.
Activities in Poland. 105.
Activities in Rumania. 103.
Capital of. 103.
Contract with Astra Romana. 103.
Equipment held in Germany by. 103.
Management of. 103.
Participation in Rumanian oil companies. 103.
Participation in oil exploitation activities in the St. Marcet area. 106.
Personnel. 103.
Shareholders in. 103.
Krauch, Dr. 2, 4, 21, 23, 31, 33, 35, 36, 45, 47, 51, 58, 65, 89, 96, 149, 152.
Kreipe. 79, 80.
Kruse. 75.
"Kybol," production of. 106.

L.

Lammers, Fritz. 64.
Landeswirtschaftsaemter. 98.
Leghorn refinery. 38.
Levant Plan. 15, 107.
Leuna. 4, 9, 46, 56, 59, 63, 64, 151, Plate 1.
Attacks upon. 51, 55, 67, 123, 124, 126, 127, Plate 6.
Capacity of. 160, Table 3.
Flak defences. 136.
Recovery of production. 66, 127.
Liberators. 53.
Lieth oilfield. 145.
Liquefied gases. 159.
Lispe oilfields. 27, 78.
Lloyd Committee. 87, 88, 90.
Lloyd, Geoffrey. 87, 115.
Lobau refinery and storage. 70.
London Naval Treaty of 1935. 84.
Low Countries—
Occupation of. 20, 21.
Depots in. 69.
Oil stocks in 1943. 81.
Lovaszi oilfield. 27.
Lubricants—
Aviation. 10.
Civil consumption. 10, 138, 169, Table 16.
Crisis in 1943. 20.
From Hungary. 27.
From synthetic plants. 46.
Production. 4, 50, 106, 138, 159, Table 1.
Regeneration of. 138.
Stocks. 5.
Storage capacity. 101.
Yield of crudes. 138.
Lubricating oil—
Plants, underground. 153.
Target system. 68.
Ludendorff, Generalfeldmarschall. 1, 56.
Ludwigshafen/Oppau. 9, 46, 151.
Capacity of. 160, Table 3.
Luetzkendorf synthetic plant—
Attacks. 55, 127.
Production. 46, 160, Table 3.
Luftwaffe. 2, 3, 21, 22, 23, 33, 35, 41, 46, 58, 65, 66, 80, 81, 82, 125.
Attacks upon Britain. 31.
Attacks upon Bomber Command airfields. 83.
Attacks upon Remagen bridgehead. 83.
Attacks upon Western Front airfields. 83.
Blending of fuel for. 69, 71, 131, 139.
Consumption of fuel. 31, 32, 49, 78, 92, 126, 135, 153.
Effect of fuel shortage. 79, 83, 85, 143, 144, 150
Fuel depots. 100.
Fuel reserves. 70, 71, 81, 82, 83.
Plans to suppress. 54.
Restrictions in training. 73, 79.
Storage of fuel. 71.
Lufttanklager. 79.
Lurgi. 113.
Lyons, attacks upon depots. 70.

M.

Macdougall, G. D. 116.
Magdeburg, synthetic oil plant at—
Attacks upon. 55.
Capacity of. Table 3, 160.
Production from. 46.
Maikop oilfield, the occupation of. 33, 34.
M.A.N.A.T. 27.
Manteuffel. 75, 76.
M.A.O.R.T. 27.
Market policy, organisation controlling. 96.
Martin. 47.
Massow van, Gen. Major, views upon Allied strategic bombing policy. 148.
Medhurst, C. E. H., Air Vice-Marshal. 115.
Mediterranean area, the closing of, to British shipping. 107.
"MEISEN," underground catalytic cracking plants. 153.

Methanol—
Disruption in the production of. 69.
Plant at Auschwitz. 9.
Production. 129.
Shortage of. 59.
Stocks of. 129.
Mexico, German transactions with. 84.
Meyer, C. E. 116.
M.I. 10 (c), War Office. 92, 117, 121.
Milch. 29, 33, 34, 35, 79, 80, 81, 149.
Views upon Allied strategic bombing policy. 147.
Military—
Consumption, British calculations of. 91, 92.
Consumption, 1940. 23.
Consumption, 1941. 32.
Consumption, 1940–44. 43.
Consumption in the Western Campaigns. 21, 23.
Consumption, diesel oil in 1940–44. 43.
Consumption, motor gasoline in 1940–44. 43.
Consumption, principal oil products in Germany in 1930–44. **Table 12,** 165, 166.
Fuel requirements. 72, 73.
Reserve stocks, plan to eliminate. 57.
Stocks of aviation spirit. 36.
Stocks of aviation spirit, the exhaustion of. 37.
Stocks of diesel oil. 36.
Stocks of motor gasoline. 36.
Consumption, stocks of motor gasoline, the exhaustion of. 37.
Supplies, distribution of liquid fuels. 69.
Training, restriction in the use of fuels for. 72, 77.
Transport and supplies, plan for the use of air power against. 56.
Mineraloeleinfuhrges. m.b.H. 96.
Mineraloel G.m.b.H., Suedost. 105.
Mineraloel Vertrieb Serbien, A.G. 106.
Mines Department, Petroleum Division. 14, 87.
Mining of the Danube. 54, 60.
Ministry Armaments and War Production (German). 96.
Ministry of—
Economic Warfare. 11, 15, 16, 38, 59, 87, 88, 90, 91, 94.
Fuel and Power. 94, 118.
Home Security, R.E. 8. 117.
Shipping. 87.
War Transport. 16.
Misburg refinery. Plate 3.
Model. 76.
Molotov. 26.
Montbartier (France)—
Air attacks upon storage. 61, 70.
Storage depôt at. 70, 71.
Moosbierbaum. 44, 46, 63, 124, 132.
Capacity of the plant at. Table 3, 160.
Morton, D. J. F., Major. 115.
Motor fuel, use of propane-butane as a. 47.
Motor gasoline—
Army supplies of. 65.
Civil consumption, 1938–44. 19.
Consumption. 124.
Consumption by the Civil Economy. 135.
Consumption by the German Army. 135, 152.
Consumption, military and civil, in 1940–44. 43.
Economies in the use of. 74, 75.
Planned production, 1942–43. 4.
Planned stocks, 1943. 5.
Production. 31, 65, 67, 124, 127, 130, 131, 132, 135.
Production, 1940–44. 43.
Production, Greater Germany in 1940–44, all products. Table 1, 159.
Requirements of the Italian Army and Air Force. 39, 40.
Shortage. 68, 76.
Stocks. 22, 23, 31, 36, 50, 53.
Stocks, 1940–44. 43.
Stocks, the exhaustion. 37.
Supplies for the armies in the field, reduction in. 75.
Munich crisis. 87.
Mussolini. 38, 40.
Nagel, Jakob. 48.
Navy—
German, consumption of oil. 32.
German, estimated consumption. 92.
German, oil shortage. 37, 85.
German, oil storage. 3, 5.
German, oil supplies. 84, 85, 145, 146.
Italian, oil stocks. 38.
Italian, oil difficulties. 39.
Naval Blockade. 11.
Naval Intelligence Division. 92.
"Navicert" System. 11.
Nesbitt, Major-General F. G. B. 115.
Neuburg, strategic storage at. 70, 71.
Neuenheerse, strategic storage at. 71.
Nicholl, A. D., Captain, R.N. 115.
Nienburg, **WIFO** depot at. 61, 70, Plate 4.
Nitrogen—
shortage of. 59, 69.
Production. 129.
North African Campaign. 40, 72.
Norway—
Invasion of. 20, 21.
Oil stocks in 1943. 81.

O.

Objectives, bombing—
Intermediate. 52.
Primary. 53.
Secondary. 53.
Oberkommando—
Der Luftwaffe. 80.
Der Luftwaffe Reserve. 81.
Wehrmacht Wirtschaftsruestungsamt. 72.
"OBOE"—
Precision ground control. 151.
Stations. 63.
"OFEN"—
Underground distillation units. 27, 78, 152.
Underground distillation units production, actual output from, **Table 9,** 164.
Production from. 154.
Office of the Four-Year Plan. 96.
Oil- and gasfields, summary of German. **111–112.**
Oilfields—
Albania. 38.
Alsace area, annual capacity at. 44.
Austrian, annual capacity at. 44.
Austria, increase in the production from. 42.
Austria, planned and actual production, 1942–43. 36.
Austria, the exploitation of. 36.
Devoli area. 106.
Dortmund area, annual capacity at. 44.
Emilia. 38.
Equipment, German supplied to Rumania. 103.
Friedrichswerk. 145.
Hamburg area, annual capacity. 44.
Hanover area, annual capacity. 44.
Heide, drilling in. 145.
Heide, mining in. 145–146.
Heide-Meldorf, contracts between Deutsche Erdoel A.G. and the German Navy for, oil from. 145–146.
Hoelle, drilling in. 145.
Hungarian area, annual capacity at. 44, 77
Tintea, plans to destroy the high-pressure wells. 18.
Oil industry—
Immobilisation. 69.
Organisation. 1–2.
Total number of persons employed in Greater Germany. 65.
Oil Plant Dispersal Programme. 27.
Oil position in Germany in July 1939. 7.
Oil problems (Vienna Conference). 28.
Oil transport. 28.
O.K.W. Reserve. 79.
Operational intelligence centre of the Admiralty. 92.
Organisation—
(Allied) set up to study the German oil position. 87.
Cl[illegible]ministration. 42.

Organisation (*continued*)—
For distribution of oil supplies for German Army. 69, 72.
Repair of oil plants. 64.
German oil administration. 96, 97, 98.
"Overlord" Plan to invade the Continent. 55.
Owtram, T. C., Major. 116.

P.

"P. 51" fighters. 54.
Paimboeuf. Tetra-ethyl-lead plant. 139, 140, 141.
Paratroops, the disbanding of German training schools. 80.
Panzer Army—
6th S.S. 76, 78.
Lack of training. 76.
Lehr Division. 76.
Pernis refinery at Rotterdam. 12.
Personalities (German). 155–158.
Personnel, aircrew, effects of oil shortage upon replacements. 143–144.
Personnel employed in the oil industry in Greater Germany. 65.
Petroleum Attaché to the United States Embassy. 88, 117.
Photographic Reconnaissance. 92, 120, 121.
Interpretation of. 94, 125.
Pipeline—
Construction. 3.
Hemmingstedt to Gruenenthal built by German Navy. 145.
Terminals, air attacks against. 60.
Pipelines. 28, 29.
Across Rhine. 29.
Planned. 79.
Ploesti to Giurgiu. 29.
Ploesti to Constanza. 29.
Proposed, Odessa to Upper Silesia. 29.
Plan—
Combined bomber offensive. 52.
For aviation spirit (Goering). 33.
For employment of Strategic Air Forces. 56.
For increased drilling and exploration. 8.
For increasing synthetic oil output, the modification of. 36.
For use of Air Power against German military transport and supplies. 56.
Four-Year. 2.
Four-Year, New. 4.
Four-Year, Office of. 96.
Four-Year, revision. 4, 33, 35.
Geilenberg, for plant dispersal and concealment. 152–154.
Geilenberg, for plant dispersal and concealment, failure of. 153.
Karinhall. 4.
Petroleum, for Europe. 24.
"Rankin." 55.
Special for aviation spirit and light metals of 23rd June, 1941. 35, 36.
The Levant. 15, 107.
To eliminate military reserve stocks. 56.
Planned—
And actual production from Austrian oilfields in 1942 and 1943. 36.
Aviation spirit production in 1942. 33.
Crude oil production in 1943. 34.
Erection of underground plants for the manufacture of ancillary products. 152–154.
Production from German raw materials, 1936–40. 4,
Planning for war. **1–5.**
Plans—
In connection with crude oil production. 21.
In connection with producer gas. 21.
In connection with synthetic oil plants. 21.
To block Danube. 17, 18.
To deny Rumanian oil to Germany. **14–18.**
To deny the use of Danube vessels to Germany. 16.
To destroy Rumanian refineries and fields. 17.

Plans (*continued*)—
To destroy the Iron Gates. 16–17.
To disrupt rail communications. 55.
To invade Continent ("Overlord"). 55.
To suppress Luftwaffe. 54.
Plant repair organisation, the achievement of. 64.
Plants—
Benzol, bombing of. 61, 68.
Benzol, priority of. 68.
Dispersed and concealed, distillation. 67.
Shale, Estonia. 58.
Underground. 65, 66.
Planungsamt. 2, 45, 47, 51, 72, 96.
Ploesti. 17.
Capture. 74.
Colombia Aquila Refinery. **Plate 8.**
Oilfields, Russian occupation. 60.
Refineries. 52.
Refineries, low level attacks upon. 53.
Russian air attacks upon. 18.
Use of smoke screens at refineries. 60.
Poelitz synthetic oil plant. 59, 63, 64, 124, 132, 136.
Air attacks upon. 55, 67, 123, 129, 130.
Capacity. Table 3, 160.
Elimination of. 134.
Production from. 46.
Production recovery. 46.
Poland—
Activities of the **Kontinentale Oel** in. 105.
Attacks upon refineries in. 59.
Crude oil production. 20.
German control of oilfields and refineries. 20.
Pollock, C. M. Major. 121.
Portland Cement Co. 113, 114.
Portugal, Allied negotiations with. 12.
Pre-emption. 14–16.
Preussag. 103.
Prisoners of war (Germany), Interrogation of. 94.
Producer gas—
British estimates of saving in gasoline by use of. 92.
Conversion of military vehicles. 58.
Conversion of vehicles. 125.
Conversion to. 47.
Inadequacy. 130.
Plans in connection with. 21.
Programme (Schieber). 48.
Use in armoured vehicles. 75, 76, 125.
Production—
Accuracy of statistical estimates of. 95.
Aviation spirit. 43, 49, 58, 63, 66, 67, 82, **123–124,** 126, 127, 128, 130, 131, 134, 135, 140.
Aviation spirit, an increase in. 83.
Aviation spirit, air attacks against. 57, 123.
Aviation spirit from underground plants. 152.
Aviation spirit in 1944. 133.
And consumption in 1944, discrepancy between. 72.
Benzol. 9, 10, 61.
Bottled Treibgas. 127, 130.
Buna. 129.
Capacity of aviation spirit in Germany. 79.
Capacity, and attack data for Greater Germany, summary of. **Table 19,** 172.
Crude oil. 1, 8, 21, 44.
Crude oil, Austrian oilfields. 78.
Crude oil, Austria, 1940–44. 44, **Table 2,** 160.
Crude oil, Austria, increase in. 42.
Crude oil, Devoli area. 38.
Crude oil, Germany, 1936. 4.
Crude oil, Germany, 1940–44. 44, **Table 2,** 160.
Crude oil, Hungary, 1936–44, **Table 8,** 164.
Crude oil, Lispe oilfields. 78.
Crude oil, Poland. 20.
Diesel oil. 31, 43, 49, 127, 130, 131, 132, 134, 135.
Ethyl fluid. **138–142,** see also under T.E.L.
Ethyl fluid, reasons for not attacking. 139.
Ethylene-dibromide. 140.
Ethylene-dibromide, from Tornesch plant. 141.
Fuel, decline in. 82.
Fuels and lubricants from alcohol. 46.
Fuels and lubricants from benzol. 46.
Fuels and lubricants from coal tar. 46.

Production (*continued*)—
Fuels and lubricants from crude oil. 46.
Germany, Allied calculation. 91, 119.
German organisation controlling. 96.
Greater Germany, 1940–44 (all processes). Table 1, 159.
Greater Germany, 1940–44, aviation spirit. Table 1, 159.
Greater Germany, 1940–44, diesel oil. Table 1, 159.
Greater Germany, 1940–44, fuel oil. Table 1, 159.
Greater Germany, 1940–44, kerosene. **Table 1,** 159.
Greater Germany, 1940–44, lubricating oil. **Table 1,** 159.
Greater Germany, 1940–44, motor gasoline. **Table 1,** 159.
High octane aviation spirit. 9.
Hungary, supplies for the German Army. 134, 135.
Jet aircraft. 82.
Immobilisation. 68.
Loss in tons per ton of bombs dropped on oil targets. Table 22, 184.
Lubricating oil. 50, 138.
Lubricating oil from underground plants. 153.
Methanol. 129.
Methanol, disruption in. 69.
Mineral oil, Central Office for. 3.
Motor gasoline. 31, 43, 67, 127, 130, 131, 132, 134, 135.
Nitrogen. 129.
Nitrogen, disruption in. 69.
Oil shales, Southern Germany. **113–114.**
Planned, 1942–43, aviation spirit. 4.
Planned, 1942–43, diesel oil. 4.
Planned, 1942–43, fuel oil. 4.
Planned, 1942–43, lubricating oil. 4.
Planned, 1942–43, motor gasoline. 4.
Planned, crude oil, 1943. 34.
Planned, from Austrian fields, 1942–43. 36.
Planned, from German raw materials, 1936–40. **4.**
Refineries and synthetic oil plants, Target Committee estimates and actual output. 122.
Rumania, decline in. 19.
Shale oil, Estonia. 45.
Summary of the oil and gas fields of Germany. 111–112.
Shale oil from underground plants. 152.
Synthetic oil, by hydrogenation. 9, 46.
Synthetic oil, by Fischer Tropsch process. 46.
Synthetic oil, expansion of industry. 45.
Synthetic oil, Germany, Allied calculation of. 119.
Synthetic oil, increase in. 9, 80.
Synthetic oil, plants. 46, 61, 80, see also " Synthetic oil plants production."
Tetra-ethyl-lead from three major plants, 1940–44. 140.
Propane-butane gas, use as motor fuel. 47, 48.
Programme—
Fighter. 80, 82.
Destruction of tactical reserves in France. 70.
Development of armaments. 72.
Reconstruction of synthetic oil plants (Geilenberg). 64.
Geilenberg dispersal and concealment. 27, 30, 45, 67, **152–154.**

R.

Raeder, Admiral. 21, 39.
R.A.F. Bomber Command. 57, 63, 64, 67, 94, 121, 151.
Weight of bombs dropped on oil targets by. 151, 173–183.
Rail—
Communications, bombing of. 30, 60, 78.
Communications, plans to disrupt. 55.
Movements to Western Front, 1944. 30.
Tank cars in use on W.I.F.O. account. 100.
Transport, difficulties in. 86.
Transport, efficiency. 30.
Railways—
Hungarian. 54.
Rumanian. 54.
Railway system, disruption of. 30.
" Rankin " Plan. 55.
Raw Materials Department and Planning Department (Germany). 42.
Refineries—
Attacks upon. 51, 129.
Attacks upon, Ploesti. 55, 57, 59, 60.
Austria, attacks upon. 60.
Available to Germany. 9.
Bombing policy. 120.
Chronological summary of strategic attacks upon. **Table 21, 173–183.**
Defence measures. 126, 149–150.
Defence measures, cost of labour and materials. 149.
France, removal of equipment from. 35, 44, 106.
Greater Germany capacities, before June 1944. **Table 4, 161–162.**
Hungary, attacks upon. 60.
Italy, production from. 40.
Lubricating oil facilities, Almasfuzito. 27.
Lubricating oil facilities, Csepel. 27.
Lubricating oil facilities, Szoeny. 27.
Ploesti. 52.
Poland, German control. 20.
Recommendations to destroy. 93.
Resistance to air attacks. 122.
Rumania, output from. 57.
Rumania, results of bombing. **137.**
Rumania, types of bombs dropped upon. 137.
Tons of production loss from, per ton of bombs dropped. 94.
Refinery—
Bari. 38, 39.
Colombia Aquila, Ploesti. **Plate 8.**
Deurag-Nerag at Misburg after attacks. **Plate 3.**
Fiume. 106.
Hemmingstedt. 145.
Hemmingstedt, bombing of. 146.
Kolin. 29.
Leghorn. 38.
Low-level attack upon Ploesti. 53.
Of the Deutsche Erdoel at Schwechat. 44.
Rhenania Ossag, A.G., attack upon. **Plate 5.**
Throughput loss in. 137.
Trieste. 106.
Trzebinia, Poland. 44.
Vienna–Lobau. 29.
Refining—
Capacity available for German and Austrian crude oil. 9.
Capacity available to Germany. 44.
Equipment plans to safeguard. 53.
Capacity, Hungary, damage to. 28.
Lispe crude oil. 78.
Regensburg, attacks upon storage at. 71.
Reichsministerium fuer Ruestung und Kriegsproduktion. 98.
Reichstelle fuer Mineraloel. 2, 84, 96, 97, 98.
Reichswirtschaftsministerium. 2, 96.
Repairs—
Synthetic oil plants. 59, 64.
Synthetic oil plants, personnel engaged in. 65.
Reserve—
" Fuehrer's." 81.
Naval strategic. 85.
Oberkommando der Luftwaffe. 81.
O.K.W. 79.
Reserves—
Depletion of. 79.
Emergency. 81.
Restrictions in flying training. 80.
Rhenania Ossag A.G. 44, **Plate 5.**
Attack upon. **Plate 5.**
Rickelsdorf oilfield. 145.
Ritter. 33.
Rivelina, Captain M. 121.
Robertelli. 33, 39.
" ROESTE "—
Underground distillation units. 152.
Underground distillation units, production from. 154.
Underground distillation units, actual output from. **Table 9,** 164.

Rohstoffamt. 2, 48, 96.
Rommel. 40, 41.
Rossi. 40.
Rouen strategic storage, attacks upon. 70.
Ruhland synthetic oil plant. 68.
Attacks upon. 55.
Capacity. **Table 3,** 160.
Ruhsert. 34, 79.
Rumania—
Activities of the Kontinentale Oel A.G. in. 103.
Allied embargo upon Rumanian imports from the United Kingdom. 15.
Astra Romana, contract with Kontinentale Oel A.G. 103.
Attacks upon Ploesti. 55, 57, 59, 60, 67, 123, 129, 130.
Attempts to block Iron Gates. **110.**
British influence in. 108.
Capture of Ploesti. 74.
Consumption, 1938–43. 25.
Consumption, German endeavours to reduce. 25.
Exportable surplus, German control over. 26.
Exports by countries, 1938–43. **Table 7,** 163.
Exports, by-products, 1938–43. **Table 6,** 163.
Exports of crude oil, 1938–43. 25.
Exports of heavy fuel oil to Germany. 84.
Exports of heavy fuel oil, reduction in. 85.
Exports to Germany. 49, 137.
Exports to Germany, 1938–43. 25.
Exports to Italy. 39. 49.
Exports to Italy, 1938–43. 25.
German domination over. 31.
German oil policy in. 25.
German organisation controlling imports from. 96.
Inflation in. 14.
Low-level attack upon Ploesti. 53.
Oilfields at Plœsti, Russian occupation of. 60.
Oilfields, proposal to " Scorch." 17.
Participation of the Kontinentale Oel A.G. in oil companies in. **104.**
Problems of oil transport. 28, 29.
Production and transport, German control of. 16.
Production, decline in. 19.
Railways. 54.
Refineries, results of bombing. **137.**
Restriction of German imports from. 14.
Special delegation to Bucharest. 26.
Supplies to the German Army. 53.
Supplies to Turkey. 13.
The Iron Gates. 16, 28.
Trade Agreement with Italy. 39.
Rumanian oil—
Industry, 1938–43, statistics of. **Table 5,** 163.
Rise in cost of. 107.
Shipped to the United Kingdom. 14–15.
The Levant Plan for the purchase of. 107.
The strategic importance of. 14.
War Cabinet policy to deny supplies to Germany. 14.
Quota system. 15.
Rumaenien Mineraloel G.m.b.H. 96.
Rundstedt, von. 73, 75, 76, 77.
Russia—
Air attack upon Ploesti. 18.
Attack upon. 31.
Battle of Stalingrad. 35.
Demolition experts sent to the Caucasus. 35.
Exports of heavy fuel oil to Germany. 84.
Participation of Kontinentale Oel A.G. in oil companies. **105.**
Russian Campaign. 32, 79.
Break out from the Baranovo bridgehead. 77.
Occupation of Ploesti oilfields. 60.

S.

Saur. 31, 65, 153.
Schell, von. 21, 48.
Schkopau. 129.
Schleswig-Holstein, development of the oil chalk deposits in. 85.
Schoerzingen shale oil production from. 45, 113.
Scholven—
Attacks upon. 57, 134.
Capacity of. Table 3, 160.
Production from. 46.
Synthetic oil plant. 124, 127.
SCHWALBEN underground hydrogenation plants. 152.
Schwechat refinery. 44.
Schweinfurt, air attacks upon. 54.
S.H.A.E.F. 54, 117, 121.
Shale oil—
Deposits, the exploitation of. 45.
Development of, in Southern Germany. **113, 114.**
Plants, construction of. 113.
Plants, Estonia. 58.
Underground plants, WUESTEN. 152.
SHELL HELLAS, Limited. 105.
Shepard, D. A. 116.
Shipping—
Allied attacks upon. 40.
Danube, dislocation of. 60.
Danube, Goeland Transport and Trading Co., Ltd., control of. 108.
Shortage of—
Aviation spirit. 37, 150.
Aviation spirit, components. 68.
Aviation spirit, crippling effect upon the Luftwaffe. 79, 81.
Aviation spirit, curtailment of flights. 125.
Aviation spirit, effects upon air crew replacements. **143–144.**
Aviation spirit, grounding of combat planes owing to. 81.
Bunkers. 85.
Buna (methanol), methanol, nitrogen. 59.
Chlorsulphonic acid. 149.
Coal. 77.
Diesel fuels. 59, 85.
Ethyl-chloride. 140.
Fuel. 58.
Fuel, effect upon Panzer replacements. 73.
Fuel, effect upon the German Navy. 84, 85, 86.
Fuel, effect upon the Merchant Marine. 85.
Fuel, effect upon the outcome of Ardennes counter-offensive. 75.
Fuel in Africa. 74.
Fuel in Italian Navy. 39, 85.
Fuel in North African Campaign. 41.
Fuel, limitation of commitments in North Sea owing to. 39.
Fuel, military effects. 72–78.
Fuel, restriction in offensive operations by U-boats. 85.
Fuel supplies to the aircraft industry. 81.
Heavy fuel oil for Naval purposes. 37.
High octane fighter fuel. 32.
Labour in synthetic oil industry. 35.
Lubricants in occupied countries. 138.
Motor gasoline. 1, 37, 68, 76.
Motor gasoline for training purposes. 77.
Oil, effects upon industrial output. 86.
Oil for national economy. 86.
Pumps. 28.
Steel. 4, 36.
Steel and labour, effects upon construction of oil plants. 33.
Substitute fuels. 60.
Simon, Captain A. L. 121.
Simon, Captain L. J. 121.
Smoke screens at Ploesti. 60.
Socony, Vacuum Oil Co. (U.S.A.). 12, 107.
Spaatz, General C., U.S.A.A.F. 95.
Spain—
Accumulation of stocks in. 12.
Allied negotiations with. 12.
Exports of wolfram to Germany. 12.
Limitation of imports. 12.
Special Commissioner for repair operations. 59.
Special plan for aviation spirit and light metals of the 23rd June, 1941. 35.

Speer. 33, 35, 42, 45, 47, 48, 56, 57, 58, 59, 60, 61, 63, 64, 65, 66, 72, 75, 76, 77, 80, 96, 150.
Conferences at Fuehrerhaupt quartier. 58, 65.
Ministry. 42.
Reports to Hitler on effects of attacks against oil plants. 123–136.
Views upon Allied strategic bombing policy. 147.
Spies, Gen.-Ing., views upon Allied strategic bombing policy. 148.
Standard Oil Co. (U.S.A.). 12.
Starling, F. C. 115.
Stassfurt, depot. 69.
Steaua Romana (Rumania). 14.
Steaua Romana (British), Ltd. 107.
Steel, shortage. 4, 36.
Steinmann. 79.
Stephens, W. D., Captain, R.N. 115.
Sterkrade Holten synthetic oil plant—
Capacity. Table 3, 160.
Attacks upon. 57.
Stock losses, 1941. 32.
Stocks—
Army. 75.
At Argenteuil for German Army in France. 69.
At refineries, attacks upon. 51.
Aviation spirit. 22, 23, 50, 66, 83, 141.
Aviation spirit, 1940–44. 43.
Aviation spirit held by Luftwaffe. 81.
B 4 fuel, depletion of. 41.
Buna. 129 .
British estimate of German. 88, 89, 91.
Captured. 22, 23, 76.
Depletion of. 79, 81.
Diesel oil. 22, 23, 49, 50.
Diesel oil, 1940–44. 43.
Diesel oil in commercial depôts. 84.
Estimates of German, prepared by Soviet Government. 89.
Ethylene-dibromide. 139.
Fuehrer's reserve. 81.
Fuel. 50, 78, 82.
Fuel in commercial depôts. 84.
Hartley Committee, calculations of. 93.
In France. 50.
In Italian Navy. 38.
In hand at W.I.F.O. depôts. 3.
Low level of. 6.
Methanol. 129.
Military, aviation spirit. 36.
Military, diesel oil. 36.
Military, motor gasoline. 36.
Motor gasoline. 22, 23, 50, 53.
Motor gasoline, 1940–44. 43.
Military, plan to eliminate. 56.
Naval diesel oil. 6.
Naval, strategic reserve. 85.
O.K.W. reserve. 81.
Of all products in July 1939. 7.
On airfields held by Luftflotte. 3, 81.
Planned, 1943, aviation spirit. 6.
Planned, 1943, diesel oil. 5.
Planned, 1943, fuel oil. 5.
Planned, 1943, lubricants. 5.
Planned, 1943, motor gasoline. 5, 6.
Set aside in 1943 in Norway, France and the Lowlands. 81.
Shortage of. 72.
Tetra-ethyl-lead fluid. 6, 139.
Three main products, 1943. 42.
Tied, at W.I.F.O. depôts. 100.
U-boat quality diesel oil in Naval storage. 86.
Withdrawals from. 32.
Storage at—
Annaburg. 71.
Argenteuil. 69.
Buchen. 70, 71.
Derben. 71.
Dresden. 70, 71.
Dulmen. 70, 71.
Ebrach. 71.
Storage at (*continued*)—
Ebenhausen. 70, 71.
Ehmen. 70, 71.
Farge. 71.
Hamburg. 70, 71.
Hitzacker. 71.
Neuburg. 70, 71.
Neuenheerse. 71.
Regensburg. 71.
Vienna Lobau. 70, 71.
W.I.F.O. depôts. 29.
Storage—
Attacks upon. 51, 69, 121.
Attacks upon the W.I.F.O. depôt at Nienburg. 61, 70, Plate 4.
Aviation fuel. 79, 81.
Baltic and North Sea ports. 84.
Bombing of major depôts in the distribution system. 68.
Capacity, Naval. 3.
Chronological summary of attacks upon. **173–183,** Table 21.
Commercial. 5.
Concealed, installations in vicinity of principal depôt at Freiham, capture of. 83.
Extension of. 100.
Government. 5.
Naval. 5.
Naval, stocks of U-boat quality diesel oil in. 86.
Strategic, construction of. 3.
Strategic, total planned capacity of. 3.
Targets. 121.
Targets, omission from bombing priority list. 68.
Underground. 59, 69.
Underground, attacks upon Montbartier. 61.
Underground, depôt at Montbartier. 70.
W.I.F.O., list of principal depôts. **101–102.**
Strategic bombing forces. 59, 62.
Strategic bombing offensive against oil. **55–71.**
Strategic reserves, attacks upon. 70.
Strategic reserves in France, programme to destroy. 70.
Strategic reserves of German Army in France. 69.
Strategic reserves of Luftwaffe in France. 69.
Student. 80.
Substitute fuels—
Inadequacy of. 130.
Shortage of. 60.
Use of, in Germany. 19, 47, 92, 170, **Table 17.**
Sueddeutsche Bau G.m.b.H. 106.
Supreme Commander Allied Expeditionary Force. 55.
Sweden—
Allied negotiations with. 13.
Imports of German lubricating oil. 13.
Switzerland, imports from Rumania, 1938–43. Table 7, 163.
Synthetic oil, attacks upon—
Bohlen. 55.
Bruex. 59.
Leuna. 55, 59, Plate 6, 67.
Luetzkendorf. 55.
Magdeburg. 55.
Poelitz. 55, 59, 67.
Ruhland. 55.
Scholven. 57.
Sterkrade. 57.
Wesseling. 57.
Plants. 57, 59, 74, 85, 126, 127, 128, 129, 130, 131. 134.
Synthetic oil—
Bergius Hydrogenation. 9, 46.
Bombing policy with regard to plants. 120.
Capacities of plants. Table 3, 160.
Change in policy regulating re-attack upon plants. 63.
Chronological summary of attacks upon. **173–183.**
Construction of plants. 58.
Construction of plants, slow progress in. 36.
Defence measures for plants. 65, 126, 128, **149–150.**

Synthetic oil (*continued*)—
Defence measures for plants, costs of labour and materials. 149.
Fighter protection at plants, decrease in. 128.
Fighter protection at plants, increase in. 58.
Fischer Tropsch production. 9, 46.
Flak protection at plants. 150.
Increase in production of. 9, 80.
Industry, development. 2, 37, 45.
Industry, man-power. 58.
Industry, shortage of man-power. 35.
Lack of protection at plants. 56.
Plans for output, modification in. 36.
Plans in connexion with plants. 21.
Synthetic oil plant at—
Auschwitz. 49, 9.
Blechhammer. 9, 36.
Bruex. 9, 29, 37, 103, **Plate 2.**
Deschowitz. 47.
Gladbeck. 36.
Heydebreck. 9, 36.
Leuna. 1, 9, **Plate 1.**
Ludwigshafen-Oppau. 9.
Zeitz. 36.
Synthetic oil plants in Silesia—
Capture by the Russians. 67.
Immobilisation of. 66, 67.
Synthetic oil production, 1940–44, from—
Blechhammer. 46.
Bohlen. 46.
Bottrop. 46.
Bruex. 46.
Gelsenberg. 46.
Heydebreck. 46.
Kamen Dortmund. 46.
Leuna. 46, Plate 1.
Ludwigshafen. 46.
Lutzkendorf. 46.
Magdeburg. 46.
Moosbierbaum. 46.
Poelitz. 46.
Scholven. 46.
Wesseling. 46.
Zeitz. 46.
Synthetic oil—
Production from plants. 61, 90.
Production from plants, British estimates. 119.
Recommendations to destroy plants. 93.
Reconstruction of plants, personnel engaged in. 65.
Reconstruction of plants. 57, 58, 59, 65, 126, 128.
Recovery from attacks upon plants. 66.
Smoke screen protection for plants. 149.
Tons of production lost per ton of bombs dropped. 94.
Weapon effectiveness in relation to plants. 151.

T.

Tanker—
Sinkings. 60.
Sinkings and North African Campaign. 40, 41.
Tankers, the *Albero* and *Coeleno* 11.
Target—
Committees. 94.
Identification. 88.
Priorities. 52, 62.
Selection. 90, 94.
Targets—
Committee. 64, 67, 68.
Tonnage of bombs dropped on oil. 57, 66.
Tar oils. 9, 10.
Tars, high temperature, the consumption of. 85.
"T.A.U.B.E." underground cracking plants. 152.
Technical Sub-Committee on Axis Oil. 87.
Technische Brigade Mineraloel (Kaukasus). 34.
Tetra-ethyl-lead. 68.
Fluid. **139–142.** See also ETHYL FLUID.
Fluid stocks. 6.
New plant at Froese. 6, 139, 140.
Tetra-ethyl-lead (*continued*)—
Paimboef. 6, 139, 140.
Plant at Doberitz. 6, 139, 140.
Plant at Heyderech. 140–141.
Production from the three main plants in 1940–1944. 140.
Texas Oil Company (U.S.A.). 12.
Thomale. 73.
Thomas, General. 6, 21, 33, 34, 35, 37.
Thomas, General, views upon Allied strategic bombing policy. **148.**
Thompson, O. F. 91, 116.
Tidewater Associated Oil Company (U.S.A.). 12.
Tintea oilfields, plans to destroy high-pressure wells. 18.
Toluol. 44.
Training—
Flying, restrictions in. 72, 79, 80, 81.
Lack of fuel for. 77.
Panzer Divisions, lack of. 76.
Paratroops, abandoning of. 80.
Pilots for Luftwaffe, decline in. 144.
Transport—
And Communication flights, restrictions in. 79.
And supply difficulties. 32.
Damage inflicted upon, by air attacks. 30, 132.
Facilities, denial to Germany of. 16.
Horse-drawn, conversion of motorised supply columns to. 75.
Mechanical, the immobilisation of. 178.
Military, conversion to producer gas. 58.
Of oil, problems of (Vienna Conference). 28.
Preservation of valuable vehicles. 74.
Rail, difficulties in. 86.
Rail, efficiency of. 30.
Rail, movements to Western Front, 1944. 30.
Shortage. 86.
Tank barges in use on W.I.F.O. account. 100.
Transportation targets, attacks on. 57.
Treasury, The. 16, 87.
Treibgas. 49, 86.
Consumption. 125.
Production. 127, 130.
Trench, B. M. C. 116.
Tripoli. 41.
Trzebinia Erdoelraffinerie G.m.b.H. 105.
Turkey, imports from Rome, 1938–43. Table 7, 163.
Turner, R. M. C. 116.
Turner, S. D. 116.

U.

U-Boat—
Offensive. 86.
Quality diesel fuel, stocks. 86.
Udet. 35.
Ueberwachungstelle fuer Mineraloel. 1, 96.
Underground—
Installations, erection. 128, 134.
Plants. 65, 66.
Plants, actual output from "ROST" and "OFEN." **Table 9.** 164.
Plants, effects of attack upon. 154.
Plants, Geilenberg, plans for dispersal. **152.**
Plants, production of J 2 light diesel fuel. 154.
Storage. 88.
Storage at Nienburg. 61, 69, **Plate 4.**
Storage at Martbartier. 70.
Storage, attacks upon Martbartier. 61.
Storage, W.I.F.O. 99.
Union Rheinische Braunkohle. 4.
Unirea. 14.
United States—
1st Tactical Air Force. 71.
8th Air Force. 54, 55, 57, 61, 64, 71, 120, 121.
8th Air Force, weight of bombs dropped on oil targets. **173–183.**
9th Air Force. 71.
15th Air Force. 53, 54, 55, 57, 60, 64, 67, 68, 69, 70, 120, 121.
15th Air Force, weight of bombs dropped on oil targets. **173–183.**

U.S.S. B.S. 151.
United States Economic Objectives Unit. 56.
United States Strategic Air Forces. 56, 59, 117.
United States Strategic Air Forces, weight of bombs dropped on oil plants. 151.

V.

Veith views upon Allied strategic bombing policy. 148.
Verteilungstelle fuer Bitumen. 97.
Vickers, C. G., Lieut.-Colonel. 115.
Vienna Conference. 28.
Vinter, S. P. 116.

W.

Walcheren, breaking of dykes. 64.
Walton, Mr. T. 109.
Wanne Eickel synthetic oil plant, capacity. **Table 3,** 160.
War Cabinet. 32, 38, 87, 90.
  Policy for denial of Rumanian oil to Germany. 14.
  Policy to purchase or charter means of oil transport. 16.
War Office. 17.
  M.I. 10(c). 92, 117.
Warlimont. 45.
Washington, liaison with. 91, 92.
Watson and Youell. 110.
Watson, M. Y. 116.
Waterways—
  Baltic. 16.
  Black Sea. 16.
  Danube. 16.
  Danube. German control over. 31.
Wehling Franz. 3, 29, 99.
Wehrmacht—
  Consumption in Western Campaigns. 21–22.
  Liquid fuel requirements. 3, 58, 135, 136.
  Requirements of J 2 fuel. 136.
  Reserves of gasoline, the exhaustion of. 37.
  Supplies from Rumania. 53.
Welheim, synthetic oil plant, capacity. **Table 3,** 160.
Wesseling synthetic oil plant—
  Attacks upon. 57, 127, 134.
  Capacity. **Table 3,** 160.
  Production. 46.
Westphal. 72, 76.
W.I.F.O.—
  Attacks upon storage. 69.
  Depôt at Nienburg, attacks upon. 61, **Plate 4.**
  Financing of. 100.
  Hauptlager. 101.
  Heerestanklager. 101.
  List of principal storage depots and capacities. 101.
  Number of personnel in organisation, 1937–43. 102.
  Organisation (Wirtschaftliche Forschungsgesellschaft). 2, 3, 29, 44, 79, 97, **99–102.**
  Properties, evaluation. 101.
  Turnover sales. 102.
  Umschlaglager. 101.
Will, cavalry major. 34.
Wintershall, A.G. 27, 103.
Wirtschaftliche Forschungsgesellschaft. See under W.I.F.O.
Wirtschaftsgruppe Kraftstoffindustrie. 2, 65, 96.
Working Committee (Oil) of the Combined Strategic Targets Committee (see also COMBINED STRATEGIC TARGETS COMMITTEE). **117–122.**
"WUESTEN" underground shale oil plants. 45, 152.
Württemberg shale oil production (see SHALE). 45, 113.

Y.

Yugoslav, activities of the Kontinentale Oel AG in. 106.

Z.

Zeitz—
  Attacks upon synthetic oil plant. 55.
  Elimination of synthetic oil plant. 134.
Zeitzler. 36.
Zeitz synthetic oil plant. 36, 46, 151.
  Capacity. Table 3, 160.
Zentrale Buero fuer Mineraloel. 2, 29, 97.
Zentrale Buero fuer Mineraloel Arbeitsgemeinschaft Schmierstoffverteilung. 97.
Zentrale Planung. 29, 32, 34, 36, 37, 85, 86, 149, 150.
  Demands made to, for aviation fuel. 80.
Zistersdorf oilfields. 8, 77.

www.ingramcontent.com/pod-product-compliance
Lightning Source LLC
Chambersburg PA
CBHW060232240426
43671CB00016B/2919

* 9 7 8 1 7 8 0 3 9 9 0 5 8 *